Principles of Vegetable Oil Extraction

This book covers the technology of vegetable oil extraction, including theory, process technologies used for various oilseeds, machinery operation and design aspects. Pertinent topics are covered in two parts: mechanical expression and solvent extraction. The importance of each preparation operation is explained as related to oilseed structure, outlining the differences in the quality of prepared material from various oilseeds. It also covers modern press designs, designs of extractors, desolventisers and distillation line, and support equipment. Effect of each unit operation on final product quality, levels of safety and precautions in plant operation, and plant automation, are discussed. Finally, the book takes a peek at possible novel and green technologies to find solutions to problems faced by the industry.

Features:

- Provides comprehensive coverage of vegetable oil extraction technology as applicable to major and minor oil-bearing materials.
- Explains the effects of each of the seed preparation steps on cell structure, and how it improves oil extraction.
- Describes mechanical expression technology in detail, including the design aspects of presses and seed preparation equipment.
- Explores the design of solvent extraction process equipment, including extractor, desolventiser, distillation and support.
- Discusses extraction plant safety, plant automation and utilities.

This book is aimed at professionals, students and researchers in chemical and food engineering.

Vivekanand Sadashiv Vadke is a leading process consultant in the vegetable oil industry. He completed his B.Tech. in Chemical Engineering at IIT Kharagpur, M.A.Sc. (chem. Eng.) at the University of Waterloo, Canada, and Ph.D. in Food Science at the University of Saskatchewan, Canada. He served for seven years with DeSmet India, where he was in charge of process design and marketing. Dr. Vadke specialises in process design of vegetable oil extraction plants and has helped upgrade many plants across continents to achieve higher process efficiencies and superior product quality, apart from designing new plants based on various oilseeds. He has several research and review papers to his credit and has been a speaker at many conferences. Dr. Vadke is currently engaged in developing novel technologies to find solutions to problems faced by the vegetable oil industry.

Principles of Vegetable Oil Extraction

Vivekanand Sadashiv Vadke

CRC Press is an imprint of the
Taylor & Francis Group, an **informa** business

Designed cover image: Er. Sanket Chakke

First edition published 2024
by CRC Press
6000 Broken Sound Parkway NW, Suite 300, Boca Raton, FL 33487-2742

and by CRC Press
4 Park Square, Milton Park, Abingdon, Oxon, OX14 4RN

CRC Press is an imprint of Taylor & Francis Group, LLC

ISBN: 9781032313832 (hbk)
ISBN: 9781032313849 (pbk)
ISBN: 9781003309475 (ebk)

DOI: 10.1201/9781003309475

Typeset in Times
by codeMantra

Contents

Foreword

In 1987, Vivek Vadke completed his doctorate at the University of Saskatchewan, in the heart of canola country. I had the pleasure to be the external examiner on his thesis, which focused on the mechanical expression of oilseeds. While preparing his dissertation, he realised that there was no readily available literature that could help an entrepreneur, a builder, an operator or a student in understanding, designing and operating a modern oil mill. The present volume, entitled *Principles of Vegetable Oil Extraction*, hopes to fill this gap.

This work is a comprehensive handbook detailing the principles and the process engineering aspects of the technology of processing sources of vegetable oils, primarily from a mechanical engineering viewpoint. Part 1 deals with mechanical pressing, covering seed preparation and pressing, the design basis of the process and the equipment, including the needs for cleaning, cake cooling, oil treatment up to but not including refining; mill layout, utilities, process control, automation and safety of both the process and the products.

While in North America and Western Europe, the oil industry is based on very large plants treating a few products such as soy or canola, Dr. Vadke is well aware of the need for smaller plants that are capable of processing a wide variety of local oil-bearing materials. This book contains specific processing and equipment information and specifications for an impressive list of starting materials: soybean, mustard/rapeseed, sunflower seed, safflower seed, cottonseed, groundnut (peanuts), copra and coconut oil, palm and palm kernel, sheanut, corn germ, castor beans, as well as sesame, niger, linseed and neem. The information can also form the basis for treating other seeds not already mentioned.

Part 2 is focused on solvent extraction technologies. It also gives the reader a brief introduction to historical practices and the evolution of the technology. It covers both the processing steps and the equipment required for feed preparation, which may include pressing, solvent extraction technology and operation and product handling, with limited discussion of soybean protein products including flour, concentrates and isolates. Chapters are devoted to quality control, safety and environmental issues.

This book is somewhat unique in detailing operation, construction and design aspects and utility requirements as well as sample plant layouts suitable for low- and middle-income locations. Both parts contain extensive appendices covering a wide range of related topics.

In summary, Dr. Vadke has assembled a very extensive review of vegetable oil processing that can be an excellent handbook for operators of small- to medium-sized plants, and can inform designers and potential investors of new plants. It is a very comprehensive introduction to this field for potential students and researchers.

Levente L. Diosady, Ph.D., P.Eng.
Professor of Food Engineering
Department of Chemical Engineering and Applied Chemistry
University of Toronto
Canada

Preface

"Can you recommend a book on extraction technology, please?" a plant engineer at a soybean crushing facility in Nigeria once asked me during a factory training session. "There isn't a comprehensive book that I may recommend to you, dear", I replied. But I promised him to write exactly such a book. I am glad that the promise has now been fulfilled.

I had noticed the unavailability of such a book first during my doctoral work at the University of Saskatchewan, Canada, during the 1980s. Later, during my years with the DeSmet group, I realised that machinery companies had kept the knowledge of technology and equipment design to themselves. The ideas for the book crystallised over the past 25 years during training sessions in India, China and many African countries.

To be fair, there are chapters and articles on extraction technology in books dedicated to specific oilseeds, for example, soybean, rapeseed, sunflower seed and cotton seed. There are also two chapters on general technology, applicable to various oilseeds, in the holy book of vegetable oil industry, the Bailey's, as well as a chapter in the *Fats and Oils Handbook* published by AOCS Press. But that is what these really are: chapters in general books that may be devoted to chemistry, processing, products and utilisation of oilseeds and oils. These, for sure, do not cover technology aspects in depth. Of the two distinct areas of mechanical expression and solvent extraction, the first one is even more neglected in the literature.

This book is written in two parts: Part 1: Mechanical Expression, and Part 2: Solvent Extraction. The first part is my area of research, and the second part is where I mainly built my career, first as part of DeSmet group, and then as a consultant to the industry. However, the first part remains my passion, and I have helped a couple of companies design large modern presses.

To compile this comprehensive handbook, I have drawn from many sources, including research papers and books. However, a significant part of the content comes from self-contemplation, training and experience.

The book covers the technology of preparation of oil-bearing materials (referred to as oilseeds hereafter) prior to both mechanical expression and solvent extraction; and then proceeds to cover actual oil extraction operations. Apart from discussion of technology for each seed, theoretical aspects of preparation, starting from the structure of oilseeds and oil cells, have been discussed. A hypothesis has been put forward to relate the preparation steps and parameters to the structure and composition of seed.

Part 1 covers the entire gamut of seed preparation, press operation and downstream operations of cake handling and oil filtration. Areas covered are process technology, equipment design and process control. It should be of help to plant operators and engineers, apart from students of oil technology. Construction, operation and design aspects of oilseed preparation machinery are discussed at length. The discussion of equipment design aspects would be of interest not only to designers but also to plant engineers.

For the mechanical expression of oil, an important parameter is the moisture content of prepared seed at press inlet. These are commonly decided based on prior experience and experimentation on the press floor; the practice may vary among different oil mills. A simple formula is suggested, supported by a logical explanation, to determine the optimum moisture for any seed and for any pressing duty, be it prepress, full press or second press.

The theory of screw pressing, and the importance of individual components on a press wormshaft are explained with the help of unidirectional pressing concepts. This will help plant engineers and operators gain insights in operation and maintenance of screw presses. Design methods are presented for the overall design of Press for specific pressing duties, as well as the design of worms and other components. These should be of direct help to mid-size press designers, especially in developing countries.

Part 2 covers all the operations and machinery in a solvent extraction plant. Once again, the preparation concepts are discussed in the light of seed composition and cell structure. Difficulties in preparation of firm flakes, or presscakes, of some oil-bearing materials, are explained based on seed structure and composition. Since most of the preparation machinery is discussed in Part 1, only a few are discussed in this part.

Discussion of processes and machinery for solvent extraction should be of help to plant operators in process optimisation. It will also help them maintain correct procedures and take adequate precautions to ensure consistency of production and product quality. A summary of trouble-shooting procedures, presented as an appendix, will be of help to plant engineers and operators.

Seed and products handling operations on both ends of the process plants have been discussed briefly, including seed pre-cleaning, drying and storage, as well as cake and meal handling and storage, and oil storage and packing.

Solvent extraction is a hazardous operation. The solvent commonly used, food-grade hexane, is flammable and explosion-prone. Safeties have to be built-in to the equipment design, and precautions followed during plant operation. Both these aspects are covered in the book. Plant automation can not only help in achieving consistent plant performance, but also may be utilised for enhanced safety. Different levels of plant automation are explained, with relative merits of each.

Chapters on utilities requirement for both types of process plants will help plant engineers not only understand and optimise utility figures, but also appreciate the quality parameters of utilities used in the plants. These aspects are often ignored, only at the plant operation's peril. Apart from the main process equipment, auxiliary equipment have been touched upon, to give a comprehensive view of a process plant to new entrants and students.

Special processes and products have been covered in appendices to both Parts. Also some appendices present material balances around crushing plants, as also typical design calculations.

Although mechanical expression and solvent extraction processes are mostly carried out together in a combined process of prepress-plus-solvent extraction, the respective plant operators, as also machinery designers, tend to operate independently of each other. The press operators are happy with presscake of certain oil content and required firmness, but will not listen to complaints of high residual oil content in solvent-extracted meal. On the other hand, the solvent plant operators do not have adequate appreciation of seed preparation steps, which not only affect the press operation, but affect also the solvent extraction and desolventising operations to a significant degree. A bridge is required to match the two processes. Determination of Milling Defect (MD), a measure of (in) adequacy of seed preparation, is that bridge. This is not a new concept, but has hardly been utilised profitably. The concept is explained, and the laboratory method to determine MD is presented as an appendix to Part 2.

Finally, I have dared peek into the future and suggested a few approaches to novel technologies to overcome problems and limitations faced by the industry today. In particular, a new press design, to help achieve lower residual oil in cake, without compromising cake and oil quality, will be of special interest to green technology warriors. A new concept of solvent extractor, to help extract oil from low-protein materials such as Rice Bran, Palm Kernel and Shea nut, will hopefully promote research in the area. I must indicate here that studies on both these technologies are currently on in the pilot plant of our engineering company.

This book will help students of oil technology, food technology and food engineering gain useful information on, and insights into, oil extraction technologies. It is my fervent hope that these insights will encourage them to go further and help innovate new, more efficient technologies. Let us all work towards that objective and together traverse the path to a greener world.

Acknowledgements

My daughters, Bhupali and Jaai, have been after me for years, "Baba, you must complete your book"! It is mainly due to their persistence that I have been able to do this. Of course, my wife, Vasanti, has been a bedrock of support through all these years and through the ups and downs of professional career; without her support, this would have remained just a dream.

Special thanks are due late Mr. Michel Knott, my Guru at Extraction DeSmet, Belgium. While my pursuit of knowledge of vegetable oil extraction started during doctoral work at the University of Saskatchewan, Canada, it was during my seven-odd years at DeSmet group that I was able to build on the knowledge and gain insights. Thanks are due Dr. Frank Sosulski, who let me work on his then newly acquired Mini-Press at the University of Saskatchewan. Thanks are also due all my colleagues at DeSmet, fellow workers, engineers and operators at various factories that I worked with, and at my engineering company, Extech Process Engineering, India.

Special thanks are also due Dr. Levente Diosady, Professor Emeritus, University of Toronto, Canada, who kindly consented to write foreword to this book. He was the external examiner for my doctoral dissertation, and he thankfully responded when I contacted him after a gap of 34 years.

Several eminent engineers and technocrats helped by reviewing the manuscript. Among them, the doyen of vegetable oil industry in India, Shri. O. P. Goenka; senior officials from Buhler group, Switzerland, Mr. Christian Ziemann and Mr. Dirk Heinrich; industry colleagues from India, Shri. Anil Modi, Shri. Pradeep Jaipuria, Shri. Sandeep Chaudhary, Shri. Manoj Peety, Shri. J. S. Parihar, Shri. Pravin Keshre, and my partner at Extech, Shri. Santosh Kumar Tiwari; my colleague, Er. Sanket Chakke put in a lot of man-days to help me with preparation of all the figures in the book; machinery suppliers who helped with sketches and write-ups of their machinery, Shri. Rabi Gandhi of United Engineering, Shri. Ramdayal Bohra of Cottor Plants, Shri. Satish Khadke of Sharplex Filters, Shri. Abhay Shah of Spectoms Engg, and Shri. C. S. Kedar of JoJon Engg. I thank all of them from the bottom of my heart.

The premier professional body of vegetable oil industry in India, The Solvent Extractors Association of India (SEA), has been very supportive of my professional journey over the past three decades. I thank the SEA, especially its energetic Executive Director, Dr. B. V. Mehta, for their continued support. I must mention the early encouragement I received from the American Oil Chemists Society, who granted me the Honored Student Award in 1987 for a research paper based on my doctoral work.

Part 1

Mechanical Expression

1 Introduction

1.1 SCOPE

This book is written in two parts; Part 1 covers the technology of, and design of machinery for, vegetable oil extraction by mechanical means. The mechanical expression plant is referred to as Oil Mill in the book. The history of extraction technologies evolved over centuries is discussed briefly, before an introduction to present-day technologies, in the first chapter. An overview of important vegetable oils consumed worldwide, along with their current production figures, is included.

Chapter 2 covers the technology of preparation of various oilseeds, as also the process parameters. It starts with the theoretical aspects, covering the oil cell structure and the resistances to oil extraction, to the structure and hardness of seeds. The importance of various preparation operations to overcome the resistances is explained. Reasons for differences in preparation methods, and process parameters, for various seeds, are discussed. A very useful formula is presented for the determination of optimum moisture content in any oil-bearing material at press inlet. It can be applied to any press duty, be it prepress, full press or second press. Technology of oil extraction from oil fruits, namely oil palm, olive and avocado, is quite different from the screw press technology; these are covered separately in Appendix 1.

Chapter 3 covers the construction features and design aspects of machinery used for oilseed preparation. Suitability for preparation of each oilseed is discussed. Layout of the machinery for ease of operation and maintenance, as also for the gravity flow of solids, is also discussed.

Chapter 4 covers the construction features and operation of screw presses. Design methods for press components, including the cage, worm shaft, as also individual worm pieces, are discussed. These may help not only machinery designers, but also plant engineers to check the design of their presses if and when faced with performance issues. Discussion of precautions during start up, and how to ensure operation at maximum press capacity, will help plant operators.

Cake cooling and breaking operations are covered in Chapter 5. Discussion of sizing and control features will help plant operators optimise these operations, which are quite often considered less important, but have direct impact on the subsequent solvent extraction process. Operations of oil filtration, drying and cooling are covered in Chapter 6. Alternative filtration techniques, along with design conditions for various equipment are discussed.

Process control parameters and plant automation is covered in Chapter 7. Different levels of plant automation, and their benefits, are discussed.

Requirement of various utilities for the operation of oil mill is discussed in Chapter 8. Differences in electrical energy requirement are explained on the basis of seed hardness. A sample calculation presented for steam requirement may be of help to plant engineers. Requirement of water for oil mill operation, while small, can be of critical importance in regions short on water. A brief discussion of the quality of

DOI: 10.1201/9781003309475-2

steam and water, usually a neglected area, may help plant operators get the best out of these utilities.

Chapter 9 discusses the effects of processing methods and parameters on the quality of products, namely cake and oil. It may help plant personnel tune the operation of plant and machinery to produce superior quality products consistently.

The last chapter of this Part is dedicated to issues of plant and operator safety and environment. The importance of constant focus on these factors can never be over-emphasised.

Apart from the main text of the Part, some important aspects of mechanical expression are covered in appendices. Appendices 1 and 2 cover the very important area of recovery of fruit oils, namely palm oil and olive oil, both very important oils in international vegetable oil basket. A few aspects of design of equipment for palm oil expression may help machinery designers as also plant engineers.

Appendix 3 covers the production of pungent mustard oil, a unique process followed widely in north India, to crush nearly 5 million ton of mustard seed. Technology challenges, and their possible solutions, are discussed.

Appendix 4 presents typical material balances for crushing of various oilseeds. This may surely be of help to plant engineers and managers.

Appendix 5 details the calculation methods for intermediate worm sizes on a wormshaft of Screw Press. This appendix is closely related to Chapter 4 which covers the design methods of Press.

Typical oil mill layouts are discussed in the last appendix, Appendix 6. Pros and cons of alternative layout philosophies are discussed.

1.2 HISTORICAL PRACTICES

Humans have extracted oil from various oil-bearing materials (seeds) for food purposes from time immemorial. While it is quite likely that the first attempts could have been to hammer seeds with stones to extract some oil, the widely used primitive method was boiling seeds in water and skim off the extracted oil (Bredeson, 1983). The latter method may be witnessed even today in parts of western Africa for extraction of sheanut butter.

In past centuries, seeds were pressed between a stationary and a moving element of a machine, forcing part of the oil out of the seeds, which then filtered out through openings in the stationary element. The ancient **stump press** consisted of a burnt-out stump which held the seeds, and a heavy pole driven by an ox to macerate the seeds and thus free some of the oil (Dunning, 1953). This method is still practised for the production of 'pungent' oil from mustard seeds in the Indian sub-continent (Appendix 3), the press is known as 'Kohlu', driven now by electric motors.

The first of **vertical presses**, in which pressure was exerted on a stationary mass by levers, screw jacks, or hydraulic cylinders, and oil flowed from the compressed mass to collecting rings below, was patented in 1795 by Joseph Bromah in England (Dunning, 1953). Many improvements were made in the first **hydraulic press** so that it was the major oil extraction unit until the early part of the 20th century.

The idea of a mechanical **screw press** was conceived by V.D. Anderson in 1876 (Dunning, 1953). In 1900, the first successful screw press, called an

Expeller, was made. It consisted of a cylindrical barrel, and a rotating shaft with helical flights. As the seed was pushed forward by the flights into a progressively narrowing space between the shaft and the barrel, it was subjected to increasing pressure, releasing the oil which then flowed out of the slots in the barrel. The partially de-oiled cake was discharged continuously through a constriction at the discharge end of the barrel. During the early years of screw press operation, raw seed was pressed and the Expeller used to be the only mechanical device in the press room of an oil mill. However, soon the benefits of seed breaking and heating were realised, and these operations were included in the seed preparation prior to pressing; also, a filtration unit was added to remove sediments from oil.

The Expeller permitted continuous operation as opposed to the batch operation of hydraulic presses, which resulted in greater capacities with smaller machines and less labour. Expellers quickly replaced the hydraulic presses as the major oil extraction units (Hutchins, 1949; Dunning, 1953). Expellers, or screw-presses, are the mainstay of oil milling at the present time.

While the screw presses offer a versatile and easy method of oilseed crushing, they leave nearly 6–8% oil-in-cake; which may be 10–20% of the original oil in seed. **Solvent extraction**, developed since the 1920s, offers an alternative to enable extraction of up to 98% of the oil, leaving only 0.5–1.5% oil in deoiled cake (known as the 'meal').

In solvent extraction, seed is brought into contact with a solvent which dissolves the oil, the oil-rich solvent phase (miscella) is then separated from the oil-free seed mass (known as the 'marc') by a simple filtration technique, and the oil is recovered by distilling off the low-boiling solvent. This method was first practised, on a large scale, in Germany shortly after World War I. In United States, solvent extraction was adopted on a major scale during the decade preceding World War II to extract oil from soybeans primarily.

The first use of a batch solvent extractor was reported in 1855 in France, where carbon disulfide was employed to extract oil from spent presscakes of olive (Kemper, 2005). The earlier solvent extractors employed the 'immersion' technique, wherein mass of oilseed was dipped in a bath of solvent, soaked for sufficient time to allow the release of oil in the solvent (Karnofsky, 1949a). The batch process was then upgraded to continuous mode. Several designs were in practice, one with a belt fitted with perforated buckets full of oilseed moving continuously, the buckets travelling slowly thru the bath of solvent, before emerging out with de-oiled seed mass; another with a multi-stage vertical extractor filled with solvent, flowing from bottom to top, and flakes travelling, by means of sweep arms attached to a central shaft, from top to bottom; and a few others.

During the late 1940s, 'percolation' type extractors were invented (Karnofsky, 1949a; LeClef, 2020). Continuous percolation-type extractors are now the norm all over the world, as these involve much less solvent hold-up, as also less solvent losses, and also produce clear miscella, compared to the 'Immersion' extractors.

Solvent extraction is now used universally either as a 'finishing' unit operation after low-pressure expellers, or as the sole process for extraction of oil, for most oilseeds.

1.3 MODERN PRACTICES

The present-day world scenario in vegetable oil extraction is dominated by solvent extraction. Mechanical expression with screw presses is certainly prevalent, but it is mostly integrated with solvent extraction facilities.

The current practice may be categorised into three processes, all related to the level of oil content in various oil-bearing materials.

High-oil content seeds, with oil content more than 35%, such as rapeseed, sunflower seed, groundnut, palm kernel, sheanut and sesame, are pre-pressed to recover between 2/3 and 3/4 of the oil, leaving about 16–20% oil-in-cake. The presscakes are conveyed to solvent extraction for recovery of the balance oil. Copra, with very high oil content, up to 67%, is mostly pressed in two stages to recover most of the oil, leaving only 7–8% oil-in-cake; the cake is either solvent-extracted, or used directly for animal feed.

Oil-bearing materials with oil content lower than 25%, such as soybean and rice bran, are subjected to direct solvent extraction.

Oilseeds having oil in the medium range, 25–35%, such as cottonseed (de-linted and decorticated) and safflower seed, are subjected to mild pressing in light press machines called Expanders, fitted with oil drain cage, to recover a small part of the oil, and to produce firm collets/pellets having oil between 20% and 22%. The collets are then subjected to solvent extraction. Drain-cage Expanders have replaced Screw Presses, for cottonseed processing, since the past two decades, although old mills continue to use the Screw Presses.

There is also the fourth category, full pressing of oilseeds, to recover maximum oil and leave just about 7–8% oil-in-cake. These cakes may or may not be sent for solvent extraction. With the advent of solvent extraction, this process has been superseded, especially for large-scale crushing operations. However, full pressing is still widely practised for small and medium-scale operations in industrially less developed countries of Asia, Africa and South America, due to low initial investment and simplicity of operation. Also, in developed countries, the growing restrictions on hydrocarbon emissions in heavily populated urban areas have renewed the interest in this method during past decades. This aspect is discussed in some detail at the end of this section.

Modern screw press sizes vary from cage internal dia 150 mm to 350 mm, cage length from 1 m to 3 m, having crushing capacities from 50 Tpd (Ton per Day) to 900 Tpd, on prepress duties. These presses are driven by motors rated from 50 kW to 700 kW. Smaller machines do exist, but these are used mostly for special applications or for on-farm crushing operations.

The drain-cage Expanders are relatively high-speed, low-power, machines. The construction is much simpler than of a screw press, and expanders cost just a fraction of screw presses for comparable crush capacities. Sizes vary from ID 200 mm to 350 mm, motor power 100–250 kW, and may process 100–600 Tpd cottonseed meat (decorticated cottonseed).

Solvent extraction plants have progressed from 'batch' type to 'continuous' process; and 'immersion' type to 'percolation' type extractors. Plant capacities which used to be 10–20 Tpd in the early years a century ago have gone up to a few thousand tons of seed per day. Batch extraction plants, of small capacities, do exist, and are utilised for special applications.

More and more plants are now automated, to enable better control of plant processes and to reduce operation manpower. Automation scope ranges from (a) small PLCs to control critical process parameters and to integrate electrical drive interlocks, thru (b) additionally to collect all process parameter data for reference and control, as also to allow operation of all drives from a central control station, to (c) additionally control all valves to eliminate the need for an in-plant operator, as also to monitor all material levels in various process vessels, and mass flow measurements, to enable real-time material balance. Safety considerations may also be incorporated into the automation systems.

Fruit oils, mainly from oil palm and olive, are either obtained by light pressing operation (palm oil) or by high-speed decantation (olive oil). Presses for palm fruit operate under very low pressures, and are of light construction. Oil palm fruit is subjected to many operations prior to pressing. Palm oil is recovered as a mixture of oil and water, which is then separated by heat treatment followed by decantation. The mixture of olive oil and water, recovered from first decantation, is further decanted without heating to separate the oil. Hydraulic presses are utilised to obtain Virgin Olive oil. The seed of palm fruit, which is recovered as a by-product of palm fruit crushing, contains the Palm Kernel, which contains nearly 50% oil. Palm Kernel (PK) is hard, unlike the soft fruit, and is crushed in high-pressure presses and then solvent-extracted like other major oilseeds.

It may be noted that all solvent-extracted oils must be refined, to be fit for human consumption. That is the law in most countries, basically to ensure that the final traces of solvent are removed. Oil refining processes do remove many other components in oil; but those matters are not in the scope of this book. Press oils, on the other hand, may sometimes be consumed directly. That, in fact, is a common practice for traditional oils in the Indian sub-continent and in parts of Africa. In other parts of the world, however, even all the press oils are also refined, together with the respective solvent-extracted oils.

In recent times, there is a niche movement towards 'virgin' oils. This urge is being met by 'cold-press' oils. True cold-press oils may be obtained from the operation of hydraulic press or the stump press (Kohlu).

There is also a growing interest in 'green' technologies, which are safe, economical and environment-friendly. This, together with the growing interest towards 'organic' food products, might see a steady move away from solvent extraction, and towards mechanical expression of vegetable oils in the future (Vavpot et al., 2012).

However, for mechanical expression to offer a practical alternative to the solvent extraction process, two major hurdles will have to be overcome. One, larger presses will have to be built. The largest presses available today, which might crush rapeseed/canola @ 900 Tpd on prepress duty, might do just about 170 Tpd on full press, and leave about 6–7% oil-in-cake. Thus, the immediate challenge might be to build a press to crush 200 Tpd which will leave only 3–4% oil-in-cake. Leaving 3–4% oil in cake may be acceptable, since the oil may be utilised fruitfully in animal feed formulations. Second, electricity consumption is much higher for mechanical expression than for solvent extraction. Thus, presses must be designed to operate with less motor power. Efforts are underway in this direction, to achieve both the objectives (Vavpot et al., 2012; Vadke, 2019).

An oilseed crushing plant was called 'Press House' in old days, because there was not much machinery other than the presses in the building. Then, with the addition of seed preparation machinery, the term Oil Mill came into vogue. Now, with the advent of elaborate preparation lines, the press section often forms a small part of the building. The plant is now often referred to as the Preparation plant, since it is commonly attached to a solvent extraction plant. However, that term can create confusion, since it is also used to refer to the preparation of oilseed for solvent extraction, without mechanical expression of oil, like in the case of soybean. To avoid this confusion, we will use the term 'Oil Mill' in this book.

1.4 VEGETABLE OILS OF THE WORLD

Nearly half a billion ton of oilseeds, and over 300 million ton of oil-fruits, are grown the world over, yielding a bit over 200 million ton of vegetable oil. The major vegetable oils, their production figures and the corresponding major producing countries, are listed in Table 1.1.

Among the listed oils, palm and olive oils are fruit oils, while others are seed oils, except for the corn oil and rice bran oil, which are derived from corn germ and rice bran, by-products of corn and rice milling, respectively. Castor oil is the leading

TABLE 1.1
Major World Vegetable Oil Production, in Million Ton

Oil	Year 2019–20	Year 2018–19	Year 2017–18	Major Producing Countries
Palm Oil	72.3	74.0	70.6	Indonesia, Malaysia
Soybean	56.6	55.6	55.1	Brazil, USA, Argentina, India, China
Rapeseed	27.3	27.7	28.1	Canada, EU, India, China,
Sunflower seed	21.2	19.3	18.5	Ukraine, Russia, Argentina
Palm Kernel	8.7	8.9	8.5	Indonesia, Malaysia
Groundnut	6.1	5.9	5.9	China, India, Nigeria
Cottonseed	5.1	5.0	5.1	India, China, USA
Corn	4.2			USA, China, EU, Brazil
Coconut	3.6	3.7	3.7	Indonesia, Philippines, India
Olive	3.1	3.2	3.3	Spain, Morocco, Turkey, Greece, Italy,
Rice Bran oil	1.6			India, China, Japan
Other oils	~3			
Total	~212			
Castor (non-edible oil)	0.8			India, China, Brazil

Sources: USDA data, June 2020; data on corn oil by Bill George, FAS, USDA; Castor oil, Rice Bran oil: SEA India.

non-edible vegetable oil, obtained by crushing the castor beans. Fruits of oil palm and olive both contain oil in the range of 20–25%. Palm oil is recovered by light pressing of the fruit pulp; while olive oil is extracted by high-speed decantation, mostly designed to keep process temperatures lower than 30C.

Soybean is the world's number one oilseed. The quantity of soybean crushed (over 300 million Ton) for oil is more than all the other oilseeds put together (less than 200m ton). Soybean oil is produced largely by solvent extraction, with a very small fraction being recovered by screw presses, for niche markets. Almost all other seed oils are recovered by a combination of mechanical expression (screw press) and solvent extraction.

Rapeseed oil is obtained from the family of rapeseed, mustard and canola. Rapeseed and sunflower seed are the major oilseeds which are pressed prior to solvent extraction. Among the two, rapeseed is easy to process, being a small and soft seed. It requires few processing steps, and less energy, compared to sunflower seed.

Production of most oils is steady over the past few years. Only sunflower oil production has been increasing at an average rate of over 4% year-on-year.

The top four oils, namely palm, soybean, rapeseed and sunflower oils, are traded internationally in large quantities. The trade is dominated by palm oil, followed by soybean oil, sunflower oil and rapeseed oil, in that order. Much of the international trade is for crude oils as extracted, or water degummed oil as in case of soybean oil, which are refined in importing countries. Some quantity of refined palm oil, and fractionated Palmolene, is also exported from Indonesia and Malaysia.

2 Technology of Preparation of Oil-Bearing Materials prior to Mechanical Expression

2.1 SEED PREPARATION – THEORETICAL ASPECTS

Oilseeds are subjected to unit operations of size reduction and steam conditioning, prior to pressing, to enhance the oil recovery. The preparation also enables the reduction in energy consumption in actual pressing operation, and helps avoid use of excessive pressures in pressing which could lead to darkening of oil and presscake.

2.1.1 Seed Structure and Cell Structure

Each unit operation plays an important role in facilitating the release of the oil globules, which are embedded within individual cells of an oilseed (Figure 2.1), and thus helps in improving the oil yield on pressing. It is clearly seen that the oil globules are dispersed in the interior matrix of an oil cell. The matrix is made up of various other bodies, of cell nucleus, protein, etc. Thus, for the recovery of oil from individual oil cell, one objective is to bring all the oil globules towards the periphery.

When the oil is near the periphery, the next, and the most important, objective is to get this past the cell wall. The cell wall is a rigid structure, and offers the strongest resistance to oil extraction. As will be seen in the following discussion, most unit operations in the seed preparation are directed towards getting the cell wall softened and ruptured. The next objective is to get the oilseed mass, which comprises numerous cells, to be porous enough to allow the oil to flow out under pressure; or to flake the mass very thin so as to facilitate the contact of the solvent with all cells (in the solvent extraction process). Cell size is typically in the range of 20 μm (Mrema and McNulty, 1984). Thus, even thin flakes, 0.25 mm, contain many layers of cells within. The flaking, therefore, has to be effective in damaging the structure of cells not only at the surface of flakes, but also within.

It may be noted that differences in seed structure impact the degree of difficulty in seed preparation. Hard seeds, such as palm kernel, are difficult to 'prepare' and require greater input of energy; whereas 'soft' seeds, such as rapeseed and groundnut, are easy to prepare and require much less energy. This aspect is explained towards the end of this section, where the proximate composition of various oilseeds and the relation to seed structure are also discussed.

DOI: 10.1201/9781003309475-3

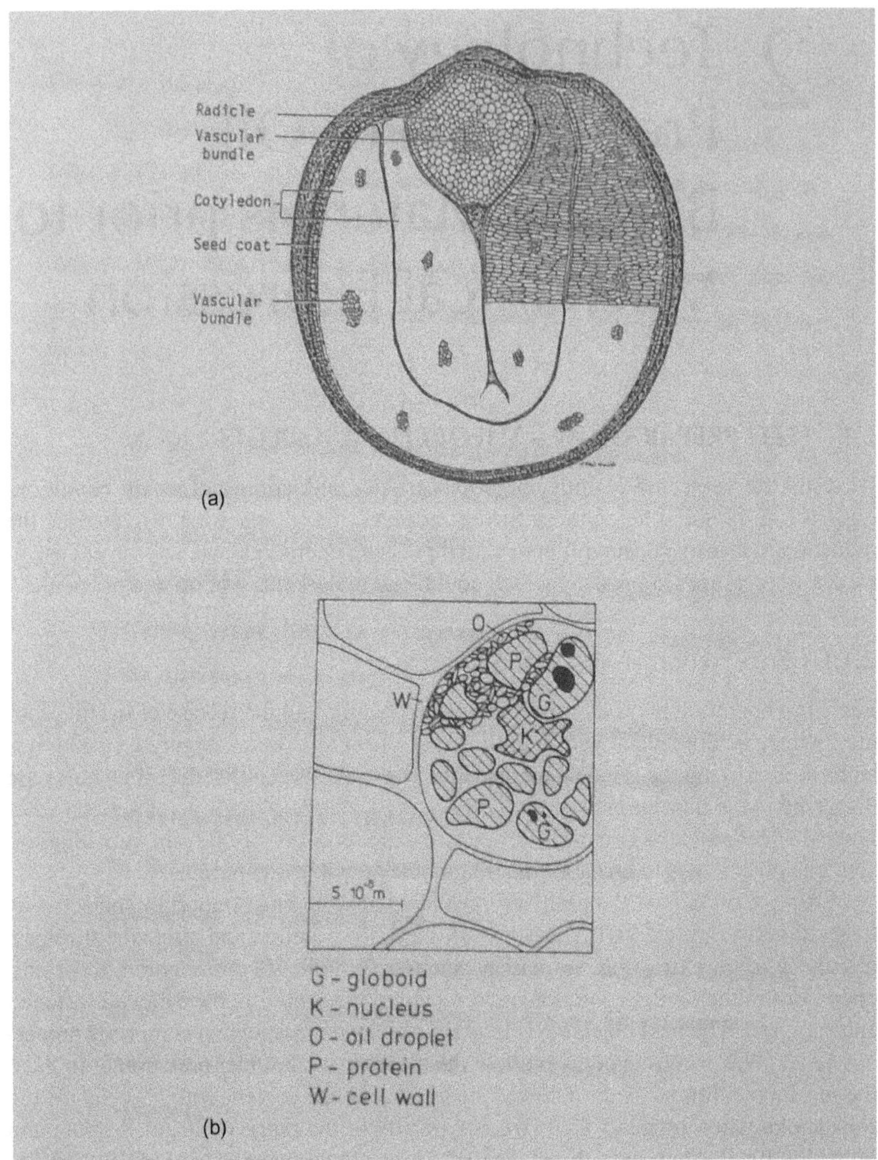

FIGURE 2.1 Oil Seed Structure (Rapeseed). (a) Seed Structure (Yiu et al., 1983); (b) Cell Structure (Eggers et al., 1985).

2.1.2 Cracking, Preheating and Flaking

The flaking operation is designed to break the seed coat and also rupture some of the cell walls (Beach, 1983; Yiu et al., 1983). The breaking of seed coat has been shown as crucial in the case of oilseeds such as rapeseed/mustard which have strongly attached seed coat and are difficult to dehull (removal of 'coat'). Preheating the seed prior to flaking avoids seed shattering by making it somewhat plastic (Beach, 1983; Ward, 1984). Preheating, with sufficient moisture in grits/seed, also helps soften the grits/seed and enhances the effect of flaking. For high-oil content seeds such as rapeseed and sunflower seed, moisture levels of 7% and 6%, respectively, are adequate for the purpose, whereas higher moisture @ around 8% is necessary for mid-oil content seeds such as cottonseed and safflower seed.

Larger seeds such as soybean or groundnut may first be broken, preheated and then flaked so as to achieve effective cell wall rupture (Norris, 1981). Another objective of flaking is to expose a greater area of oil-bearing cells to heat in the cooking operation. The pre-heating operation for larger seeds requires longer heating time, and is commonly referred to as 'Conditioning'.

Sunflower seed and cotton seed have thick and hard coat, which need to be cut open, prior to other operations of seed preparation. A portion of the hulls (seed coat) may be beneficially removed, to reduce load on presses, as also to increase the protein level in final de-oiled product. Sunflower seed may also be flaked without de-hulling.

Palm kernel and Shea nut are larger seeds and must be broken into many pieces prior to flaking. These seeds are also relatively harder than other seeds, primarily due to low protein content (details at the end of section). Copra also has low protein but is soft due to a very high oil content.

2.1.3 Cooking

Cooking is the term given to a heating operation wherein the temperature of the oilseed mass is brought up to the range of 80–115C, and the mass is held at that temperature for 20–40min. Several objectives may be achieved with proper cooking operation (Ward, 1984):

- To further facilitate the rupturing of the oil cells. This may be accomplished in part by flashing out of the intrinsic moisture.
- To coagulate the protein in the cells, to help release oil to the cell periphery. The coagulation also renders the oil separable from the meal without expelling the proteinaceous fines.
- To decrease the viscosity of the oil and so allow it to flow more readily from the meal.
- To adjust moisture content of the seed.
- To sterilize the seed to prevent growth of molds or bacteria.
- To detoxify or inactivate undesirable seed constituents, such as gossypol in cottonseed and myrosinase in mustard/rapeseed (Beach, 1983).

A microscopic examination of canola (hybrid varieties of rapeseed, with very low levels of glucosinolates and erucic acid) flakes, before and after cooking, has shown (Yiu et al., 1983) that cooking released the oil from within tiny inclusions inside individual cells, which then collected near the cell periphery. Also the numerous protein bodies in a cell fused into a large mass at the centre. Cell wall rupture also increased after the cooking operation.

The cooking operation is usually accomplished by heating the mass with indirect contact with steam. Live steam may or may not be injected, depending on the intrinsic moisture of the seed. In fact, many a times, cooker may be ventilated to reduce seed moisture. The seed moisture is adjusted in a manner that the cake emerging from Press would have close to 7% moisture. This is so because, above 7% moisture, the pressing operation is not very effective. This means that the moisture in seed at press inlet should be in the range of 4–5%; this is further explained in Section 2.1.6.

One major objective of a cooking operation, apart from the release of oil within the oil cells, and reduction in resistance of the cell walls, is coagulation of protein, and geletinisation of starch. If this not be achieved adequately, it would cause the release of the raw uncooked protein/starch with oil from the press cage. The raw, sticky 'foots', as the fines are called, would cause difficulties in oil filtration operation downstream, and as the foots are recirculated back to cooker and press, also create difficulties (reduce throughput capacity) in press operation.

Thus, cooking is a critical unit operation in seed preparation and must be controlled effectively.

2.1.4 Extrusion Cooking

Extrusion cooking has been used in recent times (Vavpot et al., 2014) in place of traditional cooking using cookers with long-residence time. This is a short-time, higher-temperature cooking process. Thanks to the intimate contact of various components, namely protein, starch, sugars and moisture, due to the high shear mixing within the extruder barrel, the cooking reactions proceed at a rapid rate.

Thus, although the contact time is very short, like only up to 30 seconds, the results are satisfactory. Protein-starch coagulation is achieved, oil collected near cell periphery, and cell walls softened and cracked. The additional beneficial effect of this operation is the creation of numerous pores through the mass, due to the sudden evaporation of moisture, as the pressure is released at discharge through the die plate. For high-oilcontent seeds, with more than 30% oil, the extruder would be fitted with a drained cage, to recover up to half the oil.

It should be noted, that for the above performance to be achieved, the flakes are to be heated close to the normal cooking temperature, in a small cooker, upstream of Extruder. This equipment may be called Pre-Heater 2.

Electrical energy requirement for this process, including that for the pre-heater 2, is several times more than that required for multi-stage cooker. However, it has been claimed that the press capacity may increase by up to 40%, for full-press operations, with this preparation technique (Vavpot et al., 2014). Also, the oil-in-cake is lower when this pre-treatment is used. This factor, of lower OIC, is important for full-press duty. This is why the extruder cooking is preferred mainly for full press duty.

2.1.5 Seed Structure and Protein Content – Effects on Processing

Protein and oil contents of various oilseeds are shown in Table 2.1. It may be seen that oilseeds which have low protein content (lower than 10%) have hard seed structure. Energy requirement during preparation steps, especially size reduction and flaking,

TABLE 2.1
Protein and Oil Content (Percent) of Various Oilseeds – Effect on Presscake Quality

Oilseed	Moisture	Protein	Oil	Remark	Protein in Presscake @ 18% Oil	Remark
Rapeseed	6–7	20–21	40–44	Soft seed	28–32	Strong cake
Groundnut	5–6	21–22	48–50		34–36	
Cottonseed Meat (dehulled)	7–8	28–30	27–34		32–35	
Castor	6–7	20–22	46–52		32–35	
Soybean	9–11	34–38	19–21		37–42 @8% oil	
Sunflower seed dehulled	5–6	17–20	50–56		28–32	
Sunflower seed whole	6–7	**14–16**	40–45	Medium	**20–24**	Medium strength
Sesame	5–6	**13–14**	52–55	Medium	**22–26**	
Copra	6–7	**9–10**	65–67	Low protein, but very high oil	**22–24**	
Palm Kernel	7–8	**8–9**	48–50	Low protein, hard seed	**13–15**	Fragile cake
Shea nut	7–8	**6–7**	48–50	Low protein, hard seed	**10–11**	Fragile cake

is high for these seeds (Palm Kernel and Sheanut). An exception to the rule is copra, with low protein content; but with very high oil content, the latter being responsible for its softness. It may be noted that as copra is crushed to leave around 25% oil-in-cake from the first press, the structure becomes harder, and is the cause for high wear on the components of the second Press.

Oilseeds such as rapeseed, groundnut, cottonseed (as 'meat' obtained after dehulling), soybean and castor, have 20% or more protein content; all these seeds are considered 'soft' seeds, and require lower energy during seed preparation, especially flaking. Sesame contains protein in the medium range, but may be considered a soft seed, thanks to the very high oil content (55%).

Seeds of sunflower are in the medium range of protein (15%), with their hard hull and soft core; and require medium energy for preparation. Dehulled sunflower seed may be considered as 'soft' material if more than 20% hulls are removed prior to flaking, wherein the protein content in dehulled 'meat' may increase up to 20%.

Another interesting inference may be drawn from the data in Table 2.1, with the numbers in the 5th column, namely, the protein content in presscakes @ 18% oil content. It may be noted that all the presscakes which have more than 25% protein content, are strong, thanks to the high level of coagulated protein. These hold very well under solvent spray action in solvent extraction plant. Those in the intermediate range of 20–25% (sunflower, sesame) are also reasonably strong, but produce fines which affect solvent percolation to an extent. But, those with less than 15% protein, such as palm kernel and shea nut cakes, are fragile, and are prone to disintegration under solvent spray. Extra care must be taken to prepare these oilseeds, with proper size reduction and longer cooking times, to improve the quality of the cakes.

2.1.6 Optimum Moisture at Press Inlet

Moisture content of seed mass entering a press is of critical importance in press performance. Quite often, oil mill operators face a question as to how one decides on the optimum moisture at press inlet, for different oil-bearing materials and different duties (prepress, full press, etc.). The answers are sought from those who might have experience of crushing a particular oilseed for a particular crushing duty. Quite likely, answers can vary among operators. So, is there a method by which the optimum moisture may be determined, apart from costly experimentation in the oil mill itself?

A simple formula is presented for optimum moisture at the press inlet. The moisture should be adjusted to such a level that the cake at the press outlet would have 7% moisture. This is so because, in the last segment of the press, if the moisture is higher than 7%, the material would be slippery and would affect oil expression adversely. On the other hand, lower moisture would make the cake hard, which would increase the energy requirement and also cause browning of the cake.

By the above formula, let us calculate, for example, the optimum moisture for rapeseed pre-press operation. Material balance around the pre-press operation of rapeseed (40% oil, 6% moisture) would show the cake (18% oil, 7% moisture) quantity to be 72% of the seed. So, the moisture at press inlet should be 0.72 times 7%, i.e. 5%. Since we had considered 6% moisture in seed for the material balance, let us correct it to 5% and recalculate. Now the cake mass would be 73% of seed, giving optimum moisture value of 5.1%. This is right within the industry practice of around 5% moisture. Some such examples are listed in Table 2.2. Readers will be able to determine optimum moisture levels for any seed and any crushing duty with this formula.

TABLE 2.2
Optimum Moisture of Oil-bearing Materials at Press Inlet

Oilseed	Oil Content (%)	Crush Duty[a]	Cake Mass Ratio	Optimum Moisture (%)
Rapeseed	40	PrePress	0.73	5.1
		Full Press	0.67	4.7
PressCake	16	Second Press	0.91	6.4
Sunflower seed (whole)	45	PrePress	0.67	4.7
Sun-seed dehulled	52	PrePress	0.59	4.1
Copra	66	First Press	0.41	2.9
PressCake	25	Second Press	0.82	5.8

[a] Oil content in PrePress cake considered 18%, and in Full Press (also, second press) cake, 8%.

2.2 PREPARATION OF MUSTARD/RAPESEED

2.2.1 Rapeseed Characteristics

Mustard and rape are two names of the same plant family. The seed is small and almost round, just about 1.5–2 mm dia (Figure 2.2), seed coat has two colours, black (or dark brown) or yellow. It is mostly the black variety which is crushed for oil, yellow is mostly used as spice. Seed coat is tightly attached to the cotyledon and is difficult to dehull (removal of coat). The cotyledon is soft and may easily be pressed with hand. Taste is bitter, due to the presence of phenolic compounds.

Rapeseed varieties have nearly 38–42% oil, 22–25% protein and 6–7% fibre. Rapeseed oil has high levels of erucic acid, which is used in the preparation of pharmaceutical products.

Rapeseed oil was thought to be responsible for some problems of heart and throat; this was a result of some studies on rats in Europe in the 1960s. The resultant brouhaha created consumer aversion to rapeseed oil in the west. In the 1970s, Canadian crop breeding scientists developed new rapeseed varieties, with very low levels of glucosinolates[a] (less than 30 micro-mole per g fat-free meal) and erucic acid (less

FIGURE 2.2 Rapeseed (Mustard, Canola).

[a] Glucosinolates are sulfur compounds present in rapeseed. Reaction by the native enzyme, myrosinase, breaks it down to produce allyl isothiocynates, compounds which give pungent flavour to the oil. These compounds are also suspected to cause 'mums' a disease of throat. However, the pungent mustard oil is a traditional dietary oil in India, and no such effect has been observed.

than 2%), the two factors thought responsible for the health issues; these varieties were named Canola. In other parts of the world, these are called 'double-zero' rapeseed. However, it may be noted that, In India, mustard/rapeseed has been one of the oils of choice for generations, with no adverse health effects reported.

2.2.2 Cleaning and Pre-heating

As the seed is received in factory, it may contain farm impurities. These are separated with a machine, fitted with deck(s) of vibratory screen, called Seed Cleaner. Magnetic separation of iron impurities is also a part of the cleaning section.

The clean seed is pre-broken in a flaker, fitted with a pair of smooth rolls. It helps to preheat the seed to around 50–55 C, to prevent the creation of fines during flaking.

In cold countries with below-freezing temperature, preheating to at least 40 C is a must to prevent shattering, as also to reduce motor load on flakers, and to help the formation of strong presscake (Unger, 1990). In cold countries, pre-heating to 40–45 C may be done in vertical seed dryers, whereas in warm countries, it is usually done in vertical steam-tray multi-stage vessels. Should the seed moisture be less than 7%, it is preferable to inject live steam during preheating, as lower moisture would cause shattering of seed.

2.2.3 Flaking and Cooking

Flaker is preferred for the rapeseed breaking operation, in place of Cracker mill with corrugated rolls, or Hammer mill, due to the small size of the seed. The process of flaking and cooking rapeseed is different for the two pressing duties, prepressing and full-pressing.

2.2.3.1 Preparation for Pre-press

The objectives of the prepressing operation are two-fold: to remove part of the oil from seed; and to produce a cake with superior solvent extraction characteristics (Ward, 1976). For rapeseed, the optimum cake quality is achieved when the residual oil content (RO) is in the range of 16–18%. This corresponds to nearly 75% oil removal. At this level of RO, most of the cell walls are effectively fractured due to the shearing action of press wormshaft, yet the cake is sufficiently porous to allow for good percolation of solvent.

To prepare rapeseed for prepress operation, the pre-heated seed is flaked hard to a thickness of 0.25–0.30 mm. This high degree of flaking is necessary to achieve rupture of more than 99% of the seed coats; this helps the press performance while utilising low pressures, as encountered in Prepress operation.

The flakes are cooked at 85–90 C, for nearly 25–30 min. The Cooker may be ventilated to remove vapours of moisture formed at cooking temperature, so that the moisture of flakes fed to the press be around 5–5.5%. Higher moisture would lead to reduction in press capacity, and lower moisture would cause unnecessarily high load

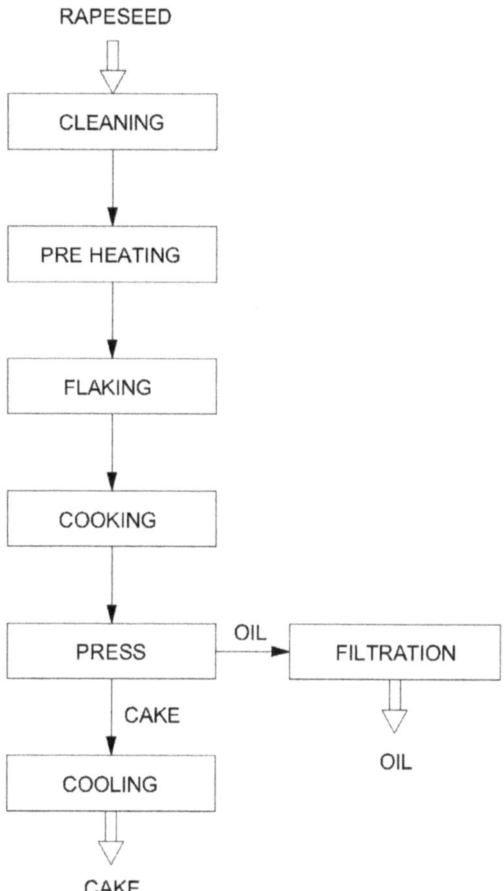

FIGURE 2.3 Rapeseed Crushing Flow Schematic.

on press motor. This level of moisture in flakes also ensures nearly 6–7% moisture in cake, after 1% flash-off at discharge and cooling, which is the optimum level for solvent extraction. Schematic of unit operations in Rapeseed crushing is shown in Figure 2.3.

It may be noted that the flakes get broken down to coarse powdery mass, by the action of the rotating arms. Thus, the material passing to the press has its seed structure totally broken. The maceration of cell wall structure, however, has to be accomplished within the press.

Common practice in China is to cook the flakes at much higher temperature up to 110C, at higher moisture 8–10%. The flakes are then dried, using hot air in slow-moving horizontal-bed Dryers, to about 5–5.5% moisture. Higher temperature increases press capacity and reduces press power consumption, but leads to dark cake and reddish oil. The red colour is removed during oil refining.

2.2.3.2 Preparation for Full Press

Full press operation is not very common and is mostly done in small capacities, for the production of 'green organic' oils and meals. The objective of a full press operation is to maximise the oil recovery. Cake extractability is of no concern, since the cake is not subjected to solvent extraction, but is utilised in animal feed formulations.

For this duty, the degree of flaking may be reduced somewhat, compared to that for prepress operation, since high pressures and high degree of shear encountered in the presses, designed especially for the full-press duty, would rupture all the cell walls anyway. The pre-heated seed is flaked to nearly 0.35–0.40 mm thickness. Cooking temperatures and time are greater than that for prepress, to help enhance press capacity. Typical conditions are, 100–105 C for 30–40 min. Alternatively, flaking may be avoided altogether, but then cooking temperatures and time need to be somewhat higher. Moisture of the cooked flakes is adjusted to 4–4.5%, nearly a percent lower than that for prepress. This is so because as the oil content in rapeseed mass reduces through the length of the press, moisture content increases correspondingly; towards the discharge, as oil content reduces below 8%, moisture content increases to nearly 7%; beyond 7% moisture, final oil release from press becomes difficult.

The steam consumption for the cooking operation, for full press, may be 50–80 kg/T seed, with seed inlet at 50 C, consisting of 50 kg for temperature rise, and up to 30 kg for moisture reduction. This is in addition to 25–30 kg steam used for pre-heating, starting from seed at 20 C. Steam requirement for Prepress would be lower by 25–30 kg/T.

Extruder cooking may also be used for full press operation. While it consumes much more electricity compared to Cookers, it offers major advantages in terms of increased press capacity, by as much as 50%, and also helps achieve lower oil-in-cake. So, while the normal OIC, with cooker, may be 7–8%, it may reduce to 5–6% with extrusion pre-treatment. It must be noted that a small cooker is still required, prior to the Extruder, just to heat the flakes to the cooking temperature, but without much holding time.

2.2.3.3 Two-stage Pressing

The full-press operation always causes higher heat generation within press, and leads to darker oil and cake. To minimise this effect, pressing maybe accomplished in two stages, first pressing to leave 15–16% oil-in-cake, a bit lower than with Prepress, and second pressing to reduce oil-in-cake to 7–8%. The first press oil is light in colour; and only the second press oil, which is just about 15% of the total press oil, is dark in colour. When the two are mixed, combined oil is much lighter than with a single-stage operation. The cake is also not as dark as in single stage. The two-stage operation entails somewhat more investment and more maintenance cost.

The current practice, the world over, is predominantly pre-press followed by solvent extraction. **Full press** operations, either single-stage or two-stage, are limited to special applications, as in the production of 'organic' oils. It must be noted that the press capacity reduces drastically, to just about 1/4th, in full press duty. Extrusion cooking is optionally practised for such operations, to enhance Press capacity, so as to reduce the number of presses required in an oil mill, and to increase the oil yield marginally.

2.2.3.4 Multiple Pressing at Low Temperature

In India, where mustard press oil is used for direct human consumption, without refining, it is important that the oil has no 'cooked' smell. The cooking temperature, therefore, is restricted to 65 C. Also, to recover oil with native flavour, short presses (barrel length 1.5 m max.) are used to prevent heat build-up.

Flaking is usually not practised. Seed is cooked at around 60–65 C, for nearly 25–30 min, with the injection of small amount of 'open steam'. The open steam helps develop a slight pungency to the oil, which is preferred by consumers. Multiple pressing is done, usually in three stages, to reduce oil-in-cake to 7–8%. Intermediate heaters, paddle screws sized for about 2–5 min retention, placed above second and third stage presses, help break the cake and heat it up only slightly; thus, the temperature of cake fed to the third press may be around 80–85 C.

Another quality of mustard oil, popular in India, is the '**pungent mustard oil**'. The pungency is developed using slow first crush in 'Kohlus' (Stump press), in presence of high moisture; this process is detailed in Appendix 1.

Typical **material balances** for prepress, full press and two-press operations are given in Appendix 4.

2.3 SUNFLOWER SEED AND SAFFLOWER SEED

2.3.1 Sunflower Seed Structure

Sunflower seed is of ellipto-conical shape (almond-like), with a spike at one end (Figure 2.4), nearly 4–6 mm wide, 8–10 mm long, and 2–3 mm thick. The seed cover (hull) is of black-grey colour, thick and hard, but the cotyledon is nearly as soft as of rapeseed.

Oil content of sunflower seed may range from 36% to 40% to as high as 45–47% in east European hybrid varieties. Protein is 14–16% and fibre is 16–20%.

The hull being hard and thick accounts for nearly 20–35% of the seed by weight. It not only causes low protein-in-meal, but also offers difficulties in pressing operation,

FIGURE 2.4 Sunflower Seed.

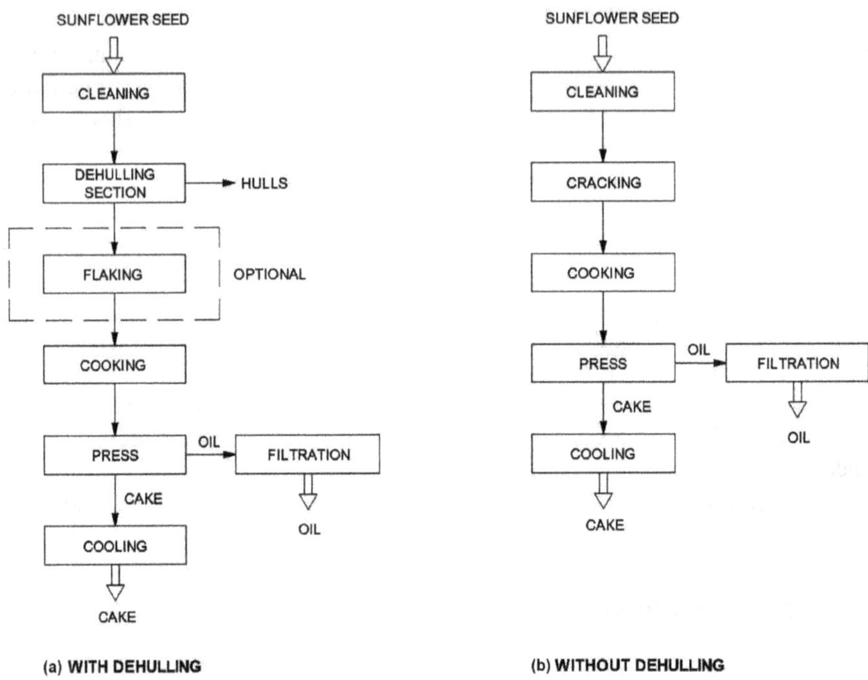

FIGURE 2.5 Sunflower Seed Crushing Flow Schematic.

namely, high energy requirement, high heat generation and high wear of press internals. Hence, it is beneficial to remove the bulk of the hulls as part of seed preparation (Figure 2.5). Separation of hulls also allows increase in crushing capacity per Press. The separated hulls are mostly burnt as fuel in steam boiler. In practice, dehulling (also called 'decortication') may or may not be adopted, depending on relative prices of meal of low and high protein content and of hulls as fuel. Both the practices are in vogue at the present time. In some instances, hulls may be separated, partially de-hulled meat crushed and solvent-extracted, to derive benefits of high capacity, and then hulls be mixed back, after grinding and heat-sterilising, with de-oiled meal.

Dehulling operation is preceded by cleaning operations, including size separation, magnetic separation and gravity separation (stones, the size of seed).

2.3.2 Sunflower Dehulling and Flaking

Dehulling, or decortication, is accomplished with Decorticator machines with a rotating blade that propels the seed against a static corrugated impact plate (LeClef and Kemper, 2015). The separated hulls are removed by size separation and/or air aspiration. Nearly 60–80% of the hulls are removed, depending on the desired level of protein in de-oiled meal. The challenge during dehulling is to not allow much meat (cotyledon pieces) to go with hulls. This is achieved by beating of the separated hulls and recovery of the meat fines by size separation. A good operation should not allow more than 1% oil, over the biological oil content of pure hulls, in the separated hulls fraction.

The dehulled meat may be flaked to nearly 0.35–0.4 mm, prior to cooking. The rolls should be driven at lower speed, compared to flaking of rapeseed and soybean, to minimise the damage to rolls due to abrasive hulls. Preheating is not required, since the meat gets preheated to some extent during dehulling. Alternatively, the dehulled meat may be directly sent to Cooker without flaking. Flaking helps increase the press capacity, but flakers add to equipment cost.

Dehulling offers another advantage; the wax content in sunflower oil reduces significantly, especially in solvent-extracted oil. This is discussed separately towards the end of this section.

The dehulling section, consisting of Decorticators, Hull Separators (vibrating screens), hull beaters and screens, along with air aspiration systems, occupies a major area of the oil mill. It is also one of the most expensive sections of the mill. For this reason, most small mills do not equip themselves with dehulling facility.

2.3.3 Size Reduction of Whole Sunflower Seed

If dehulling not be adopted, the first operation is seed cracking using pair(s) of corrugated rolls (Figure 2.5b). This breaks open the hull and makes two to three pieces (grits) of the cotyledon. The exposed grits allow for easy penetration of heat to the core during cooking.

Grits may be flaked to aid further the cooking operation, but this is mostly out of practice because of the high wear of flaker rolls due to the presence of abrasive hulls.

2.3.4 Cooking

Cooking temperature is 85–90C, only slightly higher than for rapeseed, but with a longer duration of 30–40 min, especially if grits have not been flaked. For flakes, cooking time may be reduced to 25–30 min, as used for rapeseed. Cooker is ventilated, so as to reduce moisture at outlet to around 5%. For dehulled flakes, moisture may be adjusted to a lower range of 4–4.5%, as explained in Table 2.2.

Sunflower press oil also is consumed directly in parts of the world, especially in eastern Africa. In these countries, repeated pressing, three or four stages, is done to recover more press oil, down to 8–10% oil-in-cake. For such an operation, dehulling is hardly practised, and cooking is milder.

Presscake from Prepress operation is sent for solvent extraction. The de-oiled meal, after solvent extraction, is used for animal feed.

Presscake from full press operations, with oil-in-cake less than 10%, may either be send for solvent extraction, or directly used in feed formulations for cattle and chicken.

2.3.5 Effect of Dehulling on Wax Content in Oil

Sunflower oil contains waxes, which are alcohol esters of long chain fatty acids (Carelli et al., 2002). These waxes tend to crystallise at room temperature and should be removed to maintain oil clarity. Nearly half the wax content is derived from hulls. Press oils typically have much less wax content than solvent-extracted oils.

When whole sunflower seed is crushed, wax content in press oil may be 600–1,000 ppm (mg/kg), compared to 2,000–3,000 ppm in solvent-extracted oil. Since the prepress operation recovers nearly 75–80% of the total oil, the combined oil may have 1,000–1,500 ppm wax content. For partially dehulled seed, wax in combined oil would be much lower, say just about 600–1,000 ppm.

During dewaxing of oil in Oil Refinery, separated waxes carry eight to ten times their weight of oil. Thus, lower waxes in crude oil mean lower oil losses in refining. Thus, dehulling helps reduce refining oil losses significantly.

2.3.6 Preparation of Safflower Seed

Safflower seed is similar in shape to sunflower seed, but smaller in size, nearly 3–4 mm wide, 6–7 mm long and 2–2.5 mm thick. The seed cover (hull) is of light grey colour (Figure 2.6), softer than hull of sunflower seed, and the cotyledon is as soft as of sunflower seed.

Oil content of safflower seed may range within 32–36%, protein 15–18% and fibre 21–23%. The hull accounts for nearly 35% of the seed by weight.

Since the hull is softer, safflower seed is usually processed without dehulling. Seed preparation is similar to that of sunflower (Figure 2.7). However, cracked grits may be optionally flaked as the hulls are not as abrasive as those of sunflower seed. Shorter cooking time, similar to rapeseed, is adequate for the flakes.

The safflower grits or flakes may either be pressed in screw presses, or be processed in a lighter Expander, equipped with a drain cage to allow draining of oil released in the light squeezing process. The Expander would be designed to recover nearly half the oil, leaving 20–22% oil-in-collets. The Expander, which is a much lighter machine compared to a screw press and consumes much less energy, is

FIGURE 2.6 Safflower Seed.

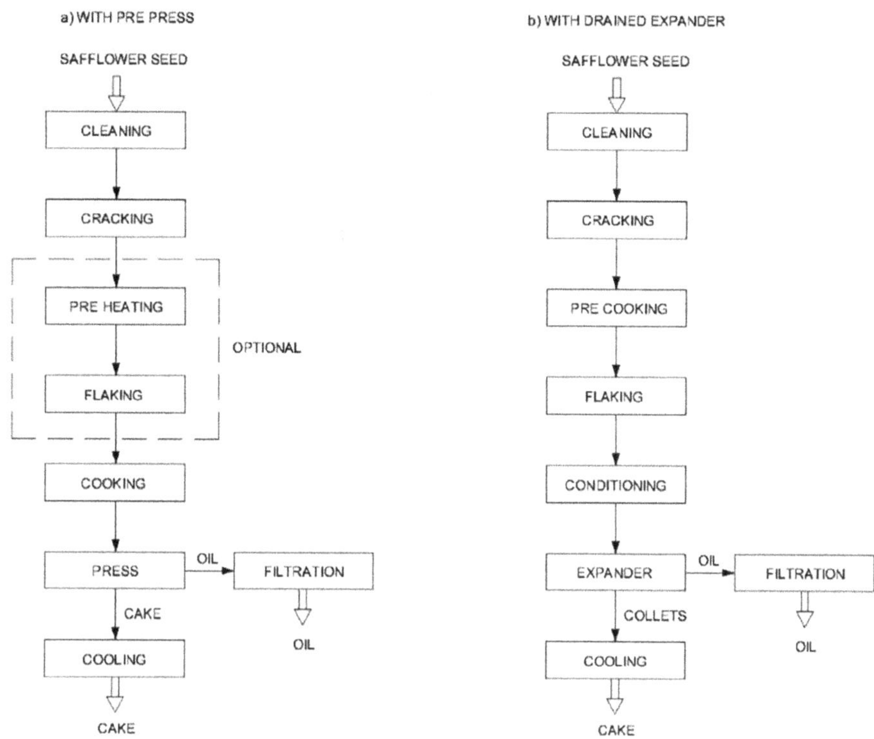

FIGURE 2.7 Safflower Seed Crushing Flow Schematic.

adequate to process safflower flakes, because of the softer hull and the lower oil content, compared to sunflower seed.

The seed preparation, upstream of expander, is somewhat different from that for prepress (Figure 2.7b). For the expander process, cracked grits may be pre-cooked at 68–70 C for nearly 20–25 min, then flaked. The flakes are then conditioned, in a small cooker, at 80 C for 5 min. The conditioned flakes are fed to the expander. This process is further discussed in Part 2, as preparation prior to solvent extraction.

2.4 COTTONSEED

Cottonseed is derived as a by-product of cotton ginning. The white cottonseed (WCS), consisting of up to 12% residual lint, has hard hull with a soft kernel inside (Figure 2.8a). Hulls are nearly 30–35% by weight of WCS. Oil content of WCS may vary from 17% to 19% for Indian, Chinese and US varieties, to nearly 21–24% for African varieties.

The white cotton seed is subjected to an elaborate line of multiple preparation steps. The first part of preparation consists of cleaning, delinting and dehulling, and the second part is the preparation of the dehulled 'meats' through the operations of flaking and cooking. The two sections are usually housed in two separate buildings.

(a) (b)

FIGURE 2.8 (a) White Cottonseed. (b) Black (Delinted) Cottonseed.

2.4.1 Delinting and Dehulling

White cottonseed contains nearly 10–12% lint. It also contains impurities, as received from a ginning house. These impurities are removed by size separation, using vibrating sieve machines.

The lint has good commercial value, and usually half the lint is recovered, using specialised Delinters, in two stages, to leave less than 6% lint on seed. First cut delinting recovers nearly 2–2.5% lint, and the second cut recovers another 4%. The number of delinters utilised for second cut is just the twice that for the first cut. Delinters may also be set to recover most of the lint, say 10%, to leave just about 2% lint-on-seed, which is then referred to as Black Cotton Seed (Figure 2.8b), but at a greatly reduced throughput. The latter process is used mainly for production of seed for planting.

The partially de-linted seed is then subjected to dehulling, where the seed is cut open between two cutting rolls. The loosened hulls are further polished using a series of slow beating rolls to free them of fine meat particles. A part of the hulls is retained with meat on purpose, to adjust the protein content of final de-oiled product. Bulk of the hulls, nearly 70–80%, is removed and may either be used as fuel in steam boilers, or be mixed with de-oiled meal for low-protein cattle feed. In the latter case, hulls should be sterilised by thermal processing (steam conditioning).

The seed house, which houses machines for cleaning, delinting and dehulling of cottonseed, is a large building, much larger than the preparation plant and oil mill. This is mainly because of the multiple delinting machines, each doing only 80–100 Tpd for the first stage and only 40–50 Tpd for the second stage. Thus, for a 500 Tpd cottonseed crushing facility, there will be two lines of delinters, a total of nearly

16–18 machines. Power requirement is also high. Thus, a cottonseed crushing facility requires much more investment compared to that for other oilseeds of comparable capacity, and also consumes much more electric energy.

The Meats, derived after delinting and dehulling, containing between 26% and 28% oil (Indian seed), or 30–34% (African), are prepared for either screw pressing or for Expander process.

2.4.2 Pre-heat, Flaking and Cooking

Prior to PrePress, meats are preheated to around 50–55 C, flaked to 0.35–0.4 mm and cooked to 80–85 C for nearly 25–30 min (Figure 2.9). Open steam may be injected only if moisture content of meat is less than 6% (African) or 7% (Indian). At start of season, moisture contents are high, and cooker should be ventilated to achieve optimum moisture content. Some smaller mills may not be equipped with flakers, and meats may directly be sent for cooking.

The cottonseed flakes contain some amount of lint, which makes the mass bulky (low bulk density). This may increase the size of cooker; these aspects are covered in Chapter 3.

The presence of lint in dehulled cottonseed (meat) poses an operation issue. As the meat is conveyed, and processed mechanically in pre-heater, flaker and cooker, some of the lint gets loose from the hulls and accumulates in corners and vapour spaces of machines and conveyors/elevators. If these are allowed to accumulate, plant capacity may get affected. Therefore, a common ventilation system, connecting all the conveying elements, as also the hopper over flaker, should be designed. Lint may be recovered from the exhaust air in air-cyclones. Even with a good aspiration system, some lint may still accumulate; thus, it is a good practice to stop the plant one day every month for cleaning of conveyors and machines.

2.4.3 Expander Process

Expander process may be considered an additional step for preparation of cotton seed for solvent extraction, rather than an oil recovery operation. A small 'drain cage' fitted on the expander barrel helps recover a small part of the oil. The objective is not to optimise the oil recovery, but to produce firm collets which will perform well in solvent extraction. Therefore, this operation, along with the preparation steps upstream, is covered in Part 2 of this book.

2.4.4 Full Pressing of White Cottonseed

There exist small mills in developing countries where whole white cottonseed is pressed to recover nearly 70% of the oil, leaving 6–7% oil-in-cake. The seed is cleaned and heated to just about 60 C prior to pressing. The energy consumption in the Press is very high, nearly 40 kWh/T. Short barrel presses are utilised to avoid excessive temperature rise with the dry fibrous material. The cake is not solvent extracted and is used directly in cattle feed.

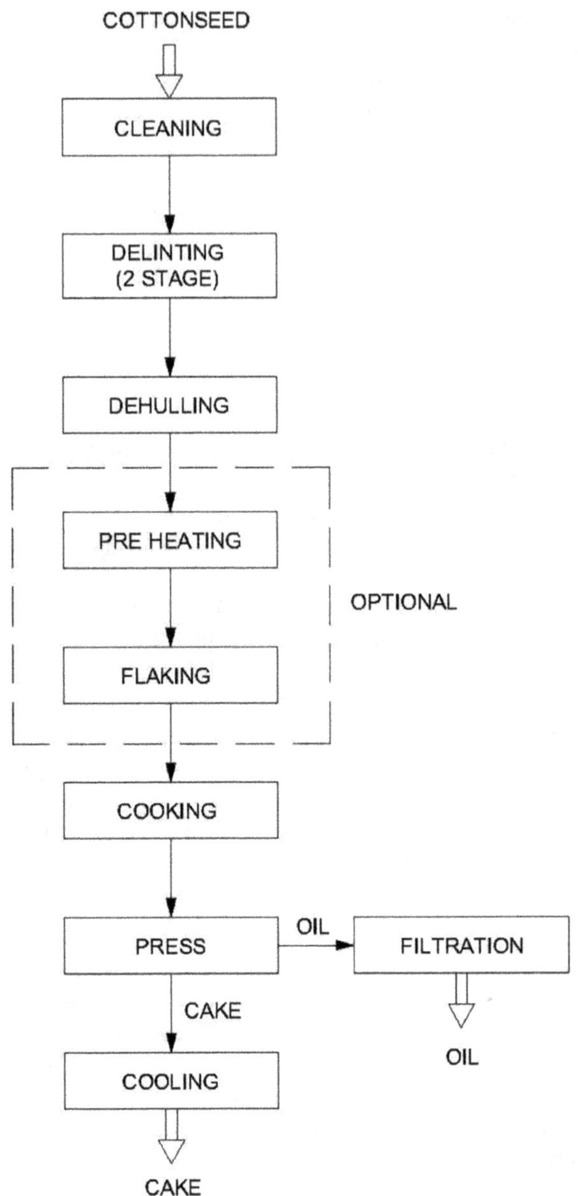

FIGURE 2.9 Cottonseed Crushing Flow Schematic.

2.5 GROUNDNUT

Groundnut is a relatively big seed nearly cylindrical in shape (Figure 2.10), around 12–15 mm long and 6–8 mm dia. Bigger nuts are found, but are usually utilised as table nuts, not crushed for oil. It is a high oil and high protein seed, with nearly

FIGURE 2.10 Groundnut.

48–50% oil and 20–26% protein. Seed coat is thin and is attached to the cotyledon, which is soft like that of sunflower seed.

Groundnut is normally received in pods in factories. The pods must be broken and removed. This is achieved using Decorticator machines, consisting of rotating cutter blades to cut open the pods, and suction fan to remove the separated light shell.

The thin seed coat constitutes only a small part of the seed, and is not removed, since the cake is fed to animals anyway.

Due to its big size, groundnut must be broken prior to cooking (Figure 2.11). This is usually accomplished in a Cracker mill, fitted with corrugated rolls. Due to the soft structure, energy requirement for cracking is quite low. Breaking the nut into 6–8 pieces allows good heat penetration during cooking.

Due to the high oil content and sticky nature of the kernel, groundnut grits are not flaked. The grits are directly cooked prior to pressing. Cooking conditions are, to raise the temperature to 80–85 C and to hold for 25–30 min to ensure heat penetration to the core. Open steam is not injected, rather cooker is ventilated to reduce moisture content to around 4%.

In view of the high protein content, it is imperative that groundnut is broken to small pieces and cooked over sufficient duration, so that the protein is fully denatured. If this is not achieved, the native form of protein may undergo coagulation during the desolventising operation in solvent extraction plant, which would lead to the formation of big lumps, causing serious operational problems.

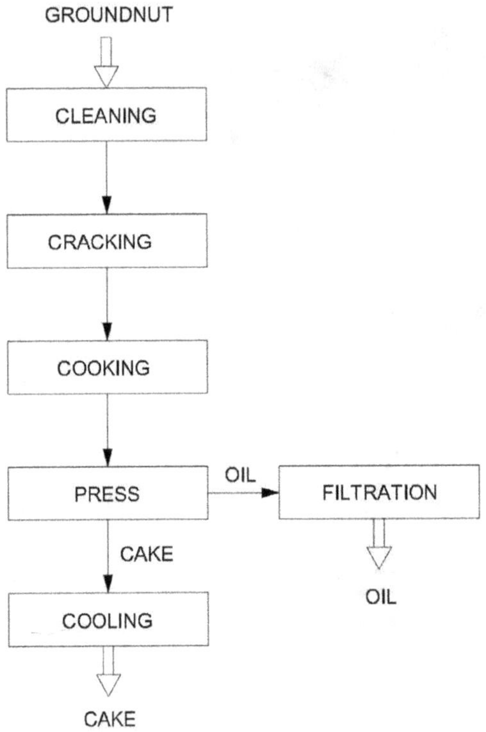

FIGURE 2.11 Flow Schematic of Groundnut Crushing.

An issue which seriously afflicts the groundnut crushing industry is the presence of aflatoxins, which come in from farm soil. This issue greatly affects price realisation for GN de-oiled meal, compared to soybean meal (which attracts the highest price among all meals, thanks to high protein content and the absence of anti-nutritional factors). The problem can be minimised with proper soil removal prior to storage of GN pods. However, the presence of even tiny amounts of aflatoxins (in many cases, maximum 20 parts per billion) is not permitted in the extracted meal in most countries. To ensure such a low level in meal, nearly all affected nuts would have to be removed prior to pressing. This may only be achieved with efficient colour sorting (since affected nuts are somewhat dis-coloured). The best Colour Sorters presently available are able to remove up to 90% of affected nuts. This would have to improve up to 97–98%, if aflatoxins standard is to be met consistently.

2.6 COPRA (COCONUT OIL)

Coconut oil is traditionally the cooking oil of choice in south-east Asian nations and in southern parts of India. It is obtained by crushing the dried cups of coconut, known as Copra (Figure 2.12). The size of half-copra cups ranges from 8090 mm for Indian copra to as big as 110–120 mm for SE Asian copra.

FIGURE 2.12 Copra Cups.

Dried copra, as received in crushing factories, typically has moisture between 6% and 8%. Copra has very high oil content in the range of 65–68%. Protein is just about 7–9%, the rest being carbohydrates, half of which is fibre. Thus fat-free copra is highly fibrous, an important consideration for mechanical expression as also for solvent extraction.

The first operation may be cleaning, to remove soil particles. For top quality copra, this may not be necessary.

To break the large copra cups into small pieces, suitable for cooking, hammer mills are more appropriate. The seed Cracker, with a small gap between corrugated rolls, would not lend itself to accommodate the cups; also the soft and sticky copra would stick to the rolls. For copra having low moisture, less than 7%, flat hammers can achieve the desired result. However, if the copra is moist, or rubbery, sharp knife edges are better suited. The size reduction, to the desirable size range of 3–6 mm, is achieved in two stages (Figure 2.13), to prevent shattering. If large copra cups are ground to the small size in a single stage, too much fines, smaller than 3 mm, are produced. These get overheated during cooking, and can impart dark colour, and burnt flavour, to the oil.

Copra grits, with high oil content, conduct heat fast enough, hence further flaking is not required.

Grits are cooked at 85–90 C for 35–40 min. Open steam is not required, since the objective is to reduce the moisture to 3–3.5%. As noted before, higher the oil content, lower is the optimum moisture level for pressing.

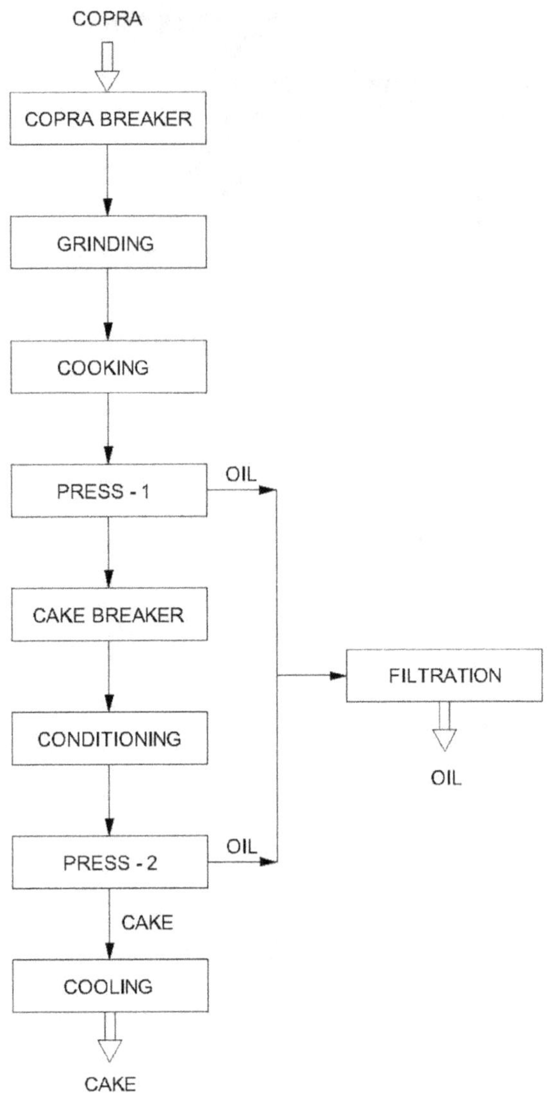

FIGURE 2.13 Copra Crushing Flow Schematic.

If the copra was dried in moist weather conditions, or if stored without adequate drying, fungus might develop and copra might become rubbery. Such copra would have to be cooked for a longer duration to achieve the desired structural changes within oil cells.

Copra is usually pressed in two stages. The cake from the first press needs to undergo additional preparation operations of grinding and cooking; these are discussed in the chapter on Press Operation.

FIGURE 2.14 Palm Kernel.

2.7 PALM KERNEL

Palm Kernel is a hard material, low in protein, with nearly 50% oil. Kernel is almond-shaped, but thicker and smaller in size, nearly 12–18 mm long (Figure 2.14). Protein content is just about 8–9%.

Kernel size varies greatly, even within a given lot; hence, it does not lend itself to proper size reduction in a Cracker mill. It is instead broken using Hammer mill to make pieces of 4–8 mm size (Figure 2.15). It helps to further pass the grits through a cracking mill to make smaller, and more uniform size, grits of 3–5 mm.

In many mills, the grits are directly cooked and sent for pressing. However, it is beneficial to flake the grits prior to cooking. Preheating to 60–65 C for 15–20 min, prior to flaking, is necessary, not only to prevent shattering but also to adequately soften the hard grits. Flaking to 0.45–0.5 mm is adequate. Producing thinner flakes from this hard seed requires too much energy, and is best avoided.

If PK is to be directly solvent-extracted, without pre-pressing, the preparation may be different. That is discussed separately in Part 2 of this book.

Cooking the grits at 85–90 C for nearly 40–45 min is required to achieve cooking to the core of the grits and to help produce a reasonably firm presscake, which would hold fairly in solvent extraction. In case of flakes, the cooking time may be shorter @ 30–35 min. Open steam is not required in either case, since the optimum moisture at cooker outlet is to be maintained low @ 4%.

Traditionally, small presses, with short barrels and high pressure, have been utilised to press PK. Since these presses are designed for high shear, the cooking times employed are relatively short, say about 20 min.

A major problem with PK processing is the fragile presscake, often quite powdery, due to the low protein content. Given the low-protein, high-carb nature of PK, may be a pre-treatment with high moisture conditioning could be tried. This could be of special interest to solvent extraction process, as a stronger cake would effectively improve the performance. This is further explored in Part 2.

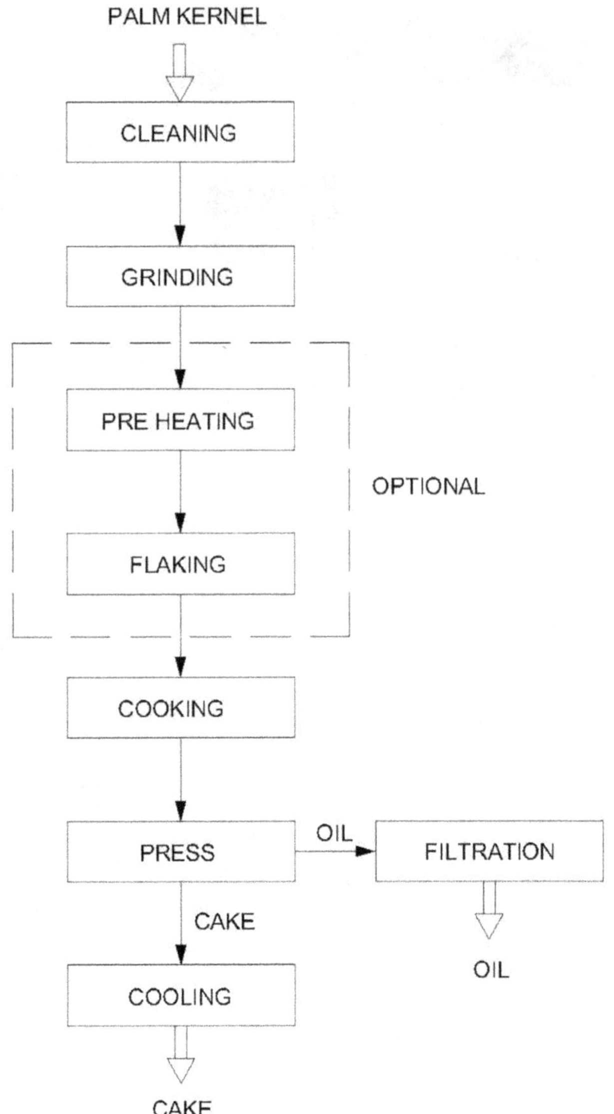

FIGURE 2.15 Palm Kernel Crushing Flow Schematic.

2.8 SHEANUT

Sheanut has the same shape as Palm kernel, but is bigger in size, typically 20–30 mm long (Figure 2.16). The seed structure is also similar to PK, as hard and low in protein. Oil content is also 48–50%, same as PK. As received, nuts moisture content ranges between 6% and 9% (beginning of season).

Sheanut is one of the most difficult seeds to crush and offers many operational challenges. Until recently, sheanut has largely been crushed in small mills using

FIGURE 2.16 Sheanut.

small screw presses. Modern milling, using larger screw presses, integrated with solvent extraction started about two decades back. Even today, only a few modern mills operate, mostly in western African countries, Ghana, Benin and Nigeria. The technology, one must admit, is still evolving.

The current practice in larger mills is as follows.

Seed is received in bags, emptied, pre-cleaned and is taken into Day Silo. Conveyor from the bottom of day silo feeds a Seed Cleaner, which are vibrating screen decks type. The clean seed is then crushed in Hammer Mills (Figure 2.17). It is a single-stage reduction to 4–8 mm size. The grits are then cooked at 85–90 C for nearly 25–30 min. Open steam is not used.

A major challenge is the handling of 'foots' in press oil. For all other oilseeds, the foots are easily circulated back to Cooker, mixed with fresh grits/flakes and re-pressed. However, the foots from Shea are very soft and sticky, and can accumulate along Cooker shell, and choke the Cooker. Hence, the foots are mostly mixed with either de-oiled meal from solvent extraction, or with husk/fibre separated in Seed Cleaner and pelletised. These pellets are mixed with press cake, cooled, and conveyed to solvent extraction plant. The mixing of foots with binding material is mostly done manually, is messy and requires manual labour. Also, the pellets disintegrate under solvent spray and cause serious problems for solvent percolation in extractor.

A plant has recently been designed in Nigeria (Extech, 2020), incorporating many significant improvements in preparation technology. First, flaking has been introduced to enhance the effects of cooking (Figure 2.18). The hammered grits are to be pre-heated to 65 C, holding time 20 min. This is to make the grits soft and aid in some cell damage in flaking. Flakes would be 0.5–0.6 mm thick. The flakes would then be sent to Cooker to ensure heat penetration to the core, with cooking time of nearly 40 min, and proper coagulation of the protein. Complete cooking is expected to reduce the quantity of 'foots' from the press. Also, the foots are expected to be less sticky, with better-coagulated and denatured proteins. These foots then would be

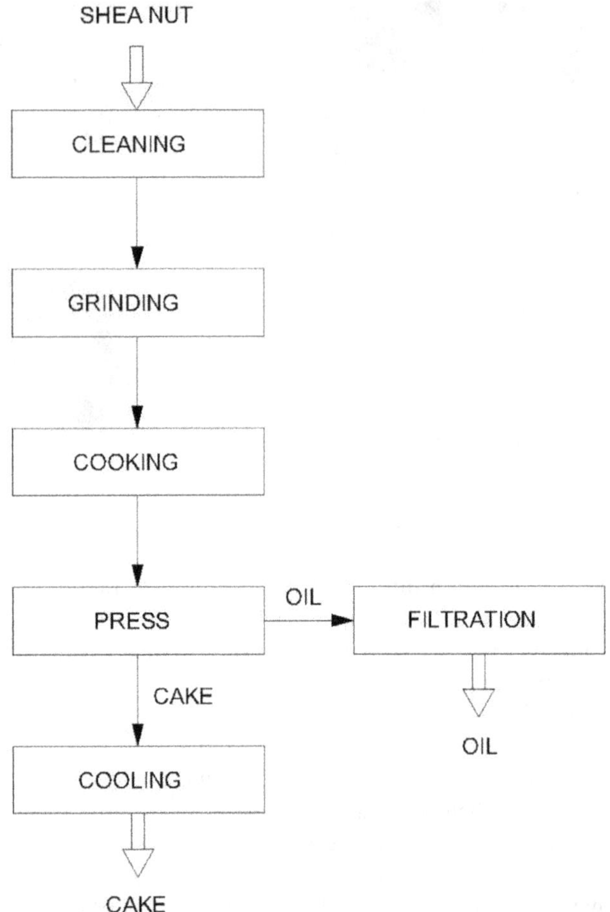

FIGURE 2.17 Sheanut Crushing Flow Schematic – Present Practice.

pressed separately in a specially designed Foots Press, with shaft speed much lower than a normal screw press, producing well-formed cake, which would not cause problems of poor solvent percolation in the extractor. Thus, the new process steps are expected to ease the problem of handling of foots, as also overcome several problems in solvent extraction plant.

2.9 CORN GERM

Corn Germ is a by-product of corn processing for the production of corn starch, sugars and corn syrup. The germ obtained from wet processing of corn, which is the most common method at present, is high in oil (42–48%), with medium protein (11–13%), and very low in ash and fibre. This is a soft material, thanks to the combination of high oil and mid-protein contents, and lends itself readily to preparation and pressing for oil recovery.

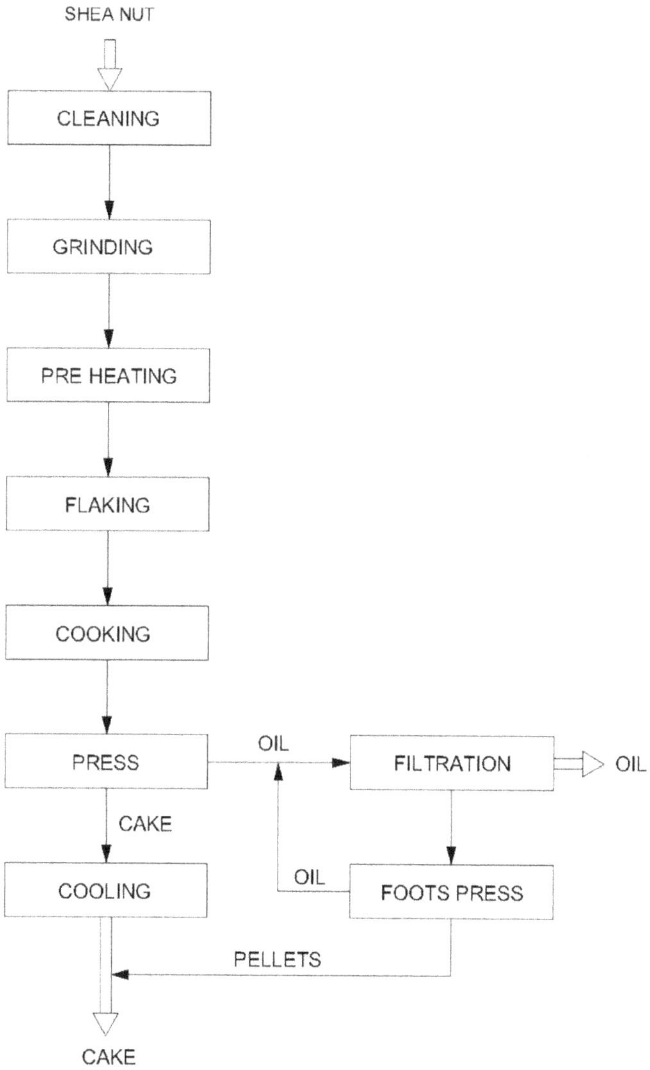

FIGURE 2.18 Schematic of Proposed Shea Nut Crushing.

The dry corn germ received at oil mils contains low moisture (3–5%). The germ may be pre-heated only in very cold countries. In tropics, with ambient temperatures around 30C, germ does not shatter on flaking. So, the first operation may be to flake to 0.3–0.4mm, followed by cooking @ 80–85C for 25–30min. Some oil mills do not even flake the germ; so, the only operation is cooking prior to pressing; however, this is not recommended, for achieving final low residual oil after solvent extraction.

Corn germ may also be directly sent to solvent extraction, without prepressing. The preparation steps are different in that case. This is discussed in Part 2 of this book.

2.10 CASTOR BEANS

Castor is the most important non-edible oilseed crop. Bulk of the oil comes from India, nearly 75% of the total world production. Castor beans generally are mid-sized, larger than soybeans (Figure 2.19), 10–15 mm long and dia 8–10 mm. However, large variations in size may be found in beans from different varieties; the length range may be 8–25, dia 5–16 mm. Seed coat is hard but thin, and the kernel is very soft and sticky. Oil content may generally be in the range of 46–52%, although lower (35%) and higher oil content (up to 55%) varieties are encountered. Protein content is around 20%.

The main challenge in the preparation of castor beans comes from the sticky nature of the kernel. Even during the cleaning operation, if some of the bean coats be

FIGURE 2.19 Castor Beans.

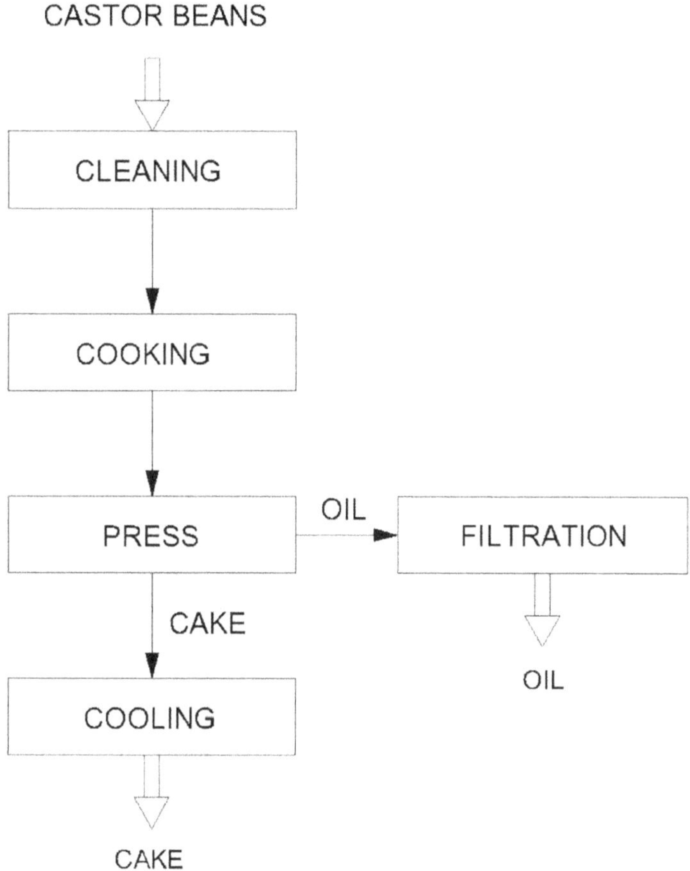

FIGURE 2.20 Schematic of Castor Beans Crushing.

broken, the kernel may stick to the screen and may have to be cleared manually. The sticky nature of the kernel does not lend the beans to operations such as cracking and flaking. Thus, whole beans are cooked and pressed (Figure 2.20). Typically, the clean beans are cooked at 80–85C for nearly 30min.

Just as with shea nut, the main problem during crushing operation is handling of foots. In some mills, which may also have a cold-press section (to produce 'cold-press' castor oil used for medicinal purposes), the foots are mixed with the cake from cold press and fed to the cooker.

It may be noted that the reason the foots are sticky, in spite of high protein content, is lack of heat penetration to the core of seed.

The problem of sticky foots may be overcome by extending the cooking time for enabling heat penetration to the core of the kernel. To avoid negative impact on colour of oil, the cooking temperature may be reduced by a few degrees. Thus, cooking at 78–80C for 1 hour may be the way to go.

2.11 OTHER OILSEEDS: SESAME, SOYBEAN, NIGER, LINSEED, NEEM

2.11.1 Sesame

Sesame is a small seed, as small and nearly as soft as rapeseed, and flat (Figure 2.21). It has high oil content, nearly @ 55%. Main producing countries are Tanzania, Myanmar, India, China and Sudan. Most of the seed in India and China goes for food purposes. Myanmar is the leading country for crushing of sesame. Tanzania is now catching up, while Sudan lags far behind.

All the mills in Myanmar are small in size, crushing typically 10–40 Tpd sesame. After cleaning, sesame is cooked to 75–80C and pressed in three to four stages, to reduce oil-in-cake to 7–8%. The operation is simple, and no major operation challenges are faced. Introduction of a flaking operation would help break the seed coat, as in case of rapeseed. That would facilitate heat penetration during cooking, help reduce oil-in-cake to the desired level just after two stages, and could eliminate the need for third and fourth stages of presses.

Introduction of large modern presses, for larger capacity mills, would require sesame to be flaked and properly cooked, with cooking conditions similar to those for rapeseed.

FIGURE 2.21 Sesame Seed.

2.11.2 Soybean

Soybean is the world's number one oilseed. However, since it is mostly extracted with solvent method, it is treated as a minor seed for the scope of this part of the book.

Soybean is mid-size seed, 7–10mm long and 5–7mm dia (Figure 2.22). Even with high protein content, it is harder than groundnut, because of low oil content of 18–22%, much lower than the latter. Beans moisture is typically 11–14% in western countries (U.S., Brazil, Argentina, Paraguay) and 8–10% in warmer countries (India, southern China, west Africa).

Soybean is by and large crushed via direct solvent extraction. However, there is a small market for presscake of soybean, which is claimed to be a better feedstock for cattle feed.

The beans, after cleaning, are cracked to grits 3–5mm size, pre-cooked to 70–75C for 15–20min, optionally flaked, and cooked to 90–95C for 20–25min. Cooker should be ventilated to reduce moisture to 6%. With proper preparation, and optimum moisture level, it is possible to recover nearly 12–14% oil, leaving 7–8% oil-in-cake, the cake being well-formed and light-colored. The oil is golden yellow, with very little red colour (Lovibond Red < 2 units in 1 inch cell).

There is now an increasing demand to produce what is called a '**Bypass Protein** Cake', which makes more protein available for nutrition of ruminant animals. The principle is to bind some of the amino acids so that these cannot be broken down in the 'rumen', and by-pass to the next part of digestive system where it is usefully digested into animal tissues. Such an objective, to bind the amino acids to make 'bypass protein', may be met by cooking the soybean grits at elevated temperature, range 110–120C, instead of the normal cooking range of 80–85C (Seaman and Stidham, 1991). The cooked grits are then pressed to recover 2/3rd of the oil. The resultant cake is high on 'bypass protein' content.

FIGURE 2.22 Soybean.

2.11.3 Niger Seed

Niger is a minor oilseed, but is important for the regions where it is grown, mostly in Ethiopia and India. Production of Niger seed oil is nearly half of Ethiopia's total vegetable oil production. Shape is long cylindrical, with a thin hard coat. In spite of the coat, the seed has overall soft structure, thanks to 15–25% protein content and nearly 40% oil.

Typical processing line consists of cleaning machinery, followed by a cooker and the press. However, it should help to flake the seeds prior to cooking, for lower press load, as also for lower RO after solvent extraction.

2.11.4 Linseed

Linseed, or flaxseed ('Ramtil' in India), oil is mostly used in paint industry for its quick drying properties, although it is a good edible oil when consumed fresh. The size and shape of the seed are very similar to sesame, but linseed is brown in colour.

The seed contains 40–45% oil and 20–25% protein. Like sesame, it is a soft seed, and is processed much the same way; cleaning, cooking and pressing. Multiple pressing is the norm, to reduce oil content in cake to 7–8% after nearly four presses.

Here too, the introduction of flaking operation would help reduce oil-in-cake to the desired level just after two presses.

2.11.5 Neem

Seeds of the Neem (*Azadirachta Indica*) tree are crushed, mainly in India, to recover oil known for its medicinal properties.

Dried seed as received at oil mill contains nearly 20–24% oil and has been obtained by removal of dry pulp by water wash (Figure 2.23). It has thick seed coat, nearly half the weight of seed. The kernel has nearly 45% oil and is soft. The seed protein content is low, just about 11–12%. Fibre and soluble carbohydrates form the bulk, almost 60% of the mass.

It is conditioned at nearly 65–70C for about 5 min and pressed to 10–12% oil-in-cake, which is suitable for solvent extraction. The cake is not strong and crumbles under solvent spray.

Should the seed be processed for the production of azadirachtin, a valuable pharmaceutical intermediate, it must be cold-pressed. In this case, seed is dehulled to leave a small fraction of hulls in meat, and crushed in a hydraulic press. The resultant cake may be heated and crushed in a screw press, prior to solvent extraction. The meal has pesticidal properties and is used with fertiliser.

A process improvement could be to first break the seed, say in hammer mill, cook at a mild temperature, say 55–60C for longer duration, say 25–30 min. Better cooking should help coagulate the protein and starch and produce a stronger cake, which would perform better in solvent extraction.

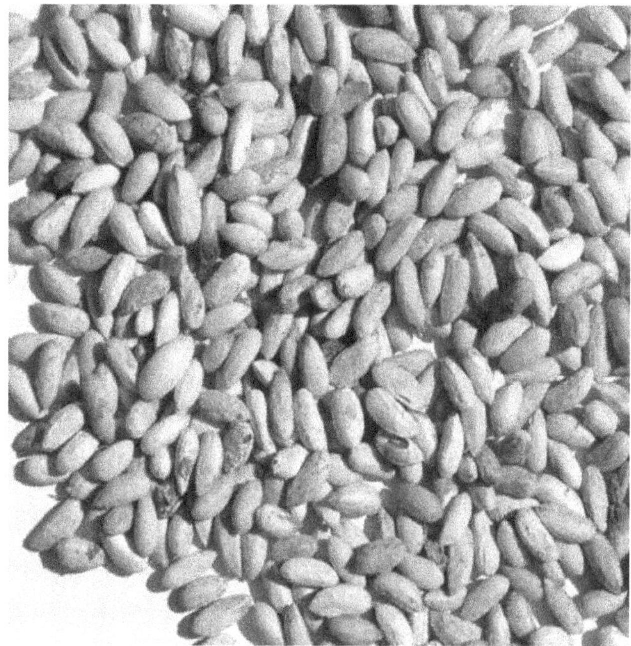

FIGURE 2.23 Neem Seed.

2.12 FRUIT OILS: OIL PALM, OLIVE, AVOCADO

Technologies utilised for the mechanical expression of oil from fruits of oil palm and olive are very different from the common screw press method. Palm oil is obtained by light pressing of palm fruit, which has been prepared suitably, and separating the oil from water. Olive and Avocado oils are obtained by high-speed decantation of the prepared mash. These technologies are covered in Appendices 1 and 2.

3 Preparation Equipment
Operation Principles and Design Aspects

3.1 UNLOADING OF SEED, PRE-CLEANING AND DRYING

3.1.1 Seed Unloading

Seed is received in either bulk containers, as in industrially well-developed countries, or in open trucks carrying bags. The bulk containers either open at the bottom or are tilted on hydraulic platforms, to dump the entire load into a large hopper below. The hopper is equipped with a screw conveyor at the bottom, to convey the seed to an elevator, which then feeds a pre-cleaner.

At port-based plants in large consuming countries, such as in Europe or China, seed may be conveyed from ships directly to the plant, either pneumatically or with bulk belt conveyors. Seed is received at high incoming rate in a buffer silo, with hopper bottom, before being sent to high-capacity pre-cleaners.

When the seed is received in bags, as in the Indian sub-continent, south-east Asia region and Africa, the bags are unloaded on a platform, dismounted manually, cut open, and the seed is dumped into an underground conveyor. The conveyor is usually a drag chain type, with protection umbrellas at feed points, to avoid shock loads on the chain (Figure 3.1). The platform can dock multiple trucks, for simultaneous unloading; and the conveyor is equipped with multiple inlets for the purpose.

The dump releases a cloud of dust which must be trapped. A suction hood is provided for the purpose. For the bulk dump, a single large hood suffices. For the bags dump, however, multiple smaller hoods are to be provided above all feed points and connected to a common duct. The dust is then sucked through a Bag Filter unit so that the outlet air is clean. The dust from the Bag Filter is periodically collected at bottom, and removed, either in bags or in small trucks.

Sampling: It is a good practice to collect a sample of incoming seed at the unloading station. An appropriate method, to collect a representative sample, is to install a small screw conveyor, say dia 250, rotating at a very slow speed like 1 rpm, below the evacuation conveyor. The medium diameter allows to intake even large chaff, and the low speed limits the sample size to about 100 kg in 20 minutes. Such a sample may be analysed for moisture, oil content, and most importantly, the amount of impurities.

3.1.2 Pre-Cleaning of Seed

The Pre-Cleaner machine is designed to remove light dust, large size impurities (overs) such as stones or farm chaff. Some machines are also equipped to remove

DOI: 10.1201/9781003309475-4

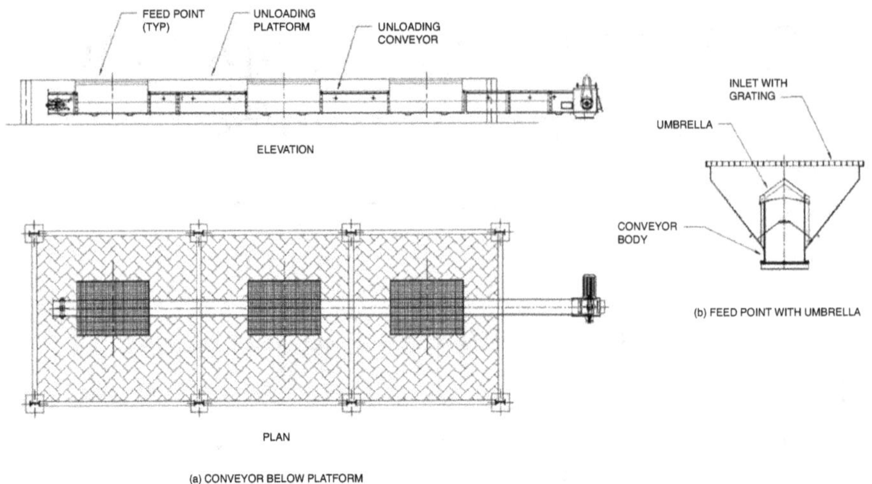

FIGURE 3.1 Unloading Conveyor. Courtesy – Eminence Equipment Pvt. Ltd.

small size impurities such as soil particles. It is imperative to remove light dust, as it can cause explosions in silos under warm conditions. Removal of large size impurities ensures smooth flow of seed out of silo.

Typically, pre-cleaners are equipped with an entry cyclone to remove fine dust, followed by multiple decks of vibratory screens for size separation (Figure 3.2). Some designs are of Rotary Drum type (Figure 3.3), where only 'overs' are removed, followed by aspiration for removal of dust. Pre-cleaners with vibratory screens are preferred over the Rotary Drum design, as the former remove more impurities, and help reduce the load on cleaning machinery in seed Preparation plant. A reduced load of impurities also helps keep the atmosphere in the preparation plant clean.

It is imperative that pre-cleaners be 'closed cabinet' type, to not allow dust to escape to atmosphere.

The separated 'dust' is collected either in an air-cyclone or an air-bag filter. Cyclones cannot collect all the fine dust and typically let out some into the atmosphere. Hence, air-bag filters are preferred for the duty, although they do increase the electrical energy requirement for air fans marginally.

The pre-cleaned seed is conveyed through an elevator to storage silo. In the case of multiple silos, a distribution conveyor is fitted on top of silos. The conveyor has pneumatic gates to individual silos, controlled from a cabin on the ground. Seed may be weighed with on-line weighers, before being loaded into silos.

Seed with high moisture content may be first fed to a Dryer before loading into a storage silo. Alternatively, should the storage be meant only for a few days, even high-moisture seed may not be dried prior to storage; these may be dried along with heat-conditioning in Dryer-Conditioners in the seed preparation plant. The latter process is more heat efficient and is discussed in detail for soybean processing in Part 2 of this book.

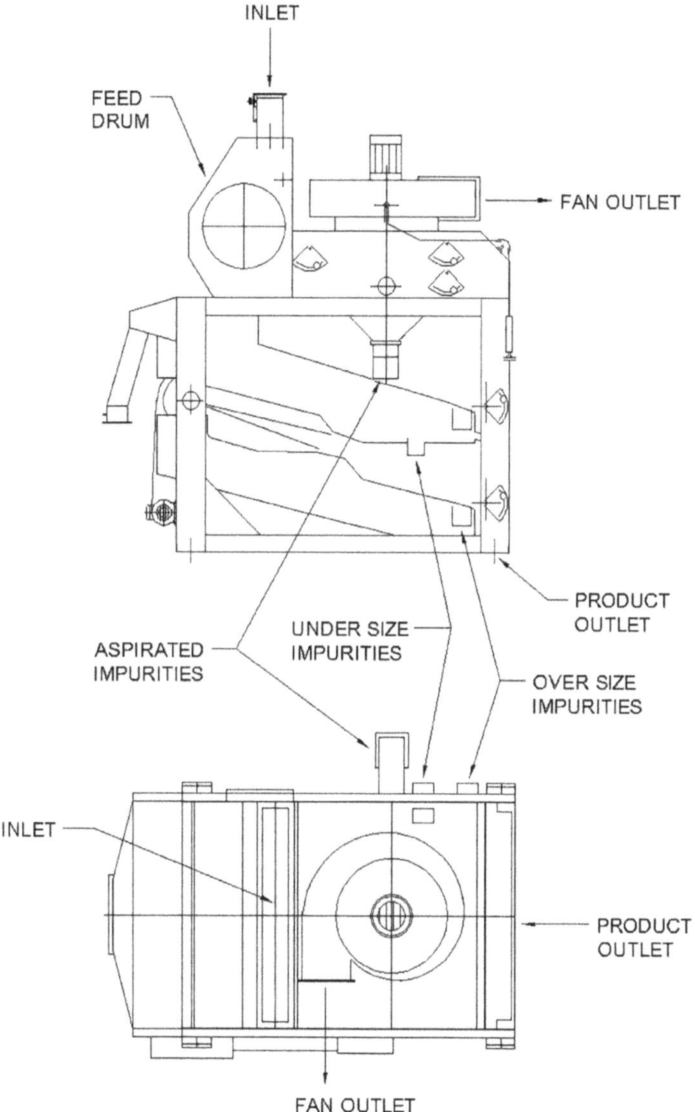

FIGURE 3.2 Seed Pre-Cleaner with Feed Drum. Courtesy – Fowler Westrup (India) Pvt. Ltd.

3.1.3 Seed Drying

Seed containing high moisture should be dried prior to long storage, to prevent damage due to respiration. The safe moisture limit may be 13% for soybean, and 9% for high-oil content seeds such as rapeseed and sunflower seed.

Dryers used for the purpose are vertical bin type. While batch dryers are used at places, continuous once-through Dryers are most common, especially at capacities

FIGURE 3.3 Rotary Drum Pre-Cleaner. Courtesy – Buhler A.G.

FIGURE 3.4 Seed Dryer. Courtesy – Neco Grain Dryer.

above 1,000 Tpd. Drying operation may usually be conducted only during the day, hence for a 1,000 T daily capacity, a continuous dryer may be designed for 100 Tph.

In a continuous Dryer (Figure 3.4), seed is fed by an elevator at the top, seed moves slowly by gravity through grids of channels, which bring in hot gases from a burner (gas or oil-fired), and dry seed is discharged by multiple long rotary valves fitted at the bottom. The speed of the RVs is varied to regulate the Dryer throughput.

Hot gas is injected at various levels, travels up through the seed bed, and is let out from a grid of channels placed up to a meter above the inlet channels. The inlet and outlet channels may be placed parallel or perpendicular to each other. The drying gases are let in at 80–90C so that the seed skin temperature does not exceed 60C. Gases leave the drying zone at nearly 65C. Thus. less than a third of the heat is utilised. The major part of the gases may be recirculated to the burner to increase the heat efficiency of the Dryer.

Soybean may be dried to 12% from a starting moisture level between 14% and 16%.

In the lower portion of the Dryer, ambient air is blown in to cool the seed to say 40C. The total residence of seed within the Dryer may be up to a couple of hours.

In countries where gas may be expensive, solid fuels may be used as heat source. In such a case, the fuel gases may not be used directly in the Dryer, as these might contaminate the seed. Rather, hot air would be produced by means of gas-air heat exchangers. When flue gases from a gas burner are used, recirculated gas from Dryer exhaust is let in to maintain the gas temperature within 90C.

Seed is preferably tempered for up to 5–6 hours in vertical bins prior to loading into silos. Tempering allows uniformity of heat and moisture within the seed. Dry-tempered seed may be stored safely for days and months without any risk of damage due to respiration.

Another kind of dryer is used when soybean is to be processed in what is known as 'Warm Dehulling'. Here, beans conditioning is accomplished along with moisture removal. This process and equipment are discussed further in Part 2 of this book.

3.2 SILOS

Pre-cleaned seed is stored in vertical silos made of corrugated galvanised iron sheets. Other treatments than galvanising are also now coming into practice. **Storage silos** are of flat bottom design, where the bottom base is of RCC (Figure 3.5). Silos constructed fully in RCC were common in yesteryears, but are now used mainly for small diameter sizes, and for special (corrosion) applications.

Silo capacities may range from 1,000 T to 20,000 T of seed, having dia up to 27 m. The silo height, typically, is a bit less than the diameter; tall silos require heavier foundations. However, should land availability be a constraint, taller silos maybe preferred.

When seed having high moisture is to be stored for long durations, say over a month, the silos are equipped with an air ventilation system. Air is blown from grates, fitted on the bottom, via underground ducts, to remove moisture accumulated on the seed surface by the process of respiration. If the moisture were not removed, lumps formation may take place, leading eventually to darkening and rotting of the seed. In large silos, temperature probes are embedded at different heights, suspended from the top cover of the silo and dispersed radially, to check for hot spots, which are indicators of moisture accumulation, since seed respiration simultaneously exhales moisture and heat. The aeration operation is automatically interlocked with the identification of hotspots, to economise on blower power.

Multiple silos may be arranged in a straight line and fed by a common overhead conveyor. Alternatively, these may be arranged in a circular layout and be fed by

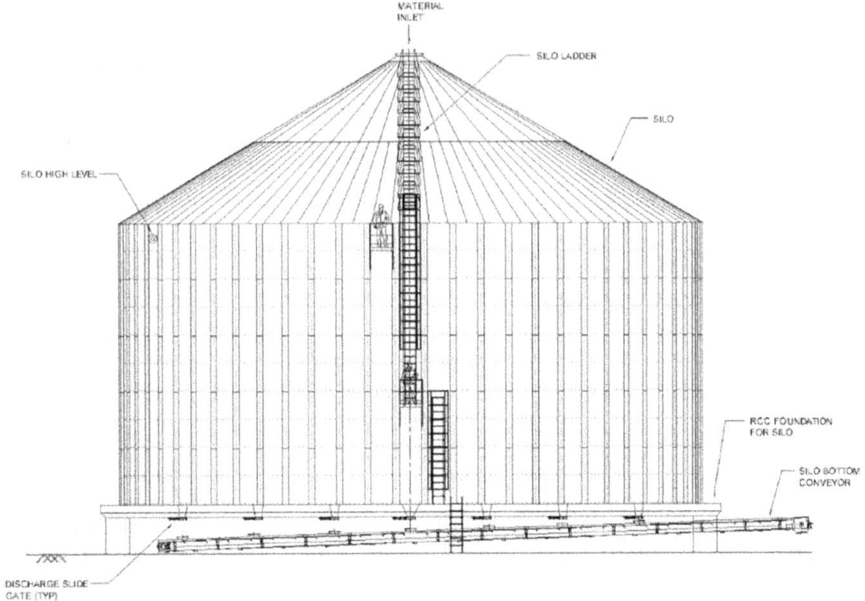

FIGURE 3.5 Flat Bottom Silo. Courtesy – Fowler Westrup (India) Pvt. Ltd.

chutes from a common elevator. In the latter case, the centrally located elevator needs to be much taller than the silo top, The former layout, a straight-line arrangement, is usually preferred, since all the silos may be fed, and discharged, with single conveyors, and the feed elevator does not have to be very tall. The overhead conveyor is supported from the silos, and a maintenance walkway is provided along the length. Discharge gate to each silo may be fitted with a pneumatic slide gate, which may be controlled from a control room at ground level.

Silos are fitted with entry doors at the lower level, and inspection windows at the upper levels. Vents are fitted on the roof. A railing is fitted on the periphery of the sloping roof for safety. A circular stairs is provided up to the roof, with intermediate rest pads. Access platforms are provided for each inspection window.

Seed is extracted from silos by conveyor, which runs below all the silos in a straight line. The multiple discharge gates at the silo bottom are regulated by slide gates, operated pneumatically or manually. Normally, only the centre gate is kept open; the peripheral gates are opened for the last discharge of seed. The conveyor is usually drag chain type, while belt conveyors are also not uncommon. Belt conveyors have wider cross-sections, requiring wider tunnels in silo foundation.

After the silo has been discharged by gravity, nearly 10–15% of the stored seed is left at the bottom, in a conical shape, sloping from high at the wall down to the centre line of silo. Large silos, dia greater than 15 m, may be fitted with a Screw Unloader (Augur), which helps extract the final seed. Prior to the operation of the screw, peripheral discharge gates are opened to discharge more seed by gravity. The unloader screw rotates and pushes the remaining seed towards the doors. It eliminates the need for manual clearing of the bottom seed, an operation fraught with safety risks.

Long seed houses may be built for larger storages, with sloping bottoms discharging seed into a conveyor running centrally along the axis. These storages may cost less per Ton of seed, since much of the load is taken by the soil below, but require more land. Short walls may support a curved roof. The seed is fed by a conveyor on top running along the length. These seed houses may be divided along the length into sections, either for seed segregation by moisture and quality, or simply for intercirculation to prevent overheating, etc.

The silo section usually includes a **day silo**, fed from storage silos, designed to hold seed for up to a day's plant capacity. This is a hopper bottom silo (Figure 3.6),

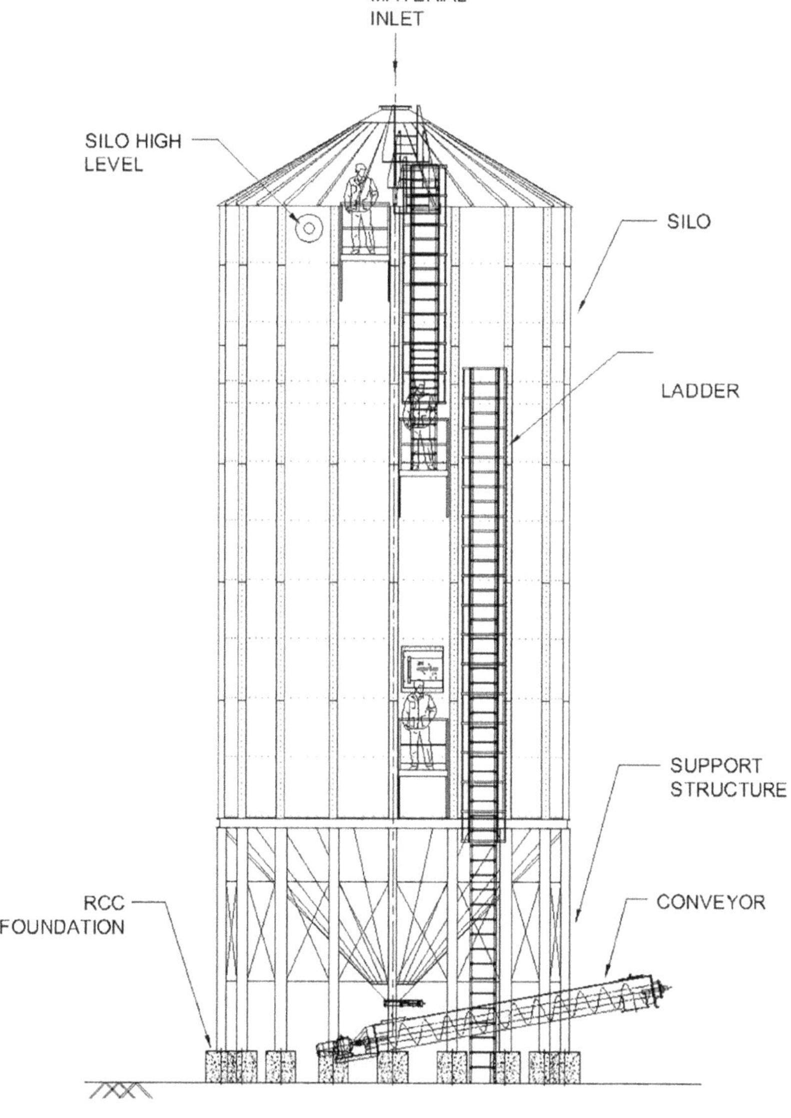

FIGURE 3.6 Day Silo. Courtesy – Fowler Westrup (India) Pvt. Ltd.

as against the flat bottom storage silos. The day silo serves the useful function of ensuring constant feed to the crushing plant. As explained in the chapter on process control, a Rotary Valve fitted below the hopper bottom, with a variable speed drive, allows to set uniform material flow.

3.3 SEED CLEANING

Cleaning of seed, to remove impurities from farm, as also metal impurities from loading unloading operations and storage, is necessary for trouble-free operation of machines in oil mill. All the machines operate in continuous mode, receiving and discharging seed continuously. Hold up of seed on the machines is very small.

3.3.1 Size Separation

Seed Cleaners are size separator machines, equipped with decks of vibratory screens. Each deck consists of two screens, upper screen with bigger opening to allow the seed and small impurities to pass through and to remove the 'overs' such as stones, farm chaff and twigs. Lower screen holds the seed and removes soil particles. The seed is discharged through an aspiration channel to remove light particles such as seed husk, small chaff, etc. Some machines are fitted with an air cyclone at top to remove the light chaff at entry itself (Figure 3.7). Each screen is fitted with a 'ball frame' below to hold 'jumping balls' of rubber or plastic, which keep striking the screen to prevent choking of the screen openings.

Slope of the screen decks may be set for optimum screening operation. Higher slope allows more throughput, but at the cost of separation efficiency.

Some machines come in 'open decks' configuration. But, in modern oil mills, 'closed cabinet' type cleaners are preferred (Figure 3.8), to avoid the escape of dust to atmosphere.

The light impurities, sucked from the aspiration channel, are separated in an air cyclone, or a bag filter, and dust-free air is let out to atmosphere. Air coming out of cyclones may contain up to 100 mg fines per m^3 of air. Bag filters can reduce the fines to nearly 20 mg/m^3 and are now preferred over air cyclones.

Screen decks, consisting of wooden frames, do need routine maintenance, and hence are kept easily accessible from the feed side. Sufficient free space must be provided for removal of the decks in the plant layout. Some newer machines are fitted with aluminium frames; these require less maintenance.

3.3.2 Magnetic Separation

After size separation, the next cleaning operation is usually magnetic separation. Magnets are of two types, permanent magnets and electro-magnets. Strips of permanent magnets may be fitted in discharge chutes of elevators, or at the inlet to heavy downstream machines such as Cracker and Flaker; these are periodically cleaned to

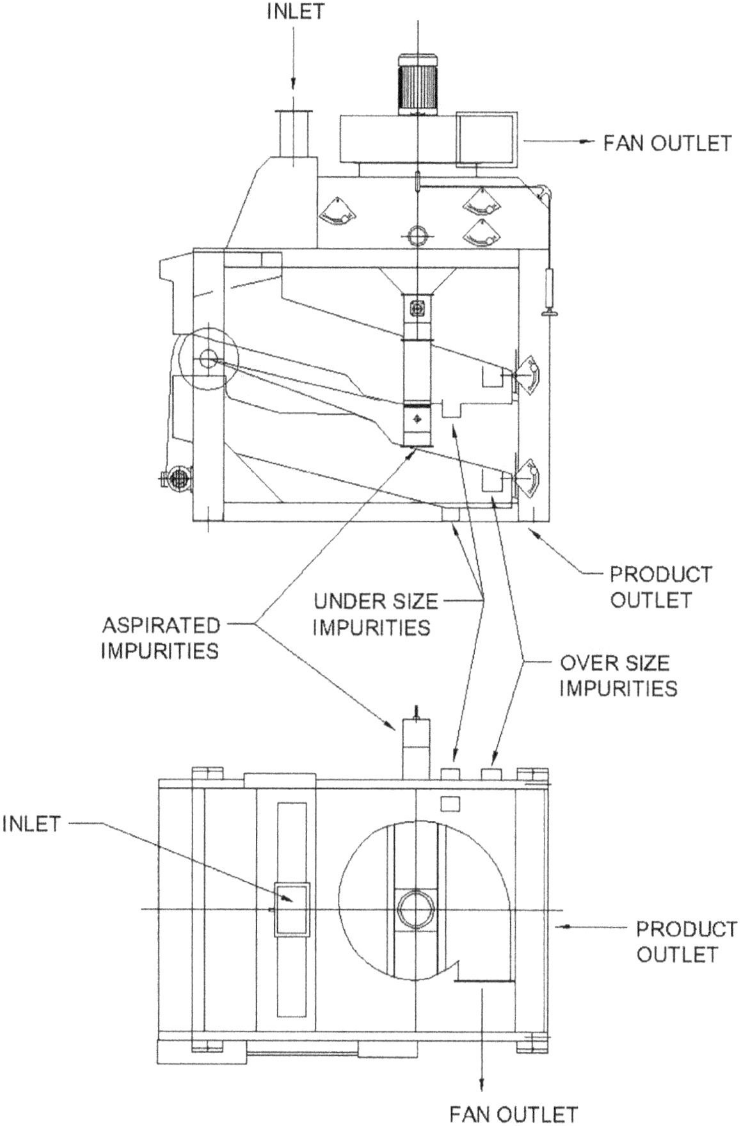

FIGURE 3.7 Seed Cleaner. Courtesy – Fowler Westrup (India) Pvt. Ltd.

wipe off accumulated iron impurities. **Rotary Drum Magnets**, fitted usually with electro-magnet, are capable of removal of iron impurities on a continuous basis and do not require manual cleaning (Figure 3.9). These are a permanent feature in modern oil mills.

FIGURE 3.8 Seed Cleaner (Closed Cabinet). Courtesy – Fowler Westrup India Pvt. Ltd.

3.3.3 Gravity Separation of Stones

Medium size seeds, soybean and sunflower seed, also require an additional cleaning step, namely Destoning. This is to remove stones, and balls of mud, similar to the size of the seed. Should these not be removed, the mud balls would break into soil particles in downstream handling, and may cause dust in the plant building; and the stones would damage surfaces of preparation equipment such as Seed Crackers and Flakers. Advantage is taken of the higher density of these stones and balls, and density separation technique is employed.

Destoner machine comprises a sloping screen, which allows air to blow through the mass of seed on screen, closed by an outer body (Figure 3.10a). The screen is vibrated to-and-fro thereby pulling the heavy particles resting on it upward. The seed, however, being lighter than stones, is fluidised above the screen, thanks to the air draft, is not subjected to the pull, and flows downward by gravity (Figure 3.10b). Thus, stone and mud balls are removed from the feed-side, and the clean seed discharges from the other end.

FIGURE 3.9 Rotary Magnetic Drum. Courtesy – Buhler A.G.

Exhaust air from destoners normally does not contain much fines. Thus, air cyclones suffice as safety equipment, and air bag filters are not necessary.

3.3.4 Layout of Cleaning Section

Seed cleaning machines come is sizes from 50 Tpd to 600 Tpd. Thus, a battery of machines is required for plant capacities of 1,000 Tpd upwards. A few suppliers, notably Buhler Inc., have now come up with a large Cleaner machine suitable for soybean up to 2,000 Tpd.

The three machines, cleaner, magnet and Destoner, are beneficially arranged one below the other, to allow flow by gravity and to avoid intermediate conveying elements. It must be noted that in a solid handling plant, much of the breakages, and hence downtimes, come from the mis-operation of conveying elements. Hence, it helps to minimise the number of conveyors and elevators.

The fans and air-cyclones, connected with the cleaners and Destoners, may be placed on the top floor of the preparation building; or sometimes may be accommodated on an extended platform just outside the building.

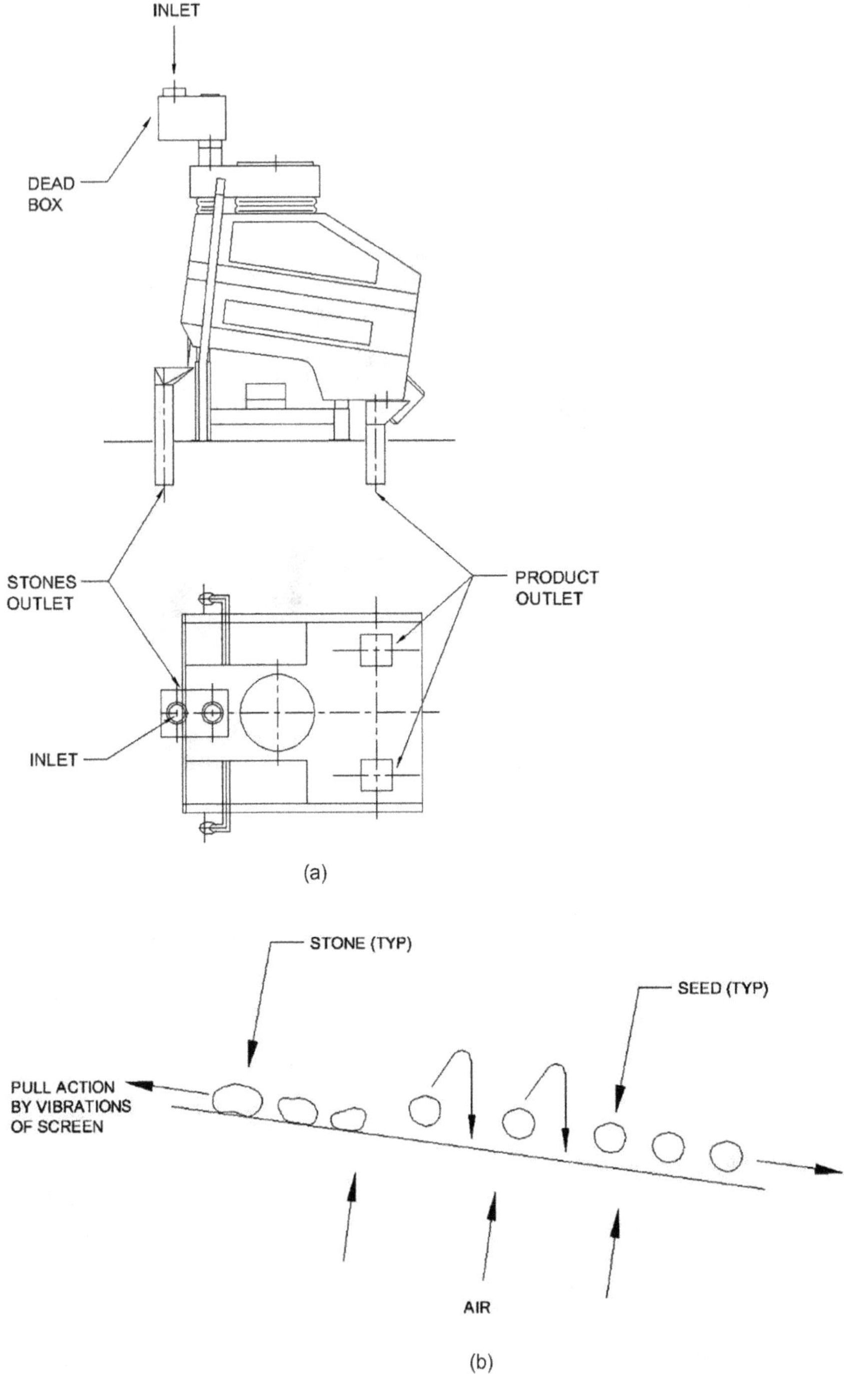

FIGURE 3.10 (a) Destoner. Courtesy – Buhler A.G. (b) Movement of Stones and Beans on Destoner Screen.

After the operations of size separation, magnetic separation and density separation, seed is thoroughly clean and is ready for unit operations of seed preparation.

3.4 ON-LINE WEIGHMENT

Seed should be weighed to keep track of material intake. Weigh-Scales are normally installed after the cleaning section, so as to avoid clogging due to chaff and twigs. High-capacity weighers may be installed in the silo section, downstream of pre-cleaning machines. In this case, the presence of some impurities, namely, small twigs and chaff, does not affect performance, since gate openings are quite large for high throughput capacity.

Online weighers are mainly of two types: belt weigher (Figure 3.11) and hopper weigher (Figure 3.12). Belt Weighers are fitted with load cells and totalisers which calculate the mass flow rate. Belt weigher is a simple and economic device. Hopper weighers are semi-continuous type, wherein seed flows in continuously to the top hopper, is taken batch-wise into the second hopper fitted with load cells, and released in batches to the third hopper. The top and bottom hoppers are buffer hoppers. The inlet to the first hopper and outlet from the third hopper are continuous, whereas the flow in between is batch-wise. The hopper weigher is more accurate than a belt weigher, thanks to the batch weighing principle. Hopper weighers are more commonly employed in modern oil mills.

A Hopper Weigher is designed for a set batch-weight. This is accomplished by initial quick intake into the second hopper to fill up to 80–90% of the set weight, followed by a slow intake up to the set batch-weight. The first hopper is fitted with two slide valves, of different openings, operated with pneumatic cylinders, for the purpose. As soon as the set weight is achieved, feed is cut off by a closing slide. Gate to the third hopper is opened, and the entire load is dumped. Gate to the third hopper

FIGURE 3.11 Online Belt Weigher. Courtesy – IPA India.

FIGURE 3.12 Hopper Weigher. Courtesy – Buhler A.G.

is closed then, and gate from the first hopper opened, for receipt of the next batch of seed into the weighing hopper.

A hopper weigher should be fitted with an air balancing line, to prevent pressure shocks caused by quick initial intake from first to the second hopper, and for the sudden release of seed from second hopper to third.

A Hopper Weigher may also be operated for Feed Control purpose, by providing a large first Hopper above, and setting the batch time. If the incoming feed rate is higher than the set value, material would accumulate in the first hopper, giving high-level alarm. That would stop the incoming conveyor and prevent over-feed.

The on-line weighers are normally fitted with a display panel showing the dynamic batch-weight, total number of batches and weight of seed for a given time, as also the flow rate. The data may also be transmitted to a remote panel.

3.5 SEED CRACKERS

Size reduction is an important step in seed preparation. Except for very small seeds such as rapeseed/mustard or sesame, all seeds are broken into pieces for better heat penetration during pre-heating or cooking operations. Cracking is also a pre-requisite for dehulling of soybean.

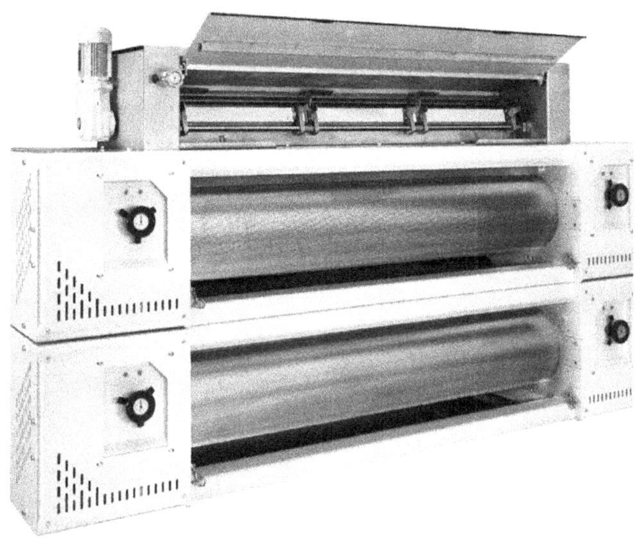

FIGURE 3.13 Seed Cracker. Courtesy – Buhler A.G.

The most widely used machine for size reduction is the Cracker mill, comprising of one or two pairs of heavy-duty corrugated rolls (Figure 3.13).

Cracker rolls are made from cast steel, and are chill-cast to achieve the hard surface depth of nearly 15–20 mm, hardness Br 500–550. Casting operation may be static or centrifugal. Roll pairs are mounted in a strong body, with suitable bearings, and are driven by one or two motors, through a set of pulleys and v-belts.

As seed is cut between the rolls, the action creates outward force on the rolls. Hence, rolls are pressed inwards with heavy-duty spring washers. In larger machines, the pressing, and shock absorption, may be effected with hydraulic cylinders, in place of spring washers.

A feed hopper is mounted on top of the mill to hold seed buffer, to ensure a steady feed to the machine. The hopper is fitted with a small-dia roll, at the bottom, which enables uniform feed over the length of the rolls. An adjustable damper helps in the regulation of flow to the mill. A level sensor may be fitted to sense the level of seed in the hopper, to help ensure that the mill does not run empty.

A full-length permanent magnet is located below the hopper, close to the curtain of seed falling into the roll pair. The magnet is fitted on hinges and can be turned aside to scrape off metal particles. As a good practice, this may be done at the beginning of every 8-hour shift.

Double-pair machines are most common when preparing the seed for direct solvent extraction. Even 3-pair designs are also available, and may be used for large-sized seeds such as palm kernel. To prepare for pressing, however, normally a single-pair action is sufficient. The rolls are corrugated to help cut the seed without too much shattering (creation of fines). The corrugation size is determined by the seed size and by the size reduction desired. A range of corrugations are in practice

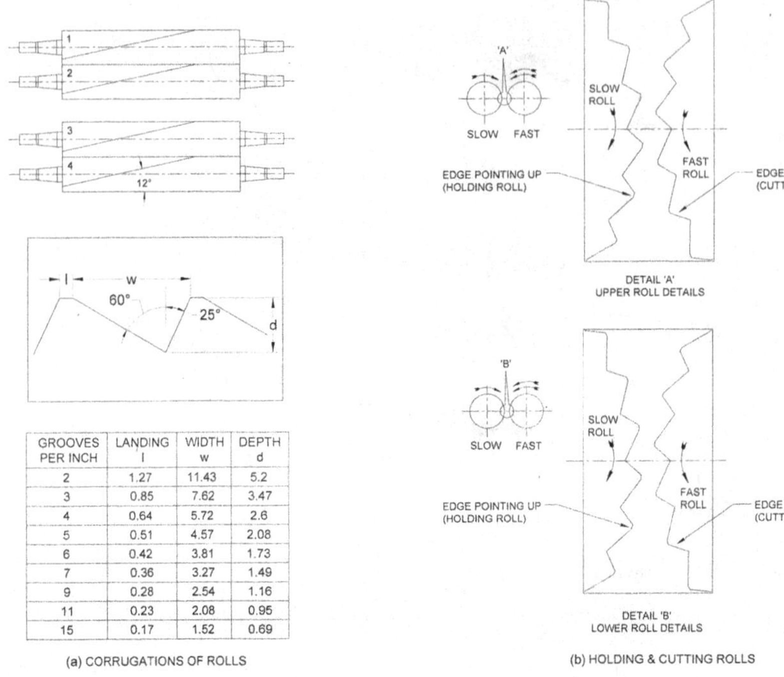

GROOVES PER INCH	LANDING l	WIDTH w	DEPTH d
2	1.27	11.43	5.2
3	0.85	7.62	3.47
4	0.64	5.72	2.6
5	0.51	4.57	2.08
6	0.42	3.81	1.73
7	0.36	3.27	1.49
9	0.28	2.54	1.16
11	0.23	2.08	0.95
15	0.17	1.52	0.69

FIGURE 3.14 Cracker Rolls Profiles.

(Figure 3.14a) for various seeds. The gap between rolls also may be set to optimise the size reduction.

Within a pair of rolls, there is a '**holding**' roll, with its edge pointing upward, to carry the seed; and the other is '**cutting**' roll, with its edge pointing downwards, to impact the seed (Figure 3.14b). As both the rolls rotate downwards, the cutting roll which is rotating at a higher speed than the holding roll cuts the seed.

A speed differential of around 1:1.3, between holding and cutting roll, suffices for effective cutting action. However, for 'Warm Dehulling' of soybean, this differential may be much higher, around 1:1.6. Should the cutting and holding configuration not be maintained while assembling the rolls, seed may be simply crushed instead of being cut; this would cause increase in load on the motor and also lead to shattering (fines creation). Individual rolls may rotate at speeds between 250 and 500 rpm.

As the seed is cut into pieces, the bulk density reduces. To account for the reduced bulk density, as also to eliminate the risk of material accumulation, the lower pair rolls (in double-pair Crackers) are rotated at higher speeds, nearly 25%, than the upper pair rolls.

Seed Crackers are set to crack the seed down only to an optimum size range, not to create very small pieces. Small pieces do not yield to effective cell-wall rupture during flaking, as described in the section on Flakers. Very small pieces, or fines, may also get overheated in cookers, and thus may impart dark colour, and over-cooked flavour, to oil and meal.

Crackers are heavy-duty machines, with strong body plates. Crackers are prone to vigorous vibrations, and noise, if proper care is not taken during machining and assembly. The rolls must be dynamically balanced to minimise vibrations. Mounting the machine on strong vibration pads helps reduce the vibrations.

As the roll corrugations wear out, the rolls may be re-corrugated. Two or three re-corrugations may be allowed until the roll diameter is down by up to 25 mm max.

Specific energy consumption of 2-roll Crackers may range between 0.6 and 0.8 kWh/T, depending on the hardness of the seed, and on the size of machine. Consumption for 4-roll Crackers would be between 1 and 1.5 kWh/T.

Seed Crackers are available in sizes ranging from 200 Tpd to 1,500 Tpd, typically with two drive motors rated at 9 kW to 50 kW each. Rolls are 1–2 m long, dia 300–450 mm. Smaller machines are available and are used for special applications.

Rolls need to be re-grooved once or twice a year, depending on usage and type of seed processed. For example, when whole sunflower seed is cracked, the grooves wear out fast and may have to be re-worked more frequently. Modern Crackers, for example the latest OLCC series from Buhler, come fitted with easy roll dismantling systems to facilitate the change-over of rolls during maintenance.

3.6 PRE-HEATERS AND COOKERS

3.6.1 Types of Equipment

Both these equipment are similar in construction. These may either be multi-stage, vertically stacked, vessels (Figure 3.15); or rotary drums. Heating medium is steam, given in jackets for indirect heating, usually at medium pressure of 8–10 BarG.

Pre-heating times are short, around 10–20 min, for various seeds (Chapter 2). Hence the **pre-heater** vessels are also relatively small. The heating temperatures are also much lower than for cooking, and no moisture removal is effected. Thus, pre-heaters are not equipped with aspiration fans. In fact, steam may be injected directly into the grits, in case of low moisture of seed.

Pre-heating may also be carried out in **Vertical Seed Dryers**, especially in cold countries in North America. The Dryer section is used for the purpose, and aspiration in Cooler section is stopped. Pre-heating temperature, in these cases, can be lower at just 45–50 C, and hold-up time be much longer, a few hours.

Cookers, on the other hand, are designed for much longer residence times, and may be larger in size. However, in most cases, where multiple screw presses are employed, each Press has its own Cooker mounted on top. Thus, individual cookers may, in fact, be smaller vessels than the single Pre-heater.

3.6.2 Rotary Drum Cooker/Dryer

At times, common cookers may be installed, instead of individual cookers on top of each press. The design of such large cookers may be the multi-stage vertical type, or horizontal Rotary Drum type (Figure 3.16). The latter is especially used in case of seed with high moisture. The shell of a rotary drum cooker may be made up of steam-heated coils, which offer large heating surface area, combined with large

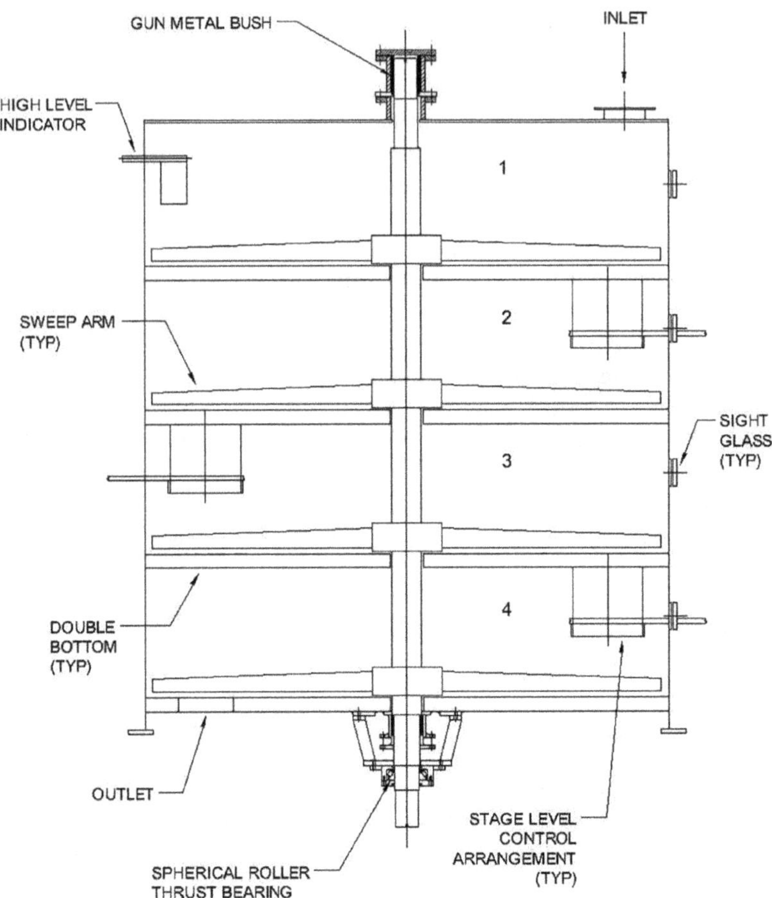

FIGURE 3.15 Vertical Cooker.

central vapour space to enable moisture removal in a short time. For applications with little drying requirement, the drum wall may be a jacketed one, in place of tube-wall. On the other hand, for extensive moisture removal duties, tube bundles may be fitted inside the drum for providing large heat transfer area. The drums rotate to lift and mix the grits/flakes. A slight inclination of the drum enables transportation of grits to discharge end. Alternative designs use stationary drums, with internal rotating shaft with paddles for mixing and transportation of seed, but these require higher energy and are mostly used for small capacities.

In view of the better heat penetration with higher moisture-in-seed, cooking times may also be shorter than in a vertical multi-stage cooker. Power requirement is also much less compared to vertical multi-stage machines, specific energy requirement being around 2 kWh/T. The major drawbacks of this design are the requirement of large floor space and difficulty in maintaining vapour seals at product inlet and outlet, which limit their usage for multiple cookers on presses.

FIGURE 3.16 Rotary Drum Cooker.

3.6.3 Vertical Cooker

These are multi-stage, stacked vessels, each stage heated with steam given to jacketed bottom. A central shaft passing through all the stages, with sweeping arms attached in each stage, sweeps the material on the bottom hot plate and also stirs the material bed. By the action of the arms, heat is transmitted to the entire material bed. Each stage has a gate for the passage of material to the stage below. Grits/flakes flow continuously from top to bottom.

Vessel size ranges from dia 2 m to 5 m, with 4–10 stages. Motor power for rotation of the central shaft may be 15–200 kW. Specific energy consumption is nearly double that for rotary drum cookers, in the range of 3–4 kWh/T, depending on the cooking time required. It may be noted that the vertical cooker helps break the flakes' structure further, thanks to the action of sweep arms. Part of energy input to the shaft is utilised for this purpose.

Individual stage height is normally 700–800 mm, with material bed height up to 400 mm (Figure 3.17). Deeper material beds are not advised, due to the limitation of heat transmission by way of conduction and slow mixing. The clear space above the material bed is the vapour space, and also provides for the fitment of level float mechanism. For cookers designed for moisture reduction, the vapour space also helps to allow the flow of air.

Cooker shells are built of mild steel, while the jacketed bottoms are preferably built with boiler-quality steel. Heavy-duty bearings are provided on both ends of the central shaft, and an additional thrust bearing is fitted at one end. Drives may be mounted on the top cover, suitably stiffened with structural members, for small cookers. However, for larger machines, drive is usually located below, supported on a separate foundation. A hydraulic coupling, fitted between motor and gearbox, helps restart a cooker in fully loaded condition. Shaft speed may typically be in the range of 10–20 rpm, lower speed for larger dia cookers. Thus, gearbox turn-down ratio is

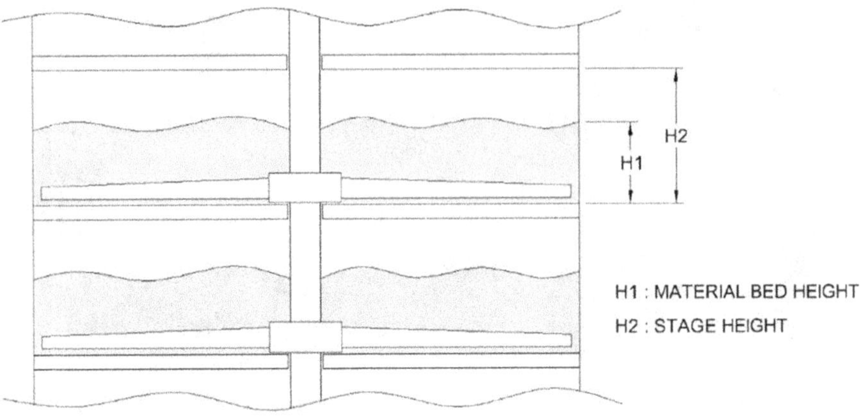

FIGURE 3.17 Material Bed Height in a Cooker.

high, @ 70–150. These are typically, three or four-stage helical gearboxes with a bevel gear for a vertical output shaft.

The **heating bottoms** are also known as double-bottoms, as these are made of two plates, with space in between for passage of steam. Should M.P. (medium pressure, 8–10 Barg) steam be used for heating, as is the normal practice, the space between the plates is kept between 30 and 40 mm. If L.P. steam (low pressure, less than 4 Barg) is used instead, the space should be increased up to 50 mm, or even 60 mm for cookers dia greater than 4 m. The upper plate should be of suitable thickness, to maintain flatness within 6 mm over the entire surface, to ensure uniform sweep of material on the plate. If the plate is uneven, pockets of material may stagnate in 'lows', would char due to long contact with hot plate, and affect product quality. To facilitate the spread of 'open steam' through the material bed, holes may be drilled in the hot plate, as described later in this section.

The **sweep arms** are thick plates, welded to the boss on the shaft at an angle, to facilitate the stirring of the entire material bed (Figure 3.18). In some designs, sweep arms may be cast together with the boss. Clearance between the lower edge of the arm and the hot plate should be kept minimum, certainly less than 20 mm, to ensure a clean sweep of the entire material on the hot plate. Angle should be steep enough for effective agitation of the bed; however, too steep an angle causes high motor load. Thus, an optimum angle range is to be determined. For most applications, the angle range of 15–20 deg is useful. For vessels dia larger than 3.5 m, it helps to have sweeping arms with inside bend (Figure 3.18b) to prevent accumulation of more material at the shell, due to centrifugal action. For such large dia vessels, arm plates are beneficially stiffened with a box to prevent deflection of the tip of the arm due to material load near the shell (Figure 3.18c).

It is useful to fit a **sight glass** on each stage to check the material level and extent of agitation. The common practice is to have round sight glasses, with a view dia 200 mm. However, it is better to have vertical sight glasses, up to 500 mm high, to provide a better view of material in case of fluctuating levels.

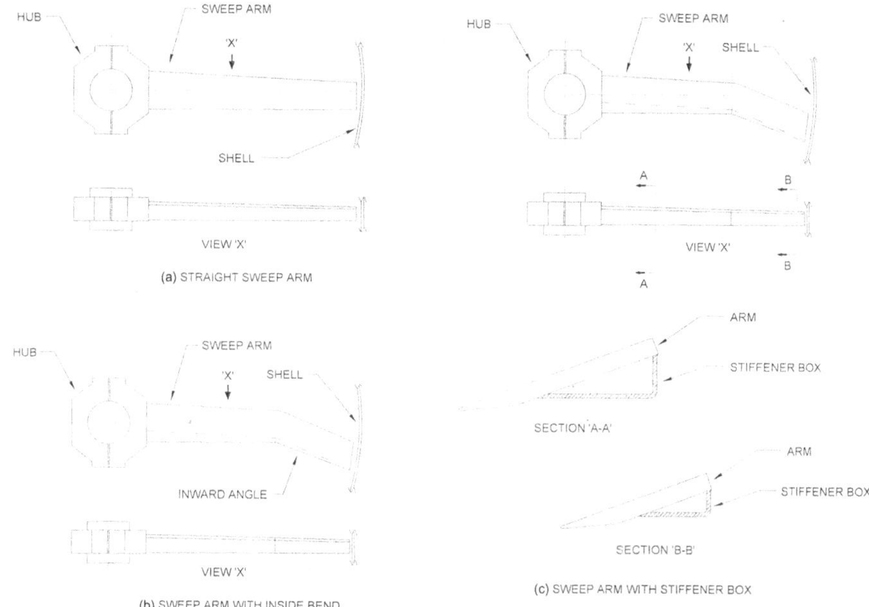

FIGURE 3.18 Sweep Arms (Vertical Cooker).

Cooker drives maybe mounted on the top cover plate for small cookers. But for larger equipment, drives also become heavy, and are preferably mounted below, supported on a floor. The main components of a cooker drive are the electric motor and a gearbox. The motor is commonly a 4-pole, giving shaft speed of close to 1,500 rpm. The gearbox has a high turndown ratio in the range of 70:1–100:1, for the output speed of 15–21 rpm. A safety component is introduced in the form of a 'fluid coupling' between the motor and the gearbox, to safeguard the main drive components against damage from shock overloads. The fluid coupling also allows re-start of the cooker in fully loaded condition. The output shaft of the gearbox is coupled with the cooker shaft with another coupling which is designed for high torque and allows minor mis-alignment, such as a 'gear coupling'.

Weight of the cooker shaft, and the material thrust on sweep arms, is absorbed by a thrust bearing mounted either at the bottom or the top of cooker. The other end of the shaft is held by a radial bearing.

In large cookers, a small **ancillary drive** may be provided to allow emptying of the cooker in case of power failure. The power rating of the drive may typically be 10% of the main drive, so that it may operate with a small diesel generator; with a higher gear ratio to run the cooker sweep arms at a very low speed (less than 1/10th of nominal speed), just to empty out the cooker within a reasonable time period. Such an arrangement prevents overheating, and charring, of material in the cooker in case of power failure. Such a drive, a gearmotor of output speed around 150 rpm, is connected to the main gearbox, through a clutch which may be put on when the drive is to be used.

3.6.4 Level Control

Material level can be known with a **level sensor mechanism** fitted on each stage. The most common mechanism consists of a float, fixed to a shaft (level shaft), resting on the material (Figure 3.19a). As the material level changes, the float moves, rotating the level shaft. The shaft rotation can be seen outside the vessel, by means of a pointer attached to the shaft. Once the position of the float is calibrated with the position of pointer outside the vessel, the pointer indicates the material level in the stage at all times.

Level sensing may also be digitised, using an electronic rotation sensor fitted on the level shaft; or be converted to a pneumatic signal with a pressure transmission valve.

It is important to regulate material levels so that heating/cooking time is maintained as desired. A simple mechanism to maintain the levels is to attach a closing plate to the level shaft, which may close the gate from the stage above when float is lifted to the required level (Figure 3.19b). This is a simple mechanism and helps maintain a pre-fixed level of grits in the stage. Pneumatic or electronic mechanisms for level control are preferred when the set levels are to be varied, either for different seed feed rates or for different cooking times.

In **pneumatic systems**, a pressure-sensing valve is used for sensing the level. Rotation of the level shaft exerts pressure in proportion to the float position inside. This pressure signal is transmitted, after conditioning, to a pneumatic actuator, either a proportional valve or a pneumatic cylinder attached to the discharge gate. The actuator opens the discharge gate in proportion to the level. Thus when the level is low, pressure signal is low, and the gate opens only slightly. On the other hand, when the level rises high, pressure signal increases, and the actuator opens the gate more, to let more material out, to bring the level down to the pre-set value. In **electronic systems**, a rotation sensor sends out signal in 4–20 mA range to a PID controller. The conditioned signal from the controller varies the speed of discharge Rotary valve. In modern plants, electronic systems are now becoming more and more common.

The control systems described above, both pneumatic and electronic, are of the **direct-acting** type. Level signal from a stage regulates the gate below, higher level opening the gate more, and vice versa. The other type is **'reverse-acting'** control. In this mechanism, the level signal from a stage would throttle the feed gate from the stage above. Thus, a higher level would open the feed gate less, and a lower level would open it more.

The **reverse-acting mechanism** is preferred for most cookers, especially those mounted on individual presses. The press load signal may regulate the final discharge gate, or screw conveyor, below the bottom stage of cooker; the level signal in the last stage would act on the gate above, and so on. If a direct-acting control were used, the action on the last gate, from the level signal of that stage, would be in conflict with the control action of the press.

The mechanical 'closing gate' mechanism described at the beginning of this section is also a reverse-acting type.

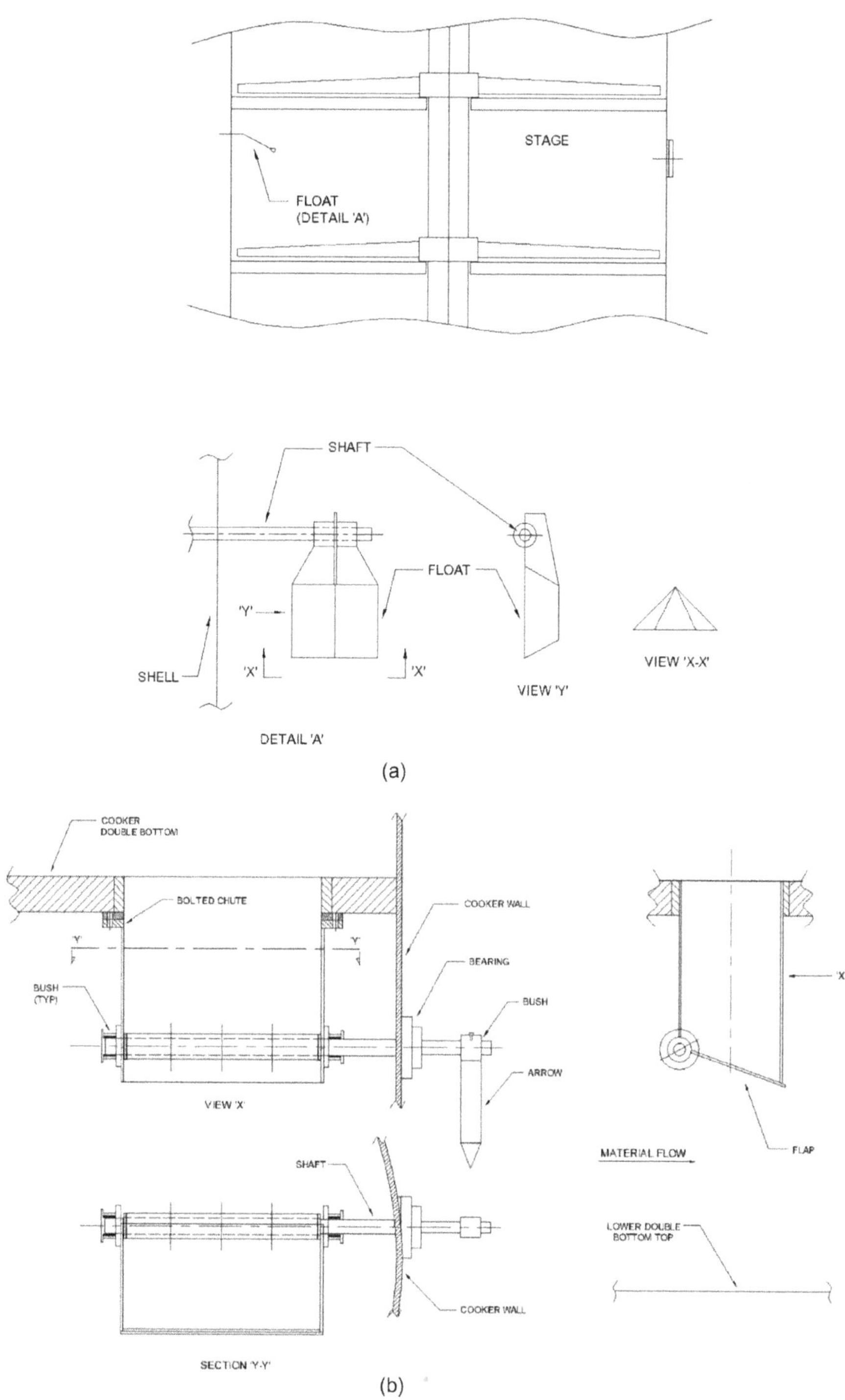

FIGURE 3.19 (a) Float Arrangement. (b) Level Flap Arrangement in Cooker.

One challenge in auto-level-control in cookers, or pre-heaters, is that the float keeps jumping up and down every time the sweep arm passes below it; this movement creates **false signal**, not related to the actual material level. This false signal, or 'noise', must be filtered out for the effective functioning of the control system. In pneumatic control systems using pneumatic valve actuator, the air buffer in the valve absorbs these momentary 'shocks', so the actuator action is not affected. In electronic systems, however, an electronic filter to remove the high-frequency 'false' signals is required to condition the input signal.

3.6.5 Temperature Control

Apart from level sensing and control, **temperature** of the material must be controlled, for desired process objective. Control system for this purpose consists of a sensor (RTD), signal conditioner (PID controller), and an actuator (Pneumatic proportional valve). Temperature of the last, or penultimate, stage of the cooker is sensed, the milli-ohms signal is conditioned in the controller as per the PID settings, and the valve opens in proportion to the signal received from the controller. The valve opening determines the steam flow to the heating bottoms of all the stages of cooker. Usually, manual valves are provided to steam lines going to all stages. It is normally a good practice to keep these valves fully open, and let the control valve regulate the amount of steam going to the cooker stages.

Open steam may be injected into the material, depending on the moisture content of seed, as described in the previous chapter. This may be required more in pre-heaters and not in cookers. For even distribution of the steam through the material, holes may be provided in the double-bottom (Figure 3.20) covering the entire cross-section, and steam be let in just below the double-bottom to let it flow upwards through the bottom plate and then through the material bed in the upper stage. Alternatively, steam may be injected through nozzles fitted along the length of sweep arms (Figure 3.21). In this case, steam has to pass through the central shaft, and connection is given to the box of sweep arm. A simpler alternative is to connect a pipe to the shaft and run it along the sweep arm (Figure 3.21b); however, this pipe is subject to wear and may need to be replaced every second year.

Open steam should be injected in upper stages, so that benefit of the higher moisture is derived in lower stages. The steam injection is also a quick way to raise bed temperature. Open steam injection may be **controlled** by sensing temperature in upper stage, where steam is being injected, and giving the conditioned signal to actuator steam valve.

3.6.6 Cooker Design

Unlike almost all other machines in seed preparation plant, there are no standard models of cookers available from machine manufacturers. A plant supplier, who aggregates all the machinery and does plant engineering, has to design the cooker for the tonnage of specific seed, or multiple seeds, and duty (prepress or full press, or preparation prior to solvent extraction).

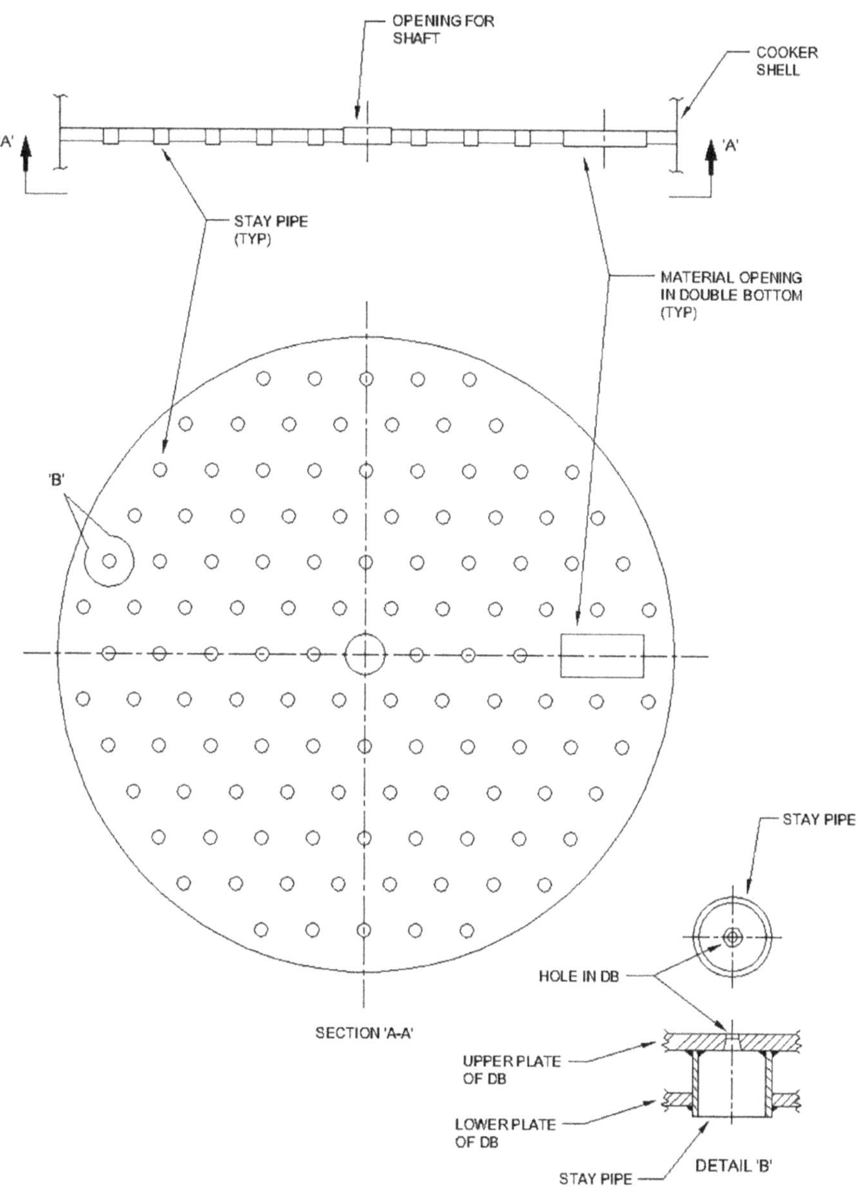

FIGURE 3.20 Holes in Double Bottom of Cooker.

3.6.6.1 Determination of Cooker Size

Two main design parameters that need to be fixed are the cooking temperature and the total hold-up time. In case of a multi-seed operation, with varying requirements of temperature and time, the maximum temperature and the maximum time need to be determined. The value of max. temperature enables the calculation of the required

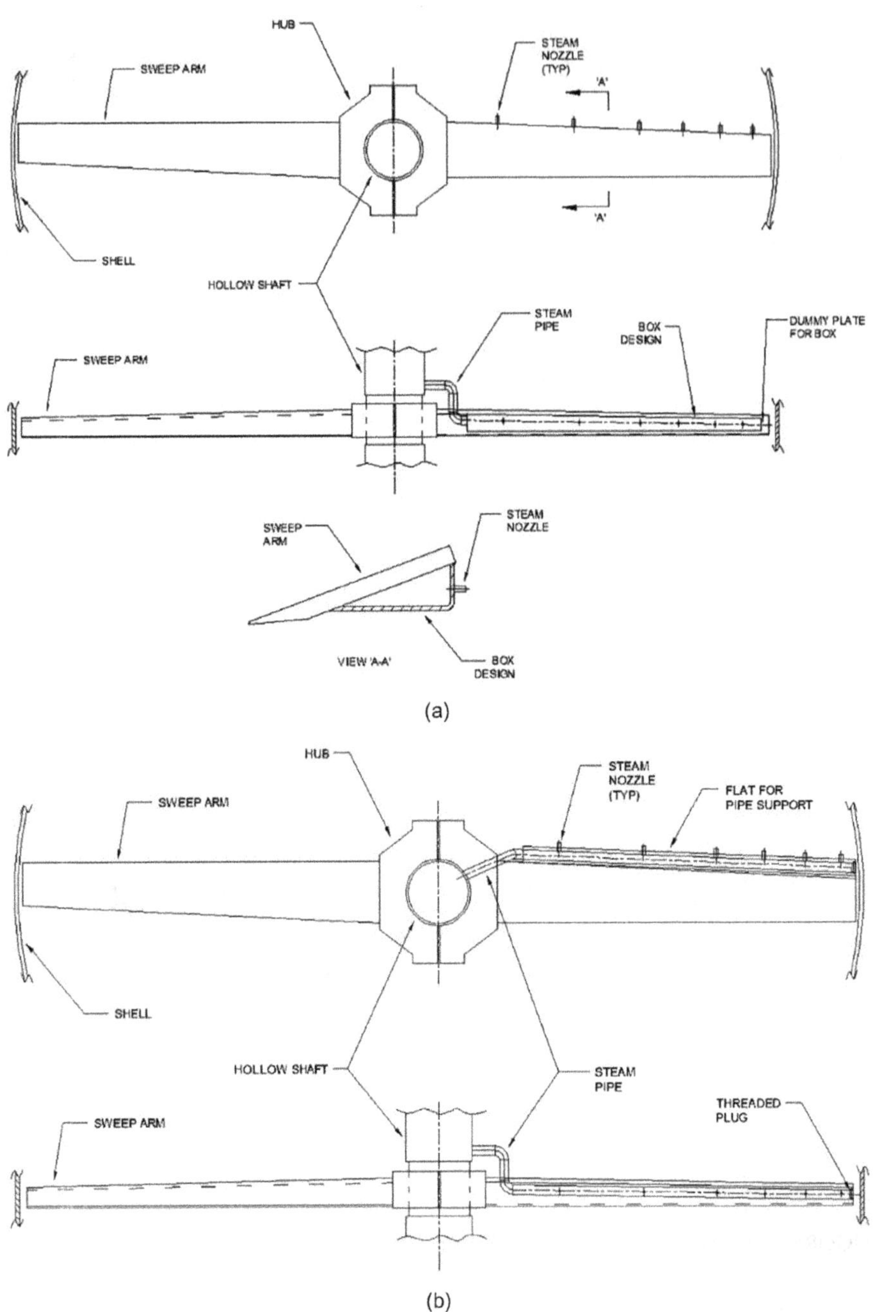

FIGURE 3.21 Open Steam through Nozzles on Sweep Arm. (a) Steam passage through Sweep Arm; (b) Steam passage via Pipe Runner.

heat load (kCal/h), and the dwell time gives the material volume (m^3) required inside the cooker.

For longer hold-up times (also called dwell time), more than 30 minutes, it is preferable to first determine the overall size for the required dwell time. Volume requirement is calculated from the dwell time, tonnage per hour, and bulk density of grits/flakes. Once the volume is known, then considering material bed depth in each stage, say, 350 mm, the total surface area is calculated (for stages where live steam is injected, bed height may be up to 500 mm). From the total surface area, multiple solutions, in terms of combinations of the number of stages and cooker dia, are obtained. From the multiple solutions, the combination which gives the ratio of cooker height:dia in the range of 1.25:1 to 1.6:1 is selected. It is to be noted that very high ratio can mean many more number of stages, and more number of level control systems to be maintained. However, very low ratio may mean large dia, higher motor power, and higher likelihood of uneven heating surface leading to non-uniform product quality.

Once the number of stages and dia are determined, possible heat transfer rate (for the available steam pressure) is checked against the required heat load. In case the possible heat transfer is less than the required rate, size may be increased as necessary. In most cases, however, the possible rate would be higher than the heat load. In such cases, the temperature control system may be depended upon to reduce steam pressure as required.

For dwell time of 25 min or shorter, it is preferable to first calculate the heating surface area requirement from the heat load. From the surface area requirement, cooker dia and number of stages is determined as explained above. Possible volume of material in the cooker is then calculated and checked against the volume requirement for the tonnage. Should the volume fall short of the requirement slightly, deeper material bed in each stage may be considered, but not more than 400 mm; otherwise, cooker size may be increased as required. Should the volume be higher than required, material height in stages may be reduced.

One important parameter required for the calculation of heat load is the specific heat of oilseed grits. The grits may be seen as a combination of three main components, namely, oil, moisture and solids. The individual specific heats of the three components may be considered as 0.5 (units: kCal/kg/C), 1.0 and 0.35, respectively. The specific heat of the grits may be calculated by the additive method.

3.6.6.2 Cooker Drive Power

Power consumption for multi-stage cookers may be computed by the following empirical formula:

$$\text{Power} = f_p * (\text{no. of stages} * \text{sweep arms per stage}) * \text{shell dia} \tag{3.1}$$

where, power is in kW, and shell dia is in meters. The power factor f_p maybe considered as 0.3 for most flakes, and 0.35 for the oily mass of ground copra, when the angle of sweep arm is not more than 20 degree, bed height is around 400 mm, and shaft rpm is around 15 rpm. For 20–22 rpm, values of 0.4 and 0.45 may be considered, respectively.

The drive motor rating may be selected at 10–15% higher than the calculated power consumption. Service factor for the selection of gearbox may be 1.6–1.7.

3.6.7 Moisture Conditioning

When moisture reduction is to be achieved in cooker, air aspiration is necessary. For small moisture reduction, say, less than 1.5%, air at ambient temperature (30C minimum) may suffice. For higher reductions, however, warm air @ 100C may be used to minimise air quantity. For very large moisture reductions, as in the case of copra where moisture may have to be reduced from 7% to 3%, hot air @ 120C may be used. The quantity of air, and steam requirement for air heating, is calculated using the psychrometric chart for air. Once the air quantity is known, the pressure drop through the cooker system is estimated. This enables finalising the specifications for **air fan**. The air ducts and fan are preferably made of stainless steel (SS), to prevent corrosion.

For seed materials containing much fines, an air cyclone (SS) may be installed between the cooker and the fan. However, for most cooker duties, cyclone is not necessary. Air is sucked from (ambient air), or blown into (warm air), preferably the lower stages of cooker. This is to allow cooking at higher moisture in upper stages, and moisture reduction in lower stages before entry to the Press. Warm air is obtained with steam radiators.

Ambient air may simply be sucked from the lower stages. Suction from individual stages may be regulated by a manual (butterfly) valve, and all suction pipes may be connected to common duct going to the fan, usually mounted on top of cooker.

Warm air, on the other hand, is blown into the stages. For proper contact with flakes, air should rise through the entire bed. This may effectively be done by blowing the air into an 'air chest' constructed below a stage (Figure 3.22). Air from the chest may rise through holes in double-bottom, and rise through the bed, before going out from suction connection. With good air distribution through the bed, and bed height of 400–500, relative humidity of outgoing air may reach up to 60%. If instead, the warm air is simply blown in above the bed at one side of cooker and sucked out from the other side, outlet humidity may just be 35–40%. In the latter case, much of the heat supplied to the air is unutilised, and air goes out at higher temperature. For outlet humidity of 60%, the steam requirement for air heating would be nearly 1.6 times the moisture removed. When air goes out at only 40% humidity instead, steam requirement maybe 15–20% higher, apart from higher electric energy for fan.

3.7 FLAKERS

Flakers tend to be one of the most expensive machines in preparation plant. They are also the heaviest, and consume more electric energy than other machines. The largest Flakers available today can flake up to 500 T seed per 24 hrs.

A feed hopper is mounted on top of the flaker mill to hold seed buffer, so as to ensure steady feed to machine. The hopper is fitted with a small-dia roll (or a paddle mixer) at bottom which enables uniform feed over the length of the rolls. An

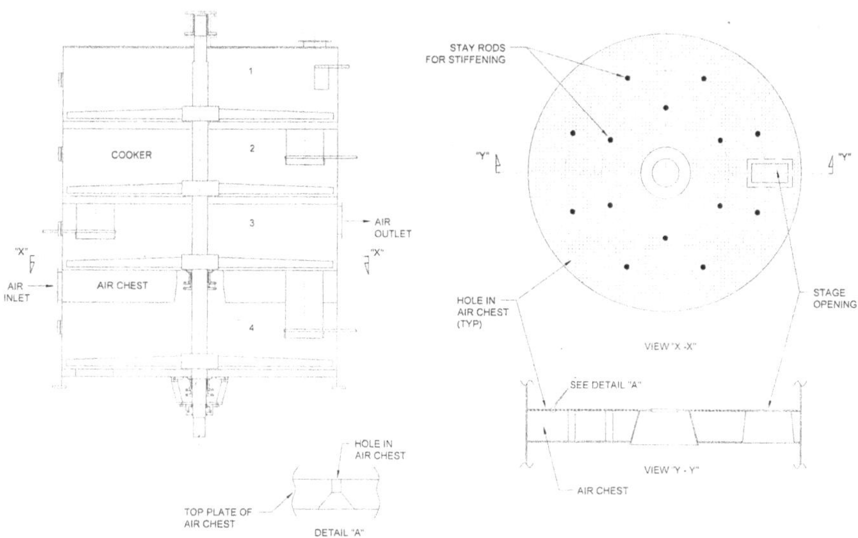

FIGURE 3.22 Air Chest below Cooker Stage.

adjustable damper helps in regulation of flow to the mill. A level sensor may be fitted to sense the level of seed/grits in the hopper, to ensure that the mill does not run empty.

A full-length permanent magnet is located below the hopper, close to the curtain of seed falling into the flaking roll pair. The magnet is fitted on hinges and can be turned aside to scrape off metal particles. As a good practice, this may be done at the beginning of every 8-hour shift.

Seed, or grits, are passed in between two rotating, heavy-duty rolls (Figure 3.23), to give instantaneous strong impact, to reduce the thickness of the grit down to 0.25–0.5 mm. The impact on the grits, in return, causes outward force on the rolls. Hence, the rolls are pressed together with great force by means of hydraulic cylinders. Pneumatic cylinders are also fitted in the line, to absorb shocks, created on account of passage of any hard material, say stones or even metallic particles.

Rotation speed of rolls is normally in the range of 280–320 rpm, smaller dia rolls speed being higher than larger dia rolls. One roll is set to rotate at 2–5% higher speed than the other, so that the rolls wipe each other, and also shear the grit to cause more rupture of cell walls. For flaking of abrasive materials such as sunflower seed, the roll speed may be reduced to around 225 rpm. Spring-loaded wipers are provided on the lower side of rolls to maintain the roll surface clean.

Rolls are made from cast steel and are chill-cast to achieve hard surface depth of nearly 15–20 mm, hardness Br 500–550. Casting operation may be static or centrifugal. Rolls are mounted in a strong body, with suitable bearings, and are driven by one or two motors, through a set of pulleys and v-belts. The flaker rolls are much heavier compared to those of Seed Cracker. Roll surface area required for equal throughput is nearly four times that required for Crackers.

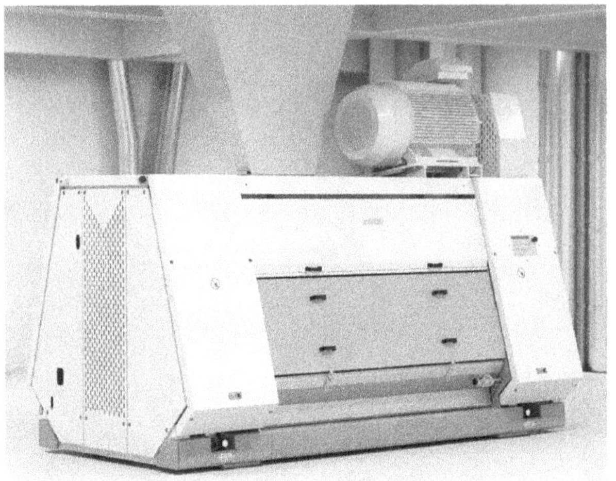

FIGURE 3.23 Flaker. Courtesy – Buhler A.G.

It is imperative that no grit by-passes the rolls and falls between rolls and the body. **Side Sealing** plates are fitted near both the edges of the rolls for this purpose. As the sealing plates prevent fall of grits on the roll edges, the edges are not subjected to wear due to flaking action. Thus, over a period, as the roll dia reduces over the working length, the side 50–80 mm edge becomes '**high spots**'. Then, there is a danger of the edge 'high spots' on opposing rolls touching each other. If machine is run in this position, the edges can hammer against each other and may develop cracks.

To avoid this problem, edges are ground regularly, and the diameter at the edge is maintained less than dia on working length by nearly 0.1 mm. **Edge grinding attachment** is provided on machines for the purpose. It is a good practice to carry out edge grinding once every week, At times, the working length of rolls also may undergo uneven wear, making the surface uneven. This can be checked visually, as also with a 'straight edge'. The indication to uneven wear can be had by checking the flake thickness on samples derived from different points along the roll length. Should the unevenness be greater than 0.1 mm, even when grits are being fed evenly along the length, it is time for grinding of full roll. **Full roll grinding attachments** (Figure 3.24) are also provided by all leading manufacturers; grinding is accomplished at low speeds (15–20 rpm), with rolls driven by a dedicated small geared motor.

A major safety built in the hydraulic power-pack, which regulates the pressure on the rolls, is to open the rolls by releasing pressure, in case of no feed which is indicated by low level in the feed hopper.

Modern flaking mills are provided with additional safeties, with display monitor on the door, such as a proximity switch, which indicates 'door is open'; it is interlocked with the main drive and does not allow to start motors in open-door condition. The motor amperages are displayed, along with the hydraulic pressure.

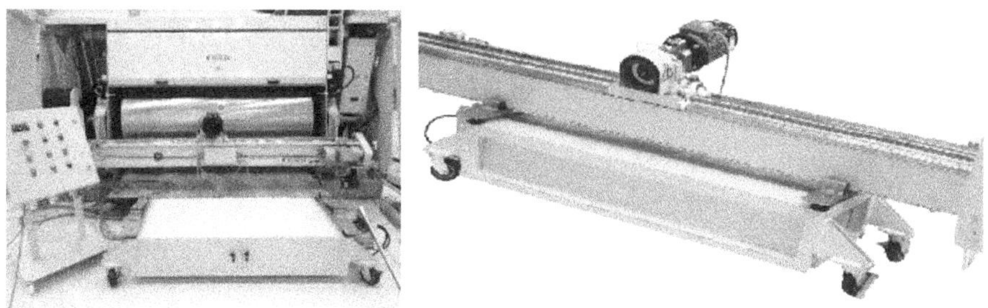

FIGURE 3.24 Full Roll Grinding Attachment. Courtesy – Buhler (India) Pvt. Ltd.

Flakers are available in capacities ranging from 100 Tpd to 500 Tpd, typically with two drive motors, each rated at 30 kW to 100 kW. Some of the modern Flakers come with a single motor. Rolls are 1,000–2,200 mm long, roll dia 500–600–800 mm. Smaller machines are available, and are used mainly for special applications, and for pilot plants. Specific energy consumption on industrial-scale machines is one of the highest among the preparatory machinery, typically in the range of 5–8 kWh/T.

Degree of Deformation

When oilseeds are first broken, using either Seed Cracker, or Hammer Mill, and then flaked, a question is often asked as to what is the desired particle size from breaking operation. Answer to this question lies in the 'degree of deformation' achieved during flaking (Figure 3.25). The degree of deformation (DoD) is simply the change in size before and after flaking. The higher the degree of deformation, the more is the work on oil cell walls. Of course, too high a DoD would mean too much power required on flakers. Karnofsky (1949b) refers to the DoD as 'distortion'.

Thus, small broken particles would be subjected to only a small DoD, and would not yield to high degree of work on oil cell walls. But, too large particles would not yield to adequate pre-heat treatment, since heat would not penetrate to the core. Thus, there would be an optimum size range of broken particles for desirable pre-heat and DoD. It is, therefore, advisable to break oilseeds to 3–4 mm size, prior to flaking.

3.8 GRINDERS

Hammer mills are commonly utilised for size reduction of large seeds, such as palm kernel and shea nut, as also large cups of copra. Hammer mills work on the principle that most materials will crush or pulverise on impact. Seed is fed through a chute into the mill chamber (Figure 3.26a). It is repeatedly struck by a number of hammers (thick metal flats) hinged on a shaft rotating at high speed inside the chamber. The seed is shattered by a combination of repeated hammer impacts, collisions with the chamber wall, and particle-to-particle impacts. A wide perforated metal screen, covering the bottom discharge opening, retains coarse material for further grinding, while allowing properly sized material to pass.

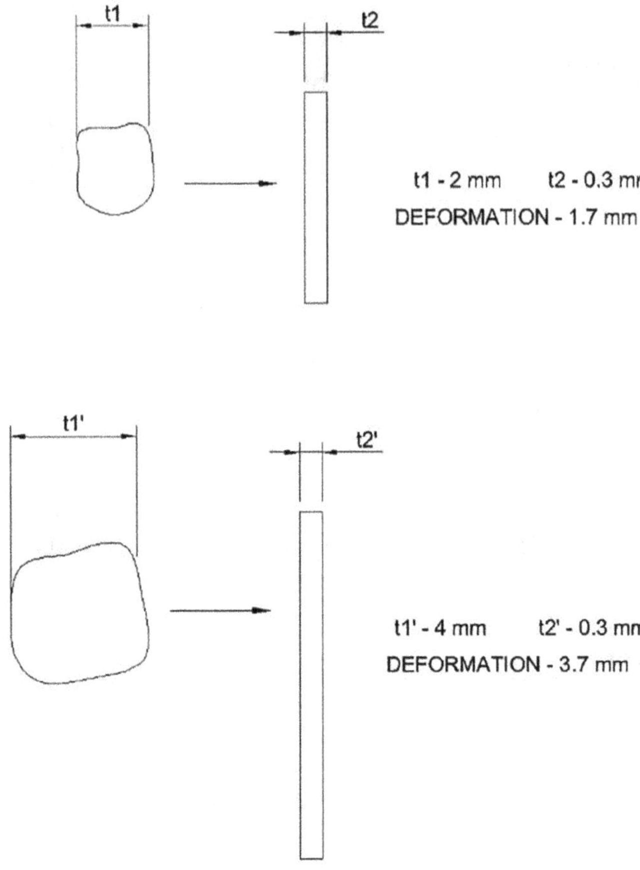

FIGURE 3.25 Degree of Deformation.

Grinders can perform well even with some fluctuations in feed rate. Thus, large feed hoppers, as necessary for Crackers and Flakers, are not required. However, a rotary feeder to help spread the feed over the width of feed chute does improve the product consistency. The spread also avoids excessive wear on only a few hammers and thus helps extend the period of change of hammer set. A permanent magnet fitted in feed chute helps guard the hammers from breakage.

Rotation speed may be selected on the basis of toughness of feed material, and the size reduction required. For fine product, high speeds are recommended. Also for tough seeds, such as palm kernel and shea nut, high speeds are necessary, in the range of 1,200–1,500 rpm. On the other hand, for soft and brittle materials, such as balls of de-oiled meal, low speed, nearly 400–500 rpm, suffices. It should be noted that higher speeds increase motor loads, and also tend to create much fines.

Grinders are heavy-duty machines, with the chamber built with thick steel plates. Hammers are made from wear-resistant materials, alloys containing chromium, or work-hardening steel like manganese steel. When the hammering face is worn out,

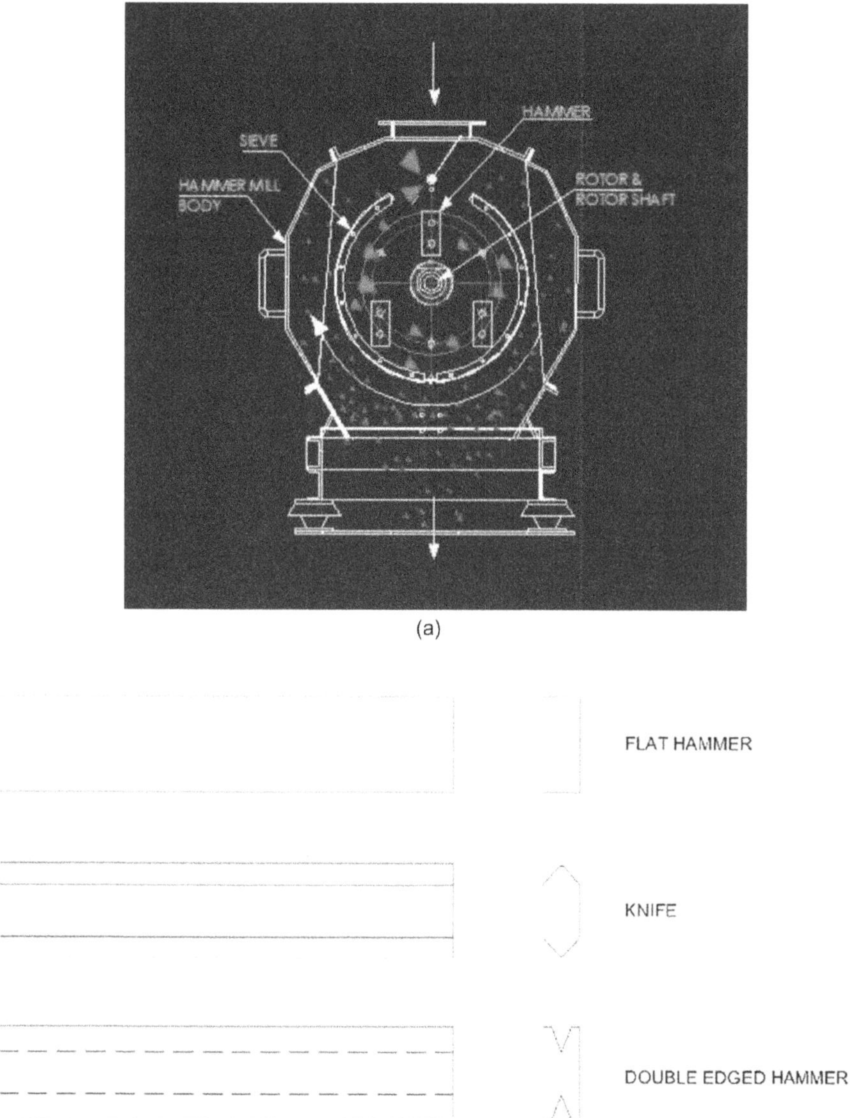

FIGURE 3.26 (a) Grinder (Hammer Mill). Courtesy – Spectoms Engineering Pvt. Ltd. (b) Knives and Double Edged Hammers.

the flat is just flipped to use the rear face. Once the rear face wears out, the hammer is flipped to hinge the worn-out end and use the intact end for hammering action.

Screen opening size is chosen as per the product size required. Typically, a 10 mm circular opening would give particle size below 6–7 mm. When bigger particle size is desired, more than say 15 mm, as in the first stage grinding of copra cups, screen may be replaced with just a bar gate, with round bars to minimise the resistance to outflow of product. Screens should have the holes closely spaced as possible, to allow maximum open area. Wide spacing of holes increases motor load and reduces mill capacity. Such 'opaque' screens also tend to increase the production of fines, since ground material does not necessarily pass the screen the first time.

Care must be taken to maintain clearance between the tip of hammer and the screen around 8–10 mm. Too close a clearance would increase motor load unnecessarily. Wider clearance would cause accumulation of material at the screen, which would eventually cause clogging of the mill.

When feed material is not brittle, but is plastic or rubbery, sharp-edged hammers, called knives, or double-edged hammers, may be used (Figure 3.26b). Here size reduction is achieved by a combination of cutting action and impact.

Light feed materials, or very fine product size, do not readily yield to gravity discharge and tend to clog the screen. Pneumatic suction discharge is preferred in such cases. Product is sucked into a bag filter, from where it is discharged intermittently. In vegetable oil plants, such a need arises while grinding hulls, or grinding de-oiled meal into a fine flour.

Hammer mills are prone to high vibration and resultant noise. These should be built robust, with dynamic balancing of the rotor. Mounting on strong anti-vibration pads would help reduce the vibrations. Machines should have a bolted cover, which would provide easy access to the rotor and the screen. The rotor is driven by a motor through pulleys and belts, which serve as the flexible component to absorb shock loads.

Hammer mills are available in capacities up to 200–300 Tpd, depending on material hardness and the amount of size reduction required, with drive motors rated up to 150 kW. Specific energy requirement is almost as high as for flakers and ranges between 4 and 8 kWh/T, depending on the hardness of material and the extent of size reduction required.

Hammer mills, together with Flakers, tend to be the most energy-consuming machines in an oil mill. So, adequate attention must be paid to energy-efficiency while selecting a hammer mill. Modern hammer mills also focus on ease of maintenance, namely ease of replacement of screen and hammers.

3.9 DELINTER AND DEHULLER (COTTONSEED)

3.9.1 Delinter

The main components of a Delinter machine are a saw drum and a grate, with the saw teeth projecting out of the grate. As the seed slides on the grate, the high-speed saw teeth shear the lint off the seed (Figure 3.27).

Feed to the machine is regulated by a magnetic Feed Roll, with a variable speed drive. Together with an adjustable damper plate, the roll regulates the feed rate closely, while also removing metal particles. The seed falls on a rotating Float Cylinder, which regulates the depth of the seed bed on the grate. Optimum bed depth ensures effective lint removal without compromising the throughput of the seed.

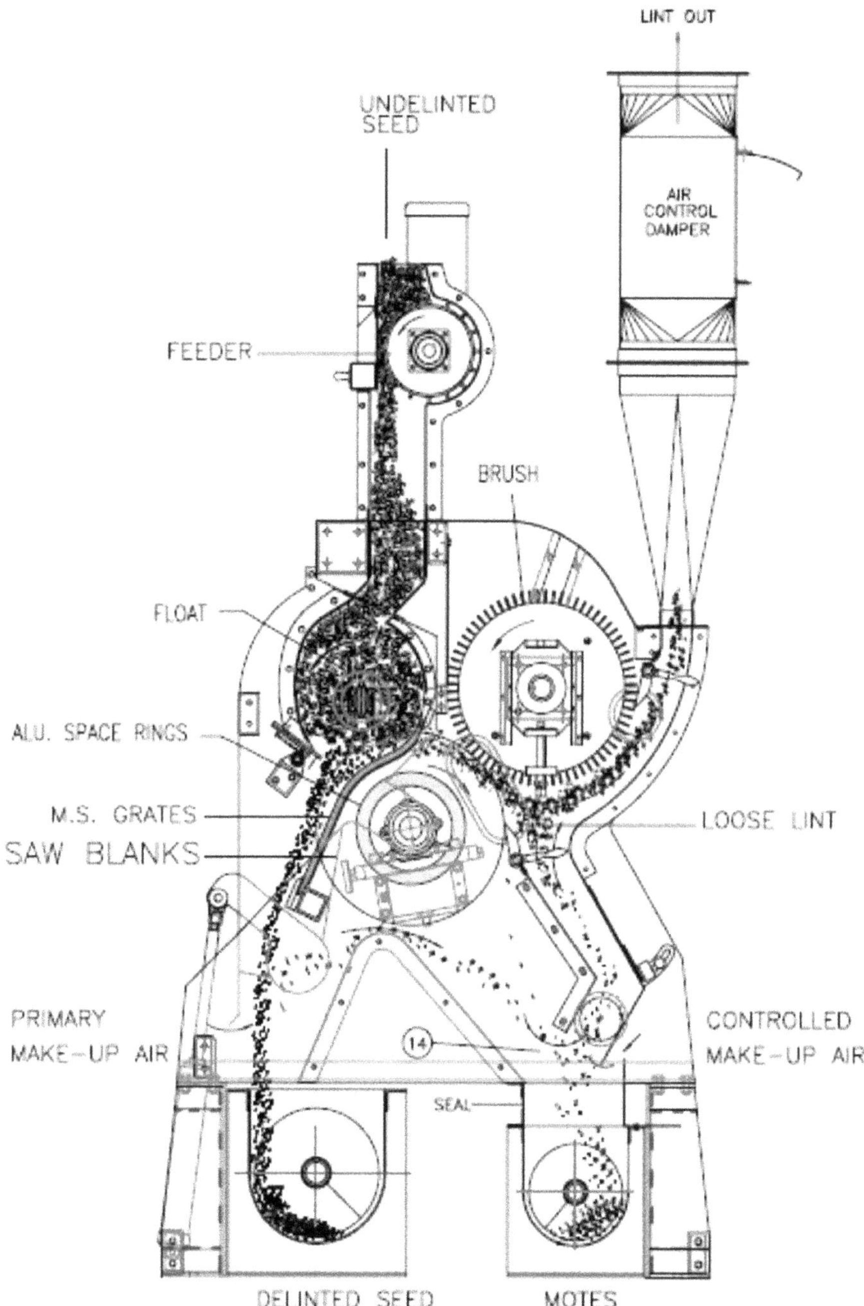

FIGURE 3.27 Delinter. Courtesy – Cottor Plants (India) Pvt. Ltd.

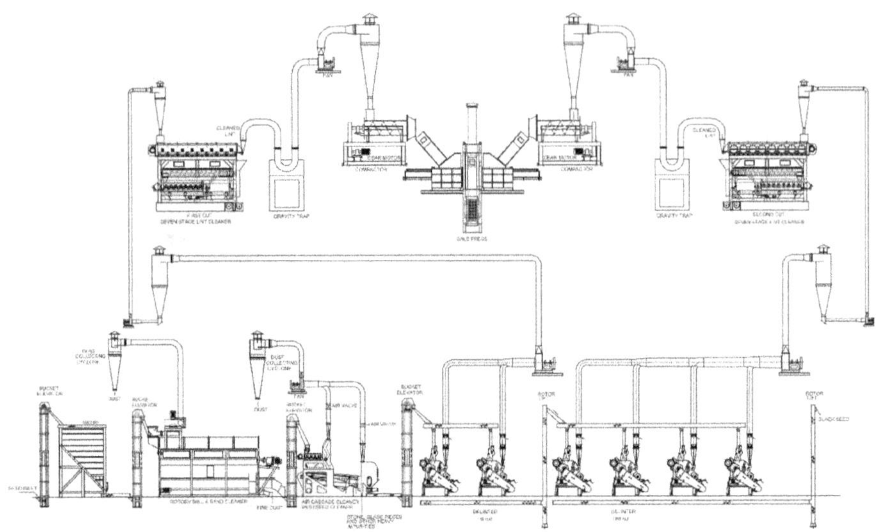

FLOW DIAGRAM FOR COTTONSEED DELINTING PLANT.

FIGURE 3.28 Delinting Flowsheet. Courtesy – Cottor Plants (India) Pvt. Ltd.

As the lint is removed from the seed by fast rotating saws, the partially delinted seed flows out by gravity, the output being regulated by a damper' rake' that is moved vertically to set the opening along the grate. The lint is wiped off the saw blades by a rotating brush. The lint is then sucked off pneumatically, cleaned and pressed into bales. The first cut delinting recovers nearly 2–2.5% lint, out of the 11–12% on white cotton seed.

The partially delinted seed is subjected to second stage delinting (Figure 3.28). Machines construction is the same as for the first stage, but number of machines is twice as much. The output damper 'rake' is pressed down to restrict the seed outlet. Here, additional 4% lint is recovered, thus leaving nearly 5–6% lint on the seed.

The first and second stage machines may also be set to recover nearly 10% lint, to produce 'black cotton seed', at a much-reduced throughput. The black cottonseed is used as planting seed.

The heart of delinting process is the rugged Saw Drum equipped with removable saws and aluminium spacer rings. The saw drum is designed and precision-machined to ensure high-speed movement between the thin clearances of the grate.

The saw cylinder requires regular reworking and sharpening of the saw edges. This is accomplished on a Saw Gummer. The unloading of the saw drum from the Delinter, fitting on the Gummer and re-loading on the Delinter, is mechanised to save time and also to minimise human error.

Delinters are low-capacity, high-energy consuming machines. Presently, the largest machines are sized to process 80–100 Tpd WCS on first stage, and just half of that on second stage. Specific energy consumption for first-stage delinters is as high as 9–10 kWh/T. Since the machine capacity halves on the second stage, the specific consumption doubles to 18–20 kWh/T.

It may be noted that until the end of the last century, first stage delinters could process only up to 20 Tpd WCS, and second stage only up to 10 Tpd. Thus, a large battery of machines was required in the seed house, resulting in a lot of manual work, apart from higher electric energy requirement. Thus, the new machines have heralded almost a revolution in the cottonseed industry.

3.9.2 Dehuller (Cotton Seed)

Twin cutting rolls are employed for dehulling of cotton seed. Impact dehullers, such as those used for sunflower seed, with a single roll and impact plate, are not preferred as these are prone to clogging by the lint present on partially delinted seed.

Feed to the Dehuller is regulated by means of a feed roll, with variable speed drive, and an adjustable damper. The uniform feed rate, together with uniform spread of seed over the entire length of the machine, ensures consistent performance of the machine. A high gauss-power permanent magnet plate is fitted just below the feed roll and allows trouble-free operation by trapping iron particles. The plate has to be cleaned periodically.

The heart of the dehulling process is the twin, rugged rolls equipped with huller knives (Figure 3.29). These twin rolls are driven at a small differential speed, to enhance the cutting process, thus minimising the amount of uncut and recycle seed. The differential speed mechanism, combined with a spiral lead on the cutting edges, allows high percentage of open 'hulled' seed, without the production of much fines.

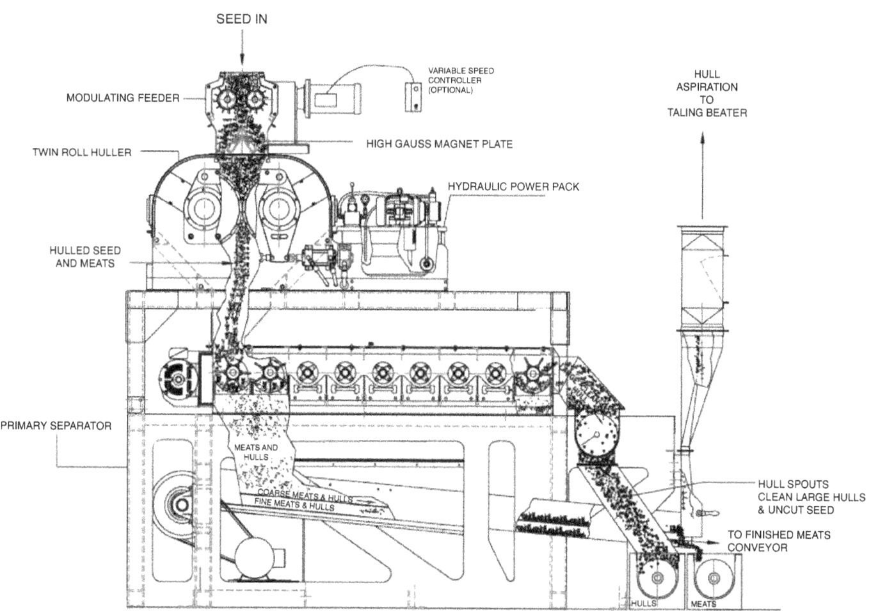

FIGURE 3.29 Cotton seed Dehuller. Courtesy – Cottor Plants (India) Pvt. Ltd.

The knives are removable, and the knife profile can be tailored to suit the specific requirement of the wide variety of cottonseed processed. Knives are made from a special tool steel.

Safety to huller knives is provided by means of hydraulic cylinders, used to maintain pressure on the cutter rolls, and relief valves that will instantly open the rolls automatically, should any metal pieces come in with the seed stream.

The Dehuller machine is mounted on top of a Primary Separator, where the meat (kernel pieces) is separated from hulls by screening, and the hulls are subjected to multiple mild beating, to recover meat fines from hulls which are collected at the end of the deck. The recovered meat fines are collected, by screening, together with the meat grits.

Hulls from the Primary Separator are divided into two parts, light hulls which are separated by air suction, and heavy hulls which have much meat attached. The heavy hulls are elevated to a Tailing Beater (Figure 3.30), where meat fines are recovered, and hulls are separated by air suction. Light hulls are subjected to slow, multiple beating (Hull Beater) to recover as much meat fines as possible. All the recovered fines are collected in a common conveyor with the meat grits, and are ready for preparation prior to oil extraction. Pure hulls, containing oil not in excess of 1% above the botanical oil content in hulls, are stored separately.

The amount of hulls separated by air suction on the Tailing Beater can be varied, depending on the protein content of final de-oiled meat desired. By throttling the suction, much hulls may be left in the meat; and by sucking more, more hulls would be removed.

Cottonseed dehulling is not as energy intensive as delinting. The machine capacities are larger, presently 200–300 Tpd, and specific energy consumption is much lower at around 2.5 kWh/T. Additional 2 kWh/T is required for the additional beaters and fans. Machine capacity reduces to 75% on white cottonseed, i.e. whole seed without delinting, and energy consumption increases by that order.

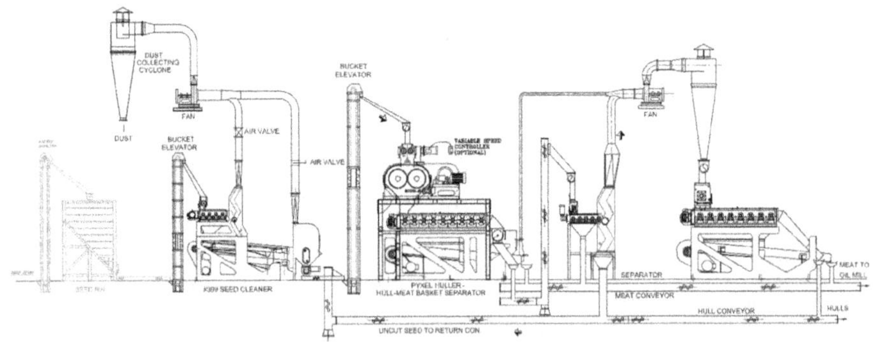

COTTONSEED CLEANING & DEHULLING PLANT.

FIGURE 3.30 Cottonseed Cleaning & Dehulling Flow sheet. Courtesy – Cottor Plants (India) Pvt. Ltd.

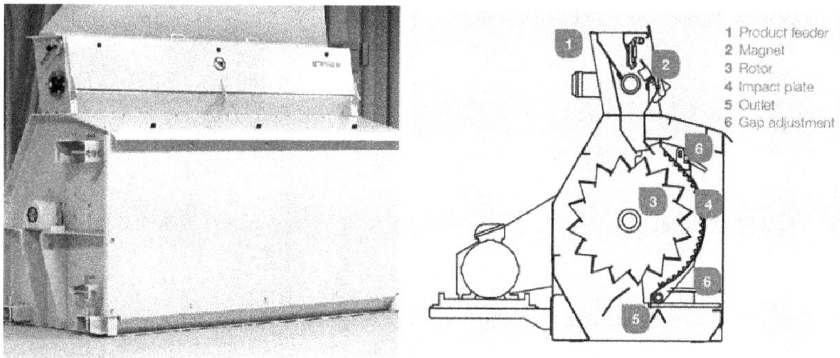

FIGURE 3.31 Impact Dehuller. Courtesy – Buhler A.G.

3.10 DEHULLER (SUNFLOWER)

Impact dehulling is practised for sunflower seed. This method gives better results in terms of less meat particles in hulls fraction and also consumes less motor power, compared to twin cutting rolls.

Impact Dehullers come in two types, vertical and horizontal (Figure 3.31). Vertical type is more versatile, can be used for other seeds, as also over wider moisture range, but requires more motor power per ton of seed, and also requires air aspiration.

Horizontal impact dehullers are sturdy, produce very little fines, do not require air aspiration; but may only be used with dry seed. With availability of large capacity seed dryers, horizontal impact machines have become the dehullers of choice for modern oil mills. These are medium-duty machines, equipped with a rotor fitted with knives that cuts open the seed and propels these against a static corrugated impact plate (LeClef and Kemper, 2015). The purpose is to detach the hulls from the meats with minimum formation of fines and minimum oil losses in the hulls fraction. This can be controlled by feed rate regulation and also by adjusting the position (inclination) of the impact plate towards, or away from, the rotor.

The detached hulls are separated by size separation, and air aspiration, on vibrating sieves (Figure 3.32). Two-stage separation removes more hulls and improves the protein content of the meats. The second stage separation may or may not be practiced depending on the protein content desired in de-oiled meal.

The separated hulls carry some meat particles, including those still attached to the hulls, and fine meat particles aspirated with hulls. Loss of meat particles means loss of protein and oil, hence these must be recovered. The recovery is achieved in two steps; (a) meat fines are recovered by size separation on vibrating sieve and (b) the attached meat is loosened, either by beating the hulls in a slow-rotating hammer mill, or by processing in second Dehuller, and then separated by screening. A good process will not leave more than 1% oil, above the botanical oil content of pure hulls, in the hulls fraction.

Horizontal impact dehullers are available for capacities up to 200 Tpd, with main motor of 20 kW. Thus, an oil mill designed to crush 1,000 Tpd sunflower seed requires five of these machines, and one or two for hulls purification.

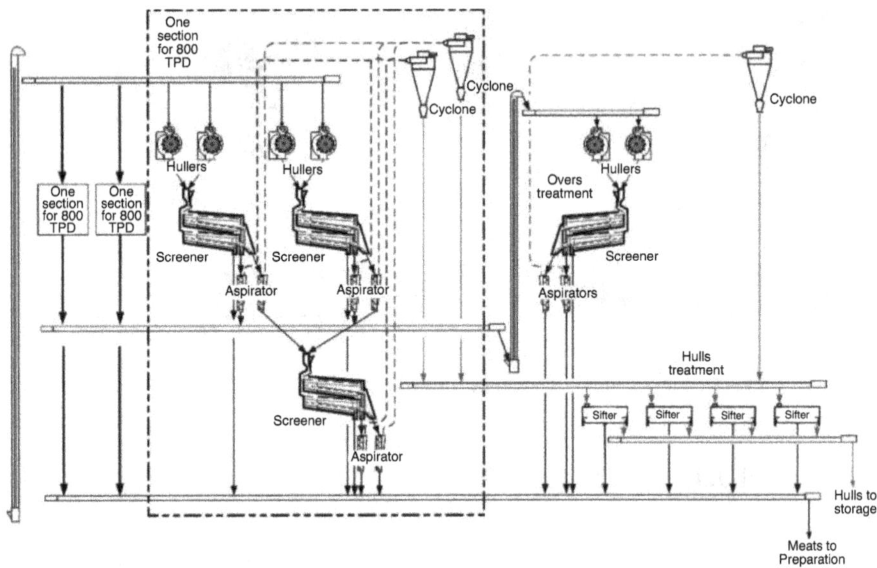

FIGURE 3.32 Sunflower Dehulling Flow Sheet (LeClef & Kemper, 2015).

With the multiple Dehullers, combined with multiple sieve separators, and hulls purification machines, the Dehulling section occupies major space in the oil mill. It is also an expensive section. For these reasons, small sunflower seed mills are usually not equipped with dehulling section.

3.11 EXTRUDER

Extruders are heavy-duty machines, consisting of a heavy-duty screw rotating inside a barrel (Figure 3.33). Material is fed at one end of the screw and discharged through a die-plate at the other end. The screw is typically supported only at the feed end, with a radial bearing and thrust bearing, and is balanced by the oilseed mass at the other end. For this reason, an extruder should never be run empty.

Extruders used in oilseed processing are high-speed machines, unlike plastics extruders which rotate at much lower speeds. This means the cooking times are very short, less than a minute. Due to the short time, external heating does not yield any significant advantage. The heat generated due to viscous dissipation, by conversion of mechanical energy, alone suffices for the required cooking effect.

Extruders are available in various sizes, with barrel dia 200–300 mm, barrel length 1.5–2 m, driven by motors rated for 100–300 kW, and may process up to 200 Tpd rapeseed. Energy consumption is almost as high as for Presses (prepress), in the range of 25–30 kWh/T. However, the main advantage of Extrusion pre-treatment is the lower oil-in-cake in Press cake, for full-press duty, apart from nearly a 50% increase in press capacity. For rapeseed, while normal OIC is 7%, it may be reduced to 5% with extruder pre-treatment.

FIGURE 3.33 Extruder. Courtesy – Anderson International.

3.12 PREPARATION PLANT LAYOUT AND CONVEYING SYSTEMS

3.12.1 Machinery Layout

Modern oil mills are organised in multi-level layouts, to benefit from gravity flow and minimise on conveying systems. It should be noted that conveyors and elevators are the most delicate equipment in an oil mill. The heavy process machinery, once erected and aligned properly, can run for weeks and months without any need for maintenance. However, conveying systems are prone to jamming, wear, overloads. Thus, minimising the conveying elements facilitates long-term trouble-free operation of the mill.

A distinctive feature of a modern oil mill is that all elevators are placed on ground and none underground. This means, there are practically no other machinery on the ground floor. All the heavy machinery are located on first and the second floors, making these the main operation areas (Figure 3.34). Conveyors receiving material from first floor machines, run just below the first floor. There are distinct advantages in this type of arrangement where the elevators are on ground, and not under-ground; it renders the elevators easy to monitor and maintain.

In old layouts, where main machines were placed on ground, conveyors tops used to be flush with the ground floor, conveyors would be in trenches, and elevator boots had to be underground. The pits used to be 2–3 m deep. Even where the pits were well-covered, it was a difficult job to keep those clean. In case of jamming of an elevator, the covers had to be opened, and the pits cleaned, before the elevator boot could be opened. This was a tedious job and took long time, often an hour or longer. For an over ground elevator, cleaning the boot is a matter of few minutes.

Having a multi-level layout means the foot print is reduced drastically. As the main machines, namely the cracker (or, dehuller), pre-heater, flaker and the Presses, are placed on the first floor, mill operation may be controlled by one operator with ease, with an assistant monitoring other machines, mainly the cookers, on the second floor. For automated operation, with remote-location control room, of course, there may not be any need for having an operator on the floor.

Since the heavy machines are elevated to upper floors, these floors must be built in RCC, to readily absorb vibrations.

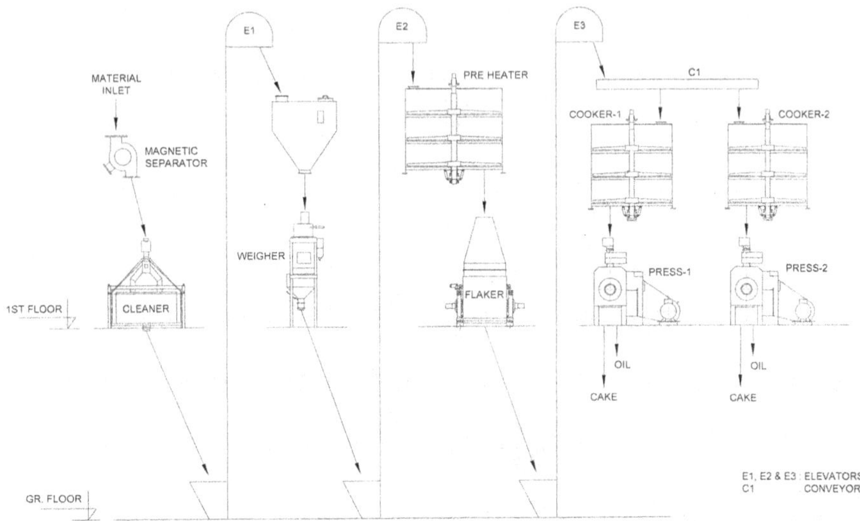

FIGURE 3.34 Preparatory Machinery Arrangement (Rapeseed).

3.12.2 Conveyers and Elevators

In small mills of yore, say with single machines of each type, conveyors used to be short. Mostly these were **screw conveyors**, which were rugged and simple to operate. With larger mills, however, the length of conveyors increased. For longer conveying, screw conveyors are not convenient. Long screw conveyor shafts must be supported with intermediate hanger bearings. These bearings obstruct the flow of materials. Thus, screws have to be designed with wider cross-section, with a low filling factor of 0.3–0.35 only. As the cross section increases, so does the power requirement, and also the wear on screw flights. Therefore, most conveyors within the seed preparation section, and overall in the oil mill, are of drag-chain type. However, screw without hanger bearing is preferably used for short length conveying and metering screws. The latter are material flow regulation devices, fitted with variable speed drives, and are usually constructed in tubular design.

Drag-chain conveyors can be built long, by bolting together successive sections. It is critical to align the sections perfectly; a task made easy with auto-drilling accomplished in laser-operated drill machines. These conveyors consist of a chain, moving within a casing of rectangular cross-section, carrying the mass along (Figure 3.35). The conveyor can have multiple inlets and multiple outlets. Casing should not be extended beyond the last outlet, to prevent material accumulation.

It is preferable to line the bottom of a chain conveyor with a soft material resistant to wear. Ultra high molecular weight polyethylene (UHMWPE) is such a material, and is widely used for the purpose, up to duty temperature of 80C. TEFLON is another such material, but is more expensive. For temperature duties more than 80C, stainless steel sheet, typically SS 304, may be used as liner. The UHMWPE liner significantly reduces friction between the chain and bottom sheet, as also between material and bottom & side plates, thus reducing the power requirement for the

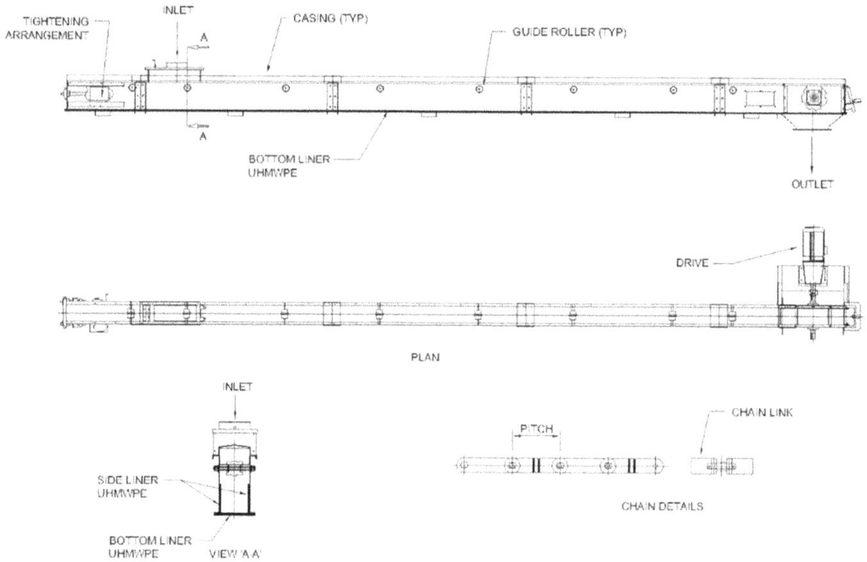

FIGURE 3.35 Drag Chain Conveyor. Courtesy – Eminence Equipment Pvt. Ltd.

conveyor, and also reduces wear of chain components. Such a liner also eliminates the risk of fire, due to sparking caused by metal-to-metal contact.

Another safety recommended for a drag-chain conveyor is an overflow switch. This may be fitted on a flap placed just above the last outlet. In case of obstruction in the outflow of material, the switch would go off, and stop the conveyor, saving it from damage to chain due to overload.

Chain conveyors may be designed for chain speeds up to 0.4–0.6 m/s. The lower speeds are used when material is to be conveyed gently to prevent friction on say, fragile cake. For horizontal conveying, for free-flowing materials like seeds, the material height may be up to 80% of width. Should the material not be free-flowing, such as flakes or de-oiled meal, the height may even be equal to the width. For straight inclne up to 15 degree, it may be reduced by that much percentage. For greaterincline, chain attachments are made taller, like an 'en-masse' conveyor. Some conveyors are configured with a bend, in between a horizontal part to collect materials from multiple machines, and an incline part to lift the material. The bend part should preferably have a long curvature, and a thin wheel should be provided to press the chain, to prevent chain lifting from the bottom plate. This is critical, to prevent material accumulation at the bend, to maintain flow capacity, and to avoid 'jerks' on the chain.

For conveyors with steep inclines, a central partition plate may be provided to make a carrying section 'box'. When a conveyor has a steep bend, the chain tends to lift and may put strong force on the pressing wheel. In such cases, a heavier chain, with inverted buckets as attachments, may be used to reduce the chain lift (Figure 3.36). Such conveyors are commonly known as 'bulk flow' conveyors.

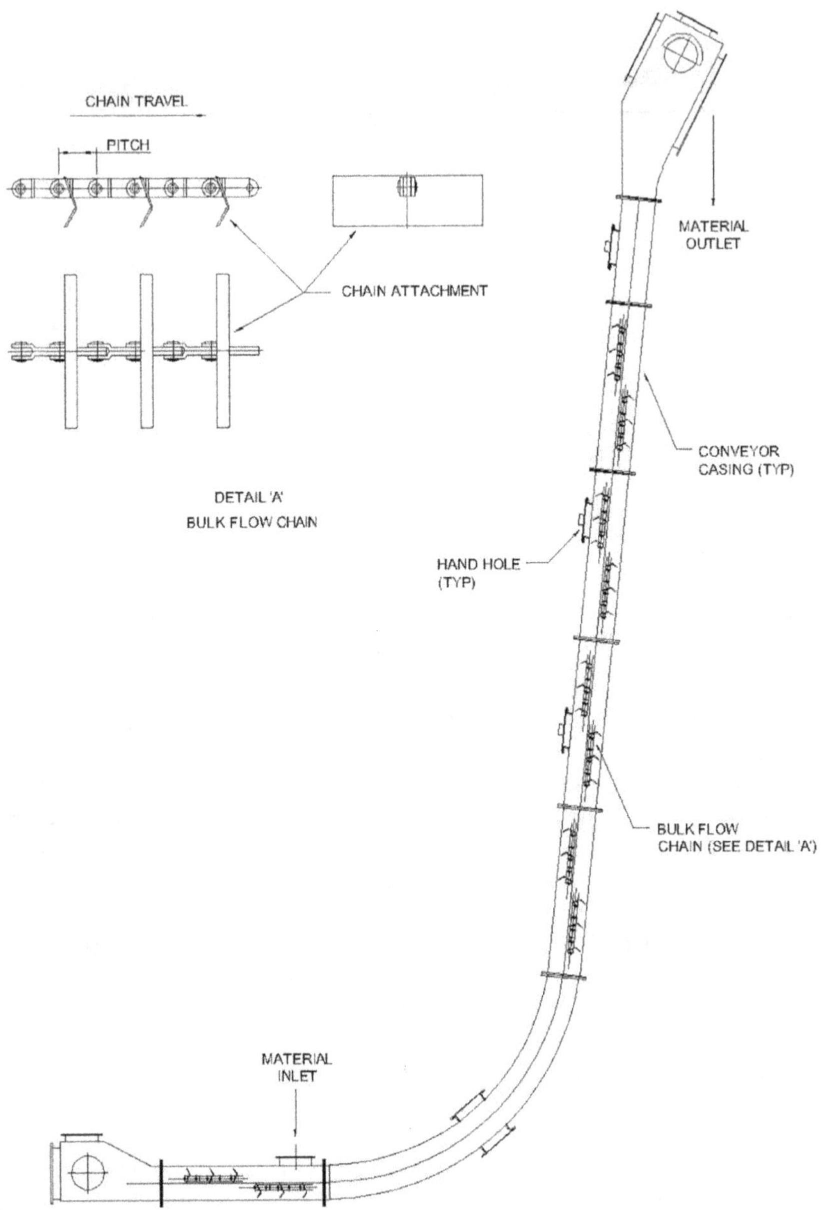

FIGURE 3.36 Bulk Flow Conveyor. Courtesy – Eminence Equipment Pvt. Ltd.

The connecting **chutes** between conveyors/elevators and machines are also prone to wear. This is especially so in the cleaning section. The seed contains soil particles which are highly abrasive. As the seed slides on the chute, it can wear out the contact side very fast. The problem may be overcome by fitting wear-resistant liner on the contact surface. This is also the case in seed unloading and pre-cleaning section.

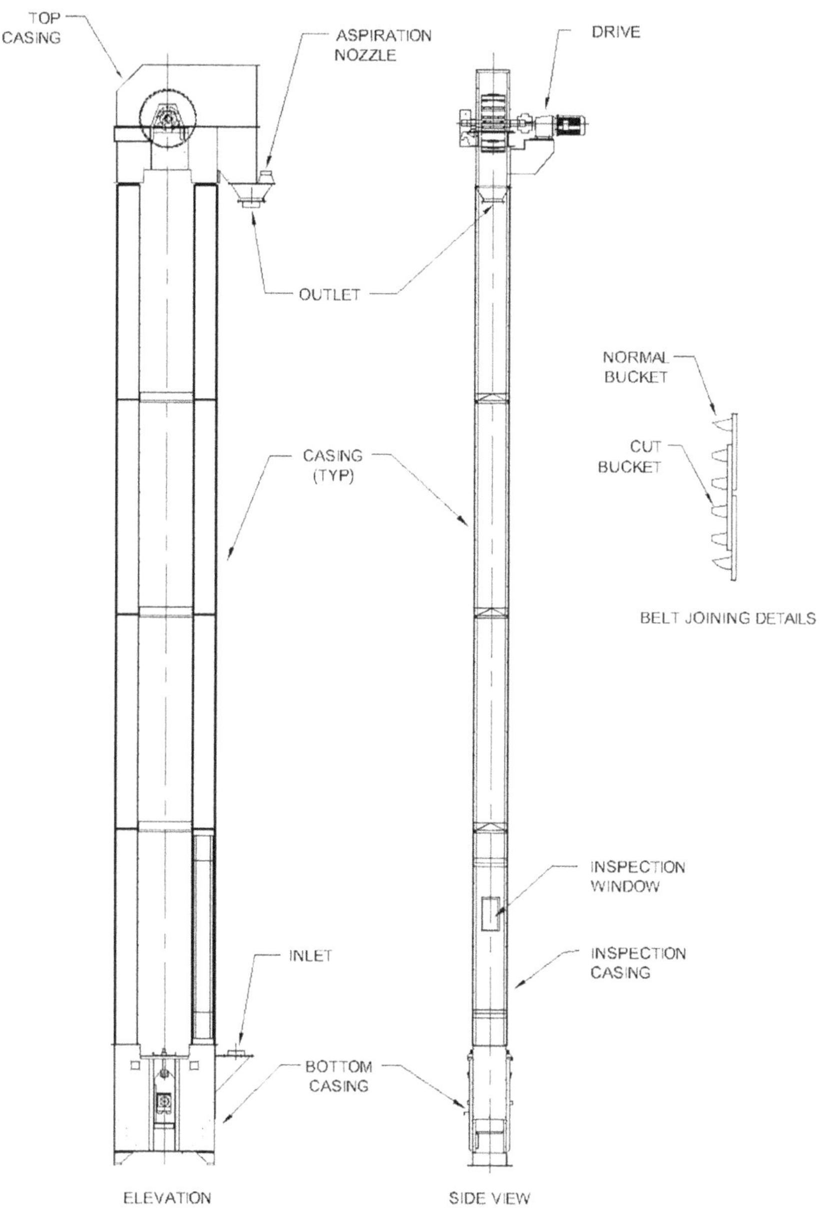

FIGURE 3.37 Bucket Elevator. Courtesy – Eminence Equipment Pvt. Ltd.

Bucket elevators (Figure 3.37) are used to lift material from ground to upper floors. These may be fitted with either moving belt or chain, on which buckets are fixed. Belt elevators are easy to maintain and also require less power, compared to chain elevators. Normal belts have limitation of operation temperature. Hence, for material lift downstream of cooker, where temperatures exceed 75C, special belts which are resistant to heat and oil, are used. During past two decades, belt speeds have increased, bucket spacing has reduced, and power requirement has reduced. Buckets with twin-leg design, with separate legs for the lifting and returning chain, require less power compared to single-leg elevators.

Bucket elevators are designed with two alternative principles, either 'centrifugal discharge' or 'continuous discharge'. The **centrifugal discharge elevators** are designed for high rotation speeds of belt drum and are used for free-flowing materials, to prevent material back-flow into the boot. Here the drum speeds may be at least 60 rpm. The linear speed of the belt depends on the drum diameter. So, while for small-sized elevators, linear speeds may be 1.4–1.5 m/s, those for bigger (high-capacity) elevators, such as those used for silo feed, may go up to 3.5 m/s. For centrifugal discharge elevators, discharge chute should be lined with a replaceable wear-resistant plate.

Continuous discharge elevators are used for non-free-flowing materials such as flour. Here the speeds are less, say up to 1–1.2 m/s, but buckets are mounted close to each other.

It is preferable to place **zero-speed sensors** on all solids moving elements, including conveyors and elevators, and interlock these in the reverse direction of material flow, so that stoppage of one machine does not cause any jamming or damages in elements upstream.

4 Screw Pressing

4.1 SCREW PRESS – HISTORICAL PERSPECTIVE

The idea of a mechanical **screw press** was conceived by V.D. Anderson in 1876 (Dunning, 1953). In 1900, the first successful screw press, called an **Expeller**, was made. It consisted of a cylindrical barrel and a rotating shaft with helical flights. As the seed was pushed forward by the flights into a progressively narrowing space between the shaft and the barrel, it was subjected to increasing pressure, releasing the oil which then flowed out of the slots in the barrel. The partially de-oiled cake was discharged continuously. The continuous operation saved a lot of manual work, which was necessary with the earlier hydraulic presses. It also paved the way for the construction of much bigger capacity presses.

As bigger presses were manufactured, for higher crushing capacities, the construction underwent major changes. In place of a single cylindrical barrel with slots, modular construction was adopted. Half-circular segments were made of flat bars placed radially, supported by a rugged frame; segment. two to four such segments were assembled to make a half-cage. Two half-cages were assembled to make a press cage (Figure 4.1). The cages had to be supported with strong frames to withstand the pressures generated within a cage. Thin shims were placed in between adjacent cage bars to create slots for the outflow of oil. Shim thickness could be varied, to optimise oil flow, without expelling too much 'foots' with oil.

The rotating shaft, with helical flights machined from the same stock, was replaced with multiple flight pieces (called 'worms') assembled on to a shaft. Such a design allowed easy replacement of individual worm pieces, without the need to replace the whole shaft, for optimising the press operation.

Knife bars were incorporated to cut the cake at different locations along the travel through the cage. This cutting action allowed for the re-mixing of cake within the chamber, and helped achieve better oil recoveries.

The 'box-frame' cage design also offered the major advantage of having to replace only the worn-out cage bars instead of the whole cylinder. Thus the frame could be used throughout the life of the press. Replaceable shims also allowed to minimise the expulsion of foots. This remained the dominant design for over half a century, and presses were used for various oilseeds, for capacities up to 50–80 Tpd.

As the need for larger capacities emerged, the cages became bulky and its assembly was cumbersome. A solution was found in single-cast half-cages (Figure 4.2). With these, there was no need to assemble the multiple frames with heavy support bars. The single-cast half-cages were sufficiently strong, and did not require external support frames. This design is now widely used for press capacities above 100 Tpd.

The largest presses built to date have an internal diameter of nearly 400–410 mm and can process up to 800–900 Tpd rapeseed, driven by motor rated at 500–700 kW. Capacity on the whole sunflower seed (undehulled) is limited to around 500 Tpd.

DOI: 10.1201/9781003309475-5

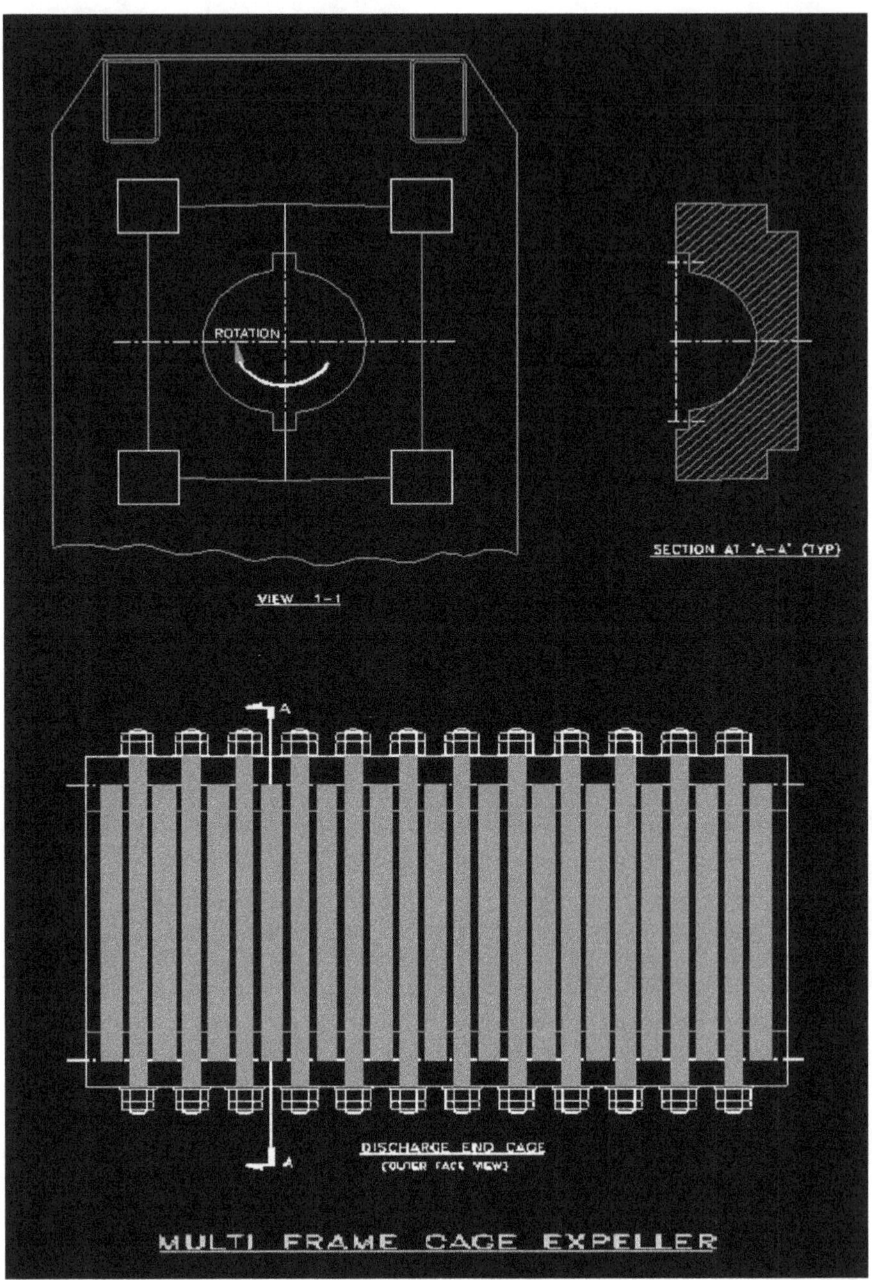

FIGURE 4.1 Box Frame Cage.

With dehulling, however, sunflower crush capacity is achieved close to that on rapeseed @ 850 Tpd, but with motor power higher than for rapeseed (it is to be noted that after dehulling, nearly 150 T of hulls is separated from the 850 T sunflower seed; so only about 700 Tpd dehulled meat will be crushed in the press).

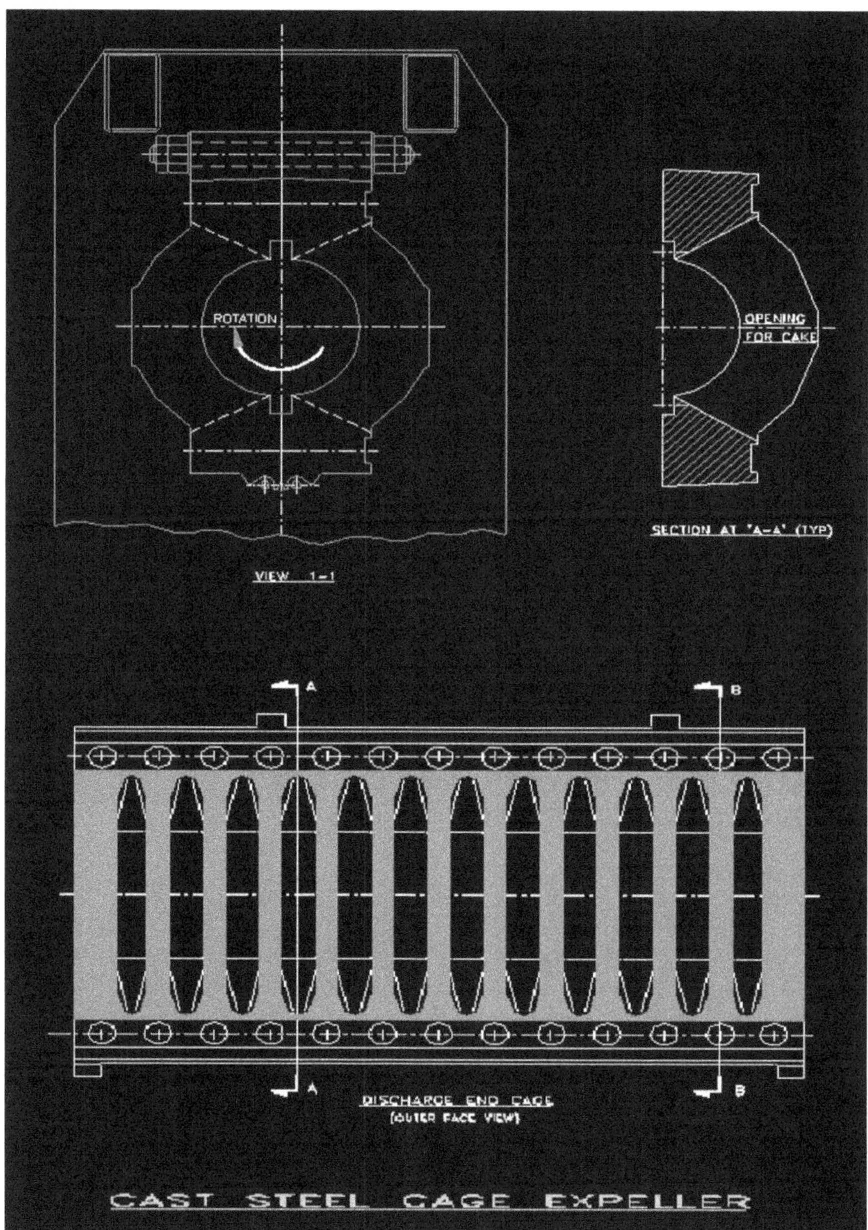

FIGURE 4.2 Single Cast Cage.

It may be noted that the box-frame design is still widely used in African nations, as also in south and south-eastern Asia. Press capacities are limited to mostly 80 Tpd, chamber ID up to 230 mm, 2.5 m long, driven by motors up to 100 kW. As special wear-resistant materials are not used for the wear parts, these machines are priced much lower than those of single-cast design. Thus, typically, four such machines of 50 Tpd each, cost much less than a single machine (single-cast design) of 200 Tpd. Although the operation and maintenance costs with the former design are much higher than with the latter, as may be seen in the following pages, and also requires much more manpower, the former is quite often preferred for the lower initial investment.

Modern presses are designed for high throughput, and work with low internal pressures compared to smaller presses. In smaller presses, thanks to higher working pressures, and to narrow clearance between screw shaft (known as the 'worm' shaft) and circular cage, the oilseed may be broken and subjected to substantial heat. Thus, whole oilseed may directly be fed to such small presses. On the other hand, large modern presses must be accompanied with elaborate preparation equipment upstream, for seed breaking and cooking.

Many advances have been made in materials of construction of presses, over the decades. Notable among these are the materials used for wormshaft and wear components. Worm shafts are now heat treated to a high degree to increase the shear strength. This allows the use of a smaller dia shaft for a given torque duty, which helps in the selection of worms with deeper flights for high press throughput. Wear parts such as worm pieces and cage bars are either made from special alloys which offer high wear resistance, or are hard-faced with very high wear-resistant steels, to increase their life. While the use of such special materials increases the cost of a press, it ensures maintenance-free, continuous operation over a long period, increasing the productivity of an oil mill. This factor is of critical importance for modern high-capacity mills.

4.2 COMPARISON BETWEEN BOX-FRAME AND SINGLE-CAST CAGE DESIGNS

4.2.1 Principles

There are significant process benefits to be derived from the single-cast design over the box-frame design. These benefits are derived thanks mainly to two construction differences. One, as the single-cast cages are through-bored, commonly on a horizontal bore machine, the cage bars segments line up perfectly linear. This is not the case on box-frames, where errors creep in while assembling the frames. The unevenness of cage bars alignment causes unnecessary resistance to the flow of oilseed mass within the cage, which affects the throughput capacity of the press. Less resistance to flow also results in lower electricity requirement for single-cast designs.

Second, in single-cast design, since there are no holding box-frames, the oil flows out unhindered. In box-frame designs, foots (fines expelled with oil) accumulate on the frames and hinder the oil outflow. Foots must be cleared periodically and manually. The single-cast designs, on the other hand, are 'self-cleaning'.

Since the single-cast cages are self-cleaning, these presses are operated with closed covers, with only a couple of windows for occasional inspection. Since the box-frame cages need to be cleaned manually all the time, covers cannot be kept closed. Typically, the box-frame cage presses are provided with a screen at the base to retain coarse foots, which are manually scraped and fed back to the cooker. The screened oil flows by gravity through an opening in the base.

TABLE 4.1
Box-frame vs. Single Cast Cage Case Study 1: Copra Crushing

Parameter	Box-frame	Single-cast	Remark
Crush capacity per press	18 Tpd	24.5 Tpd	Increase 35%
No. of Presses required to crush 120 Tpd	7	5	
Motor Amps, each press	40 A	45 A	
Combined power, kW	180 kW	145 kW	Savings 20%
Specific Power, per T copra	36 kWh/T	29 kWh/T	
Oil loss, w/w copra	0.3%	0.1%	Oil saved; cleaner plant

4.2.2 Case Study 1: Copra Crushing

A comparative study was made on presses used for copra, first press, by replacing box-frame cages with single-cast cages (Vadke, 2007). These were small presses, ID 180 mm, 8 nos., to crush 120 Tpd copra to leave 22–25% oil-in-cake. Process parameters were monitored closely before and after the replacement, for a month each. There was a substantial increase in crushing capacity of each press (Table 4.1). This resulted in utilisation of only five of the presses, in place of seven earlier, to achieve 120 Tpd capacity. Load on motor did increase, but not to the extent of capacity increase; hence, specific electricity consumption per ton of copra reduced significantly. Other significant benefit accrued was that the accumulation of foots on cages reduced drastically, which allowed a closed-door operation. As manual cleaning was not required, workmen strength in the oil mill could be reduced. Also, due to closed-door operation, splashing of oil stopped, hence loss of oil to atmosphere stopped. The oil loss, which had been estimated at nearly 0.3% (copra weight basis), reduced to just 0.1%.

Another major benefit accrued with the single-cast cages was that the press operation stabilised, sudden increases in motor load (current shocks) reduced greatly. This resulted in reduction in break-downs of press components, especially the press shaft, and the gearbox input shaft and gear. Also, since two presses became stand-byes, these could be maintained and kept ready. Thus, mill down-time reduced significantly, and mill productivity increased.

4.2.3 Case Study 2: Mustard Seed Crushing

Study was conducted at a 400 Tpd mustard seed crushing facility in north India (Vadke, 2007). There were two lines of presses. One line had a single machine of single-cast cage design, nominal capacity 300 Tpd (Cage ID 280 mm), and the other line had five machines of box-frame design, capacity 35 Tpd each (cage ID 230 mm). Since both lines had a common seed preparation section, the performance of the two lines could be compared directly. Mustard seed was cleaned, pre-heated, flaked and cooked at 75–78 C for 30 min. The process was designed to produce 'mustard expeller oil' for direct human consumption, while leaving nearly 18% oil-in-cake, which was conveyed to solvent extraction plant after cooling.

TABLE 4.2
Box-frame vs. Single Cast Cage Case Study 2: Mustard Seed Crushing

Parameter	Box-frame	Single-cast	Remark
Crush capacity per machine	35 Tpd	250 Tpd	
No. of Presses	5	1	Less space, less maintenance
Power incl. conveyors	190 kW	175 kW	
Power per T of mustard seed	26 kWh/T	17 kWh/T	35% saving
No. of workers per shift	1 operator, 3 helpers	1 operator	Less manpower cost
Material Loss, w/w seed	0.3%	0.1%	Low material loss to atmosphere
Process Cost, incl. wages	INR 245/T	INR 155/T	>35% saving overall

Note: Value of Indian Rupee was INR 50 = 1 USD.

Comparison between the two lines of presses is shown in Table 4.2. Crush capacity achieved by the single-cast design press was 250 Tpd, and the five presses in the second line together could crush only about 175 Tpd. There was huge difference between the electricity consumption of the two lines, although a part of it could be attributed to the presence of multiple machines in the second line.

The cages of box-frame machines had to be cleaned frequently, hence 1 helper had to be assigned for two machines, apart from the line operator. Since the cage of single-cast machine self-cleaned, no helper was required there. The self-cleaning cages also ensured that the door of the press could be kept closed, thus saving on material splashing to atmosphere. Table 4.2 shows cost savings, due to savings in electricity, wages and material loss, to be more than 35% in the case of the single-cast cage design press.

4.3 MODERN PRESSES – CONSTRUCTION FEATURES

4.3.1 Overall Construction

Core of a screw press consists of a sturdy cage and a worm shaft rotating inside the cage. The cage is held in between strong body plates (Figure 4.3). Most modern presses have two cages, with the one at feed end having slightly larger ID than the second. However, some presses are also built with a single cage. There is either a small Feed Cage separate from the main cage, or feed inlet may be given into the first cage itself. Cake is discharged through a constrictive annular opening at the other end, in to a small cake chamber.

The configuration of worms, the discrete flights, is such that the space between the worms and cage bars narrows progressively from feed end to discharge end. As the prepared grits/flakes are fed continuously to the press at feed inlet and are pushed forward in to the narrowing space, the worm shaft exerts high pressure on the material, causing oil to squeeze out. The released oil is pushed out through thin slots between cage bars. The partially de-oiled solids exit through an adjustable annular port in the form of flat cake (Figure 4.4).

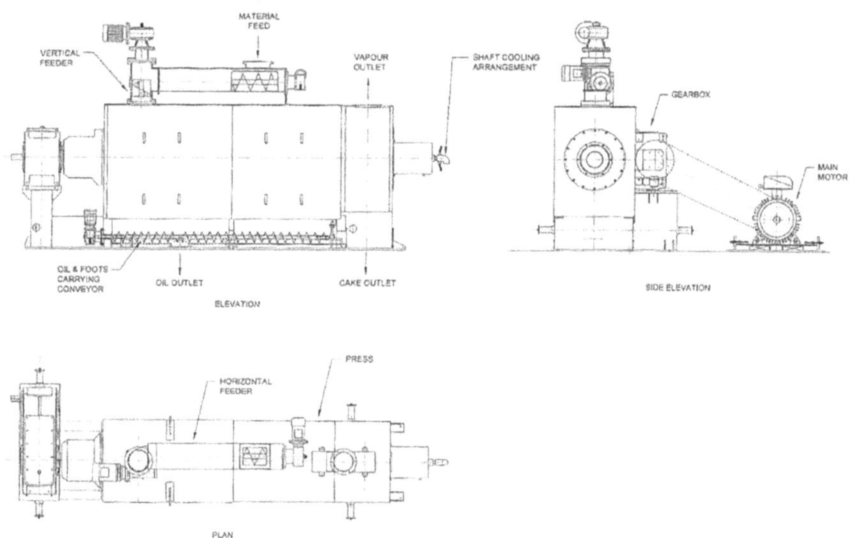

FIGURE 4.3 Press Assembly. Courtesy – United Engineering (E) Corporation, New Delhi.

FIGURE 4.4 Cake Emerging from Choke.

The annular port, for exit of presscake, consists of a fixed outer ring, called 'choke ring' and a movable cone. As the cone moves in and out of the choke ring, the annular width varies and so does the thickness of cake. Such an adjustable choke allows adjustment of final pressure on the mass, to ensure the desired level of residual oil. Some presses are designed with 'fixed' choke. Both the systems are discussed in detail in next section.

Most presses are fitted with a feed system consisting of (a) a Horizontal Feeder screw, with variable speed drive; and (b) a Vertical Feeder screw, which push-feeds the prepared grits/flakes into the press cage. For free-flowing materials, vertical feeder may not be required, and the screw may be removed to save on its motor power. In some designs, the vertical feeder is used to force feed, together with initial compression to remove air within the oilseed mass (Vavpot et al., 2014). The body of the vertical feeder is bolted on to the feed chamber, and the horizontal feeder body is usually bolted to the inlet flange of the vertical feeder. Since the vertical feeder is a compressing screw, its construction is relatively heavy-duty compared to the horizontal feeder, which is constructed as a well-machined screw conveyor.

The body plates which hold the cages are mounted on a **base box**, which may hold a tapered oil collection tray inside. A screw, fitted at the bottom of the tray, discharges the oil and foots to evacuation conveyor. The base is extended to support the main drive gearbox, placed usually at feed end.

Each section of a press, comprising a cage and the supporting body plates, is referred to as a chamber. Thus, a press typically consists of a feed chamber, the main chamber consisting of the main cage and bottom screw, and a cake chamber. As mentioned above, in some designs, with a single cage, the feed and main press sections may form a common chamber.

4.3.2 Cages

Feed Cage: Usually a separate short cage, made of a thick schedule pipe piece, machined on the inside. The lower half may be made of cage bars, with spacer shims. Shims may be thick, say 1 mm, to let out the air, which escapes from the initial compression of oilseed material.

Main Cage(s): The main cage, where oil is expressed, usually comprises two cages, the first cage with a marginally higher bore. A transition taper ring is fitted between the two, for smooth transition from bigger bore to smaller bore. The difference in bore of the two cages is around 25 mm (1 inch, traditionally) in most cases, but may even be up to 40–50 mm in some designs. Each cage is made of two halves bolted together. Each half is single-cast and through-bored. The inner face of the cage-half is lined with cage bars in a semi-circular segment (Figure 4.5). Thin spacer shims are placed between adjacent bars to provide slots for the outflow of oil. Segment length, or the length of cage bar, is typically 250–300 mm, and typically three or four segments fit into a cage. In case of presses with a single cage, the cage may be five to eight segments long. In each segment, the cage bars are held tight by means of holding bars on either end; these are bolted to the cage-half. One of the holding bars is plain-faced, just like cage bars, and is commonly called a shoe-bar. The other holding bar has protruding 'knives', meant to cut the presscake inside the cage, and is called the 'knife bar'. The two cage-halves, fitted with bars, are bolted together to form a cage, which is supported on vertical body plates at both ends. The bolting may be done with pneumatic wrench, so as to save time, and to ensure the tightening of each bolt with a preset torque.

FIGURE 4.5 Assembly of Cage Bars in a Half Cage & Worm Shaft. Courtesy – United Engineering (E) Corporation.

In large presses, each half-cage may weigh up to 3 T. Thus proper arrangement, as an overhead monorail, should be provided for the handling of the cages. For dismantling of the cage, first all the bolts are removed, using pneumatic wrench. Each half-cage is then held to a hook, attached to the top frame of the press, by a chain pulley-block. The cage is then lowered slowly using a telescopic hydraulic jack, to finally rest in the horizontal position on the base bed. In this position, the cage bars may be inspected and replaced conveniently.

Cage lining bars are long flats, typically 250–300 mm long, 10–12 mm thick and around 25 mm high. These are either made of wear-resistant alloy steels or are hard-faced. The latter are more expensive, but last much longer and are most common in practice. The hard-face layer is 3–4 mm deep and maximum wear up to 3 mm is allowed. With hard face material of chromium-nickel or high-chromium, the life of bars may be 8,000–10,000 hours on prepress duty for soft seeds such as rapeseed. For prepress of undehulled sunflower seed, the life may just be 1/3rd, say 3,000–4,000 hours.

Since the pressures inside the press are not uniform along the length of cage, all cage bars do not wear out at the same time. Thus, after the bars in high-pressure areas have worn down by 2–2.5 mm, these are beneficially interchanged with those in low pressure areas, to get more life out of the set of bars.

In some designs, cage bars have longitudinal reliefs (Figure 4.6), to help the flow of oil and foots. Such relief asks for extensive machining, hence the bars are more expensive, but help make the cages better 'self-cleaning'.

Spacer **shims** are set between adjacent bars, to make gap for outflow of oil. Typically, shim thickness ranges between 0.3 and 1 mm. Shims 1 mm thick are used for the feed chamber segment, where air is to be driven out with first compression. Subsequent shims are not thicker than 0.8 mm. In large presses, where there are multiple high-pressure areas, oil flow from high-pressure areas up to middle of the cage is high and shims may be 0.5–0.6 mm thick. In high-pressure areas towards the discharge end, thin shims of 0.4 mm may be used. For a press configured for second press duty, the oil recovery is to be quite low, hence, shims near the discharge end might be very thin, say 0.3 mm.

Knife bars are much thicker than cage lining bars (Figure 4.7), and only two of these are used in a single segment. Thus, on a half-cage, as the lining bar segments are set, only 1 knife bar is placed at one end. It is also utilised to tighten the segment of bars. At the other end, a bar of similar thickness as knife bar is used, but without the protruding 'knives'; these are often called '**shoe bars**'. Knife bars profiles are discussed in the section on 'Design aspects of Screw Presses'.

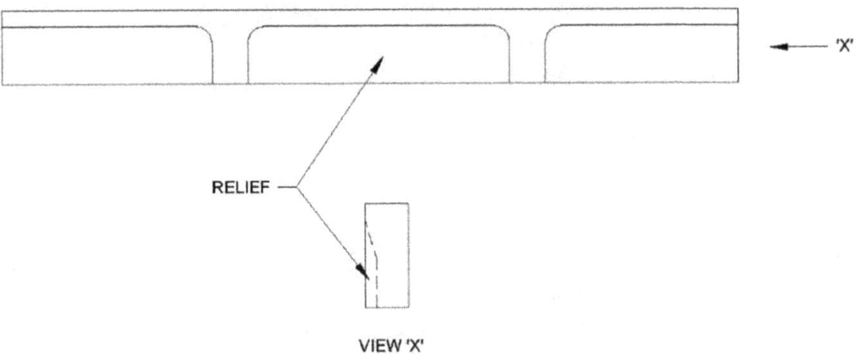

FIGURE 4.6 Cage Bars with Longitudinal Relief.

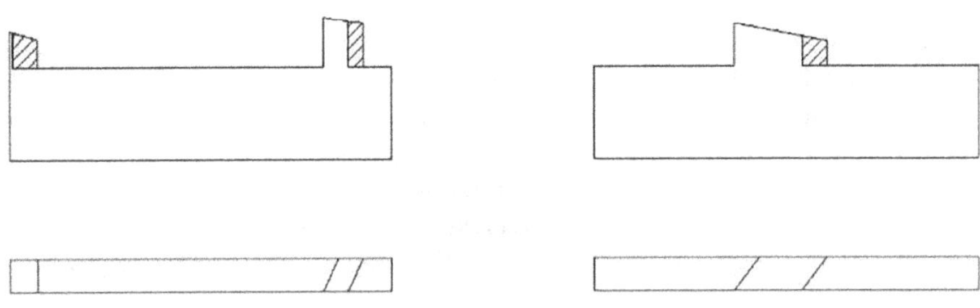

FIGURE 4.7 Knife Bar.

Choke: The choke provides the annular space for exit of the press cake, at the opposite end of the cage. It comprises an outer fixed ring and an inner cone. The ring is held within a body plate, and the cone is held on to the worm shaft. The gap between the ring and the cone determines cake thickness. Three different choke designs are practiced by different press manufacturers. One is a fixed choke, second the variable choke, and the third a 'floating' choke.

In the fixed choke design, positions of both the outer ring and cone is fixed, giving a constant width annular space for exit of the cake. It is claimed that since the press has many intermediate compression zones along the length of worm shaft, final 'adjustment' of exit pressure is of little concern.

Variable choke offers a movable cone, which enables the adjustment of the choke gap. The movement of the cone may be by lateral slide or by rotational progression on screw thread. An adjustable choke presents a means to assure desired residual oil-in-cake, even when worms and/or cage bars undergo some wear. As the worms and bars wear down, there is a progressive reduction in pressure exerted by worm shaft, resulting in slow increase in residual oil. Tightening of the choke allows a final extra squeeze to adjust the RO, although at the cost of press throughput.

The 'floating' choke incorporates a movable cone, held in position by hydraulic pressure. During normal press operation, there is always some variation in exit pressure, due to several factors including variation in feed rate, moisture content, degree of cooking, etc. As the pressure varies, the cone self-adjusts its position so as to balance the exit pressure. This mechanism ensures uniform exit pressure and uniform quality of cake.

Out of the three choke designs, the second, variable choke, is the most common. It allows reasonable control on exit pressure, and hence on RO, with a simple adjustment mechanism. The floating choke does offer somewhat better control, but involves an elaborate mechanism.

Cake Chamber: The worm shaft extends beyond the choke and is supported by a bearing on the other end, which in turn is supported by a body plate. The space between the choke and the bearing constitutes the cake chamber, where the cake is discharged into a chute below. The shaft may be fitted with cutter blades, to cut cake into small pieces, which are easier to convey downstream.

As the cake is discharged in hot condition, some moisture flashes off. These vapours should be evacuated to prevent water condensation, which would cause corrosion of the body plates. Normally, a suction fan is provided, on top of the chamber, for the purpose. The vapours are discharged into atmosphere outside the oil mill building.

Along with the moisture flash, some fine particles get entrained in the flash vapour. At times, the flash is quite forceful, due to sudden 'shocks' in exit pressure of cake. In such a condition, even with air suction, some fines may ingress into the bearing. A good oil seal is provided on the face of the bearing to prevent such ingression.

Cake chamber is normally closed with bolted covers on both sides. Inspection windows are provided in the covers, to enable routine check on cake quality. Covers, or 'shrouds', may be opened for adjustment of choke or for removal of worm pieces for maintenance.

Discharge chute, below the cake chamber, may normally be divided into two separate outlets, one for 'reject' cake, produced at press startup; and the other for the 'normal' cake, which is sent for further processing. A flap, adjusted with a lever, allows opening of one or the other outlet. In recent times, over the past decade, with the advent of variable frequency drives (VFDs), which allow the adjustment of press-shaft speed during operation, cake quality may be controlled with low speed at startup, to avoid the production of 'reject' cake. In this case, the dual chute arrangement may not be necessary.

4.3.3 Press Covers

Some press designs have full shrouds covering all three sides over the base (Vavpot et al., 2014). Most other designs have vertical covers on both sides, while the top is covered with thick plates, either bolted or welded to the top frame which ties the body plates. In some designs, the vertical covers are split into two, with the upper part bolted to the body plates, and the lower part hinged to the base, to allow inspection of oil drainage from the cages. The upper part is fitted with inspection windows to check for clogging of cages. The lower covers usually have sloping flats fixed inside near the bottom, to prevent oil leakage. Edges of the covers may be lined with rubber gaskets to prevent the escape of oil mist from the press chamber. The covers should be made of stainless steel, to prevent corrosion due to condensation of moisture and corrosive gases.

In large presses, the press chamber(s) may be ventilated, to prevent condensation of moisture. As the vapours, consisting of air, water vapour and oil mist, are vented out, these would cause air pollution in the surroundings of the oil mill building. It is, therefore, a good idea to fit a baffle assembly and a demister pad in the duct. The oil collected at the baffle assembly is in small quantity, say 1–4 kg per 100 T seed crushed, and could be mixed with crude solvent-extracted oil, or utilised for soap-making. The demister pad must be cleaned at a regular interval to avoid excessive pressure drop in the vapour line.

4.3.4 Worm Shaft

This is the most important part of a screw press, which majorly determines its crush performance. It comprises a base shaft on which discrete worm pieces are mounted (Figure 4.8). Worm pieces come in various types – worm with flight, worm without

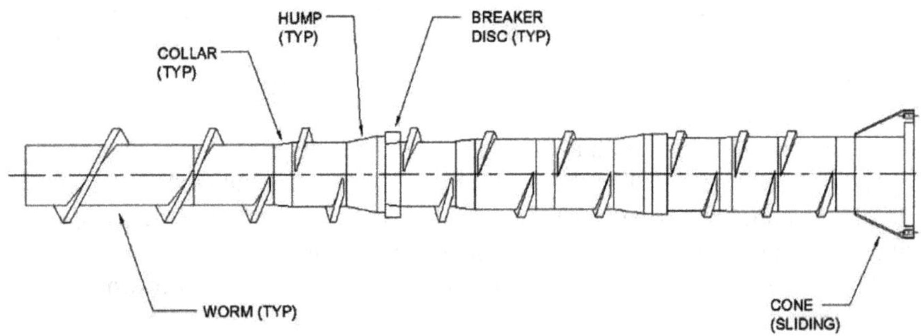

FIGURE 4.8 Worm Shaft Assembly.

FIGURE 4.9 Pressure Worm. Courtesy – United Engineering (E) Corporation, New Delhi.

flight, worm with straight boss, worm with taper boss, breaker disk and finally, the cone. The worm pieces are held on the shaft with full-length keys. To avoid hot oilseed mass getting in between adjacent worms, and then between worm pieces and shaft, a metal ring is fitted into grooves made on worm matching faces. Metal rings have replaced rubber O-rings, which used to harden over time.

Worms with flights may either be just to convey the material, e.g. the feed worm, or to compress the mass, called 'pressure worms' (Figure 4.9). In most designs, a short distance piece or a 'collar' (worm without flight) is provided in between two pressure worms. The presence of a collar allows the fitment of a 'knife', on the knife bar, at the location of the collar, to cut and re-mix the mass. The collars may be of constant dia or be tapered, to match the boss dia of the following worm.

In large presses, there are multiple compression zones. At the end of each such zone, a high-taper collar, or 'hump' is provided to act like an intermediate 'fixed cone' (Figure 4.10), to create major compression. It is desirable to cut the mass and re-mix it well, after such major compression; a specially designed breaker disk is fitted, just after a hump, for the purpose.

In older press designs, 'reverse worms' used to be provided, to ensure complete maceration of grits/seeds. Now, when the oilseed is subjected to elaborate 'preparation', and the seed structure is already broken down, there is no need for reverse worms. Reverse worms create sudden pressure peaks, which cause high wear of its flights as also of cage bars in the segment. Such sudden pressure peaks also affect the press throughput adversely; this is discussed in the next section.

Within a cage, the OD of flights on all worms is the same, very close to the ID of the cage bar circle. The clearance between the flight and the bar should be as small as possible. Usual clearance is around 1–1.5 mm. As the worm flights and bars wear down, clearance increases, which results in reduction in pressure exerted by the wormshaft. Clearance beyond 6–7 mm is detrimental to press performance; and should trigger replacement of bars and/or replacement of worms.

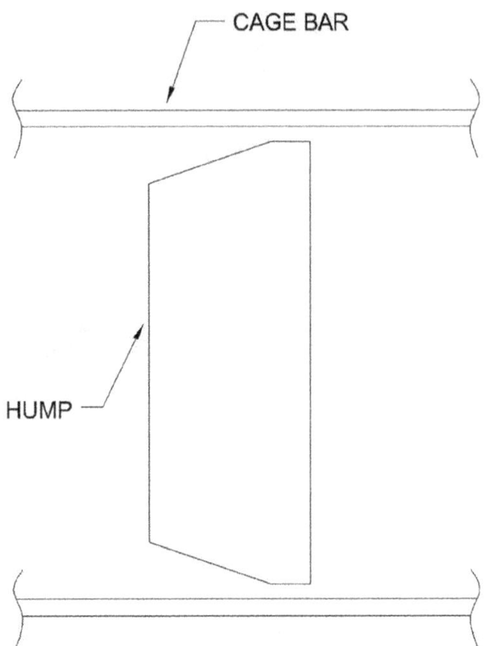

FIGURE 4.10 Hump.

The tips and the compressing face of a worm flight are hard-faced, usually to a depth of 4–5 mm. The main property of the hard-face material is high wear resistance and not merely the hardness. **High chromium** or **chromium-nickel steels** are preferred for the purpose. Nickel is a more expensive material, but its presence makes the alloy heat-resistant, and is good for full-press duties.

The core **shaft** is made of high-tensile steel, to provide good shear strength. Special heat treatments help increase the shear strength and allow the use of smaller dia shafts for a given duty. **En 24 T**, the last letter denoting the type of heat treatment, offers high shear strength. A smaller dia shaft offers flexibility in design of worm pieces for higher press throughputs.

4.3.5 Worm Shaft Cooling

In most modern press designs, a provision is made to cool the worm shaft (Figure 4.11). Such a provision helps to keep the temperature of oilseed mass, and the discharge cake, moderate, and avoids charring of cake due to overheating. It also helps in final oil recovery, to a degree, in the last segments of the cage, as can be seen from the discussion in the next section on 'operation practices' of screw press. The cooling water inlet and outlet pipes are connected through a double rotary joint.

In some press designs, even the cages are water-cooled (Vavpot et al., 2014). It is claimed that such a provision helps reduce the temperature of press oil and preserve its quality.

FIGURE 4.11 Cooling Arrangement of Worm Shaft. Courtesy – United Engineering (E) Corporation.

4.3.6 Drive and Bearings

The shaft is extended on both sides of the cages. At discharge side, it extends beyond the cake chamber and is supported by a radial bearing, which in turn is supported by a body plate. In some press designs, a simple **bush-bearing** is provided, to avoid maintenance issues on account of ingress of cake fines. When a radial ball-bearing is provided, it must be protected with a good-quality oil seal. The feed side shaft extends through a **radial bearing** and a **thrust bearing**, and connects to the gearbox through a coupling. In some press designs, the shaft extends through a hollow-shaft gearbox. Thrust bearing has to be adequately designed to take full thrust created due to the compressive action of the shaft, and also be able to absorb 'shock' loads. A heavy-duty spring is often provided, with thrust bearing, for the latter purpose.

The **gearbox** is an important, and expensive, component of the press. In modern presses, these are typically three-stage reduction, parallel helical type. The gearbox may be air-cooled or water-cooled. Considering that most prepress machines are operated at shaft speed within the range of 30–40 rpm, the gearbox reduction ratio is typically around 30:1. Further reduction can be achieved by change of pulley, or by turning down the speed with a VFD. Full press machines are typically operated at 18–25 rpm, hence the gearbox ratios have to be higher. Alternatively, the additional reduction may be achieved with pulleys.

Some press manufacturers offer direct coupling of drive motor with the gearbox. However, most others offer **pulleys** between the two. Pulleys not only offer a flexible component for the safety of the motor, but also allow for speed reduction when necessary. Such flexibility is also useful to allow the utility of a press for either first press or a second press duty, or for multiple oilseeds.

Some of the press designs offer an option of '**barring motor**'. This is a small geared motor, coupled to the other end of input shaft of the gearbox, with a clutch, and can operate the press at a very low speed, say 2–3 rpm. In case of any interruption in main power supply, as the main motor stops, it is important to empty the press of the hot oilseed mass, which may otherwise solidify into a hard cake within about 10 minutes. In such a situation, the barring motor is started with an auxiliary power supply, say a diesel generator, to run the press slowly to empty it out within a few minutes.

In recent times, with the availability of rugged VFDs at reasonable prices, these are increasingly being put into practice. A VFD offers multiple advantages for press operation. One, it allows for speed reduction without having to change pulleys. Second, it obviates the need for a barring motor, since in case of power disruption it may be re-started at very low rpm and very low amperage. Third, it allows the startup of the press at low rpm to prevent high peak currents. Finally, it presents an option of small speed variation, during press operation, for fine-tuning the oil content in cake. This effect is explained in the section on 'operation practices' of screw press. Even when VFD is used, pulleys may be retained as a flexible component for the protection of drive motor.

4.4 OPERATION PRACTICES

4.4.1 A Modern Oil Mill

The practice of screw pressing has undergone a sea change from the days when the first screw press was operationalised around 1900. Not only are the presses now much bigger, but are also supported by a lot of preparation equipment. Thus, as against a long line of small presses in old mills, maybe with small heating kettles on top, a modern oil mill incorporates only a few presses, but with a series of other machines, arranged over two or three operation floors. Even though the mill crushing capacities have increased manifold, the workmen per shift have reduced drastically. Also, where there always used to be a number of presses opened for maintenance, and maintenance teams working continuously, there are none now. A modern oil mill is a clean environment, with only a moderate sound of the drive motors and humming of machines. From where people had to work bare-bodied to withstand the heat emanating from the 'open' kettles and presses, and do a lot of heavy manual work, a modern mill is operated by uniformed workmen who do practically no manual labour, but are the 'operators' who control the processes and the quality of cake and oil.

4.4.2 Prepress and Full Press

Modern mills are designed for two types of duties, pre-pressing or full pressing. The main objective of the former is to produce presscakes @ 16–20% oil content, the cake being well-formed yet porous enough to allow solvent percolation in the solvent extraction plant located in the adjacent building. While meeting the main objective, nearly 34th of the oil is recovered from the press. Full press machines are operated to maximise oil recovery.

Since pre-pressing is basically a preparation step for solvent extraction, the quality of cake has to reflect in its performance in the latter. A quality parameter, which may be determined in laboratory, is the '**Milling Defect**'. If the preparation, including prepress, were perfect, an ideal extractor would be able recover all the oil, leaving no oil at all in the meal. A laboratory extractor is considered an ideal extractor. Thus, when a cake is extracted in laboratory, for say 3 hours, the oil remaining in the extracted cake is considered the defect in preparation, or the milling defect. This concept is elaborated in Part 2 of this book, on solvent extraction. In practice, an MD of 0.3% is considered as a good quality index for most oilseeds.

For full-pressing duty, the objective is to recover as much oil, leaving as little oil as possible in the cake. In practice, most presses are operated to leave about 7–8% oil-in-cake. Oil-in-cake as low as 5% has been reported, but mostly at the cost of cake and oil quality. The feed to full press machines may be oilseed grits/flakes, or presscake from first press (like a prepress), or collets from extruder treatment.

The peak pressures employed in full pressing are much higher than those for pre-pressing. Medium pressures in the range of 30–40 MPa (300–400 kg/cm^2) are adequate for the purpose of pre-pressing, which recovers over 70% of the oil (Norris, 1981; Ward, 1984). However, for the additional recovery of further 15% of the oil, in full pressing duties, pressures required are nearly three to four times, in the range of 100–140 MPa. Thus the machines built specially for full press duties may be of heavier construction. As a result of the high pressures employed, full press cakes are discharged at higher temperatures, 110–120 C, compared to those from prepress @ below 100 C (with shaft cooling).

As discussed before, full press machines are operated at much lower shaft speed (18–25 rpm), as compared to the prepress machines (30–40 rpm). Prepress machines are designed for high throughput, while those for full press are designed to give maximum oil recovery.

4.4.3 Effect of Press Operation Parameters

Prepress machines are operated mostly at shaft speed range of 30–40 rpm, whereas the full press machines are operated at lower speeds, in the range of 18–25 rpm. The lower speed of wormshaft creates higher pressure on the oilseed mass, and also allows longer press time, resulting in lower residual oil (RO). This effect has been explained with the help of a mathematical equation (Vadke and Sosulski, 1988).

$$Q = G_1 * N - (G_2/\mu) * P \tag{4.1}$$

where, Q is the mass flow rate (press throughput); G_1 and G_2 are geometric parameters of the press comprising terms representing cage width and depth of worm flights, the helix angle; μ is the viscosity (apparent) of oilseed mass, and P is the exit pressure. G_1 and G_2 are also affected by parameters related to the seed mass, namely, the power-law index of viscosity and density; however, these parameters may be considered as constants for a given oilseed.

The first term on the right-hand side of Eq. (4.1) represents the flow due to the conveying action of rotating shaft and is called 'drag flow'. The second term can be seen as a back flow, caused by the pressure at choke. In actuality, there is no back flow

TABLE 4.3
Summary of Patterns of Change in Press Performance, Corresponding to Press Settings and Seed Pretreatments

Variable	P	Q	RO	T
Press Parameters				
Choke Opening ↓	↑	↓	↓	↑
Shaft Speed ↓	↑	↓	↓	↓
Seed Treatments				
Heating ↑	↑	↑	↓	↑
Flaking	↑	↑	↑	↓
Flaking + Heating ↑	↑	↑	↓	↑
Moisture Addition ↑	↓	↓	↑	↓

Source: Vadke and Sosulski (1988)
(Symbol ↑ denotes increasing in value, and ↓ denotes decreasing in value)

in the press; the material churns within the channel instead of advancing uniformly forward, which causes a net reduction in the throughput.

It was shown that the lower shaft speed resulted in higher viscosity of the pseudoplastic mass, which in turn caused higher pressure. The lower speed also reduces the drag flow, causing reduction in throughput.

The overall effects of press parameters, as also of pretreatments, are shown qualitatively in Table 4.3 (Vadke and Sosulski, 1988). It should be noted that these effects were observed on a small laboratory-scale press, with a continuous flight screw shaft, yet are valid for large presses as well.

Tightening the choke would have similar effects as of reducing the shaft speed, but with one difference, the cake temperature would rise instead of falling. Also, the effect of tightening the choke, on reduction in RO, may only be marginal on large presses which have multiple compression zones inside. The press throughput would get affected significantly, however, by tightening the choke. Hence, the choke is only to be used for a small adjustment of RO. Better option is to reduce the shaft speed, using the VFD.

4.4.4 Effects of Seed Preparation

The beneficial effects of the flaking and cooking operations, on higher press throughput and lower RO (Table 4.3), also have been explained with the help of the same equation as cited above (Vadke and Sosulski, 1988). These operations cause agglomeration of protein and starch, which causes increase in viscosity of the mass. As explained earlier, in the discussion on the effect of shaft speed, higher viscosity increases the pressure exerted on the mass and increases oil recovery.

Higher viscosity also reduces the churning within the press and increases net axial flow, or the throughput, as seen from Eq. (4.1). One may argue that simultaneous increase in pressure and viscosity would keep the second term in the equation constant, and thus would keep the throughput constant. However, the increase in pressure is not to the same degree as the increase in viscosity; hence, the net effect is some reduction in the second term. This is the explanation for the observed increase in throughput with the flaking and cooking treatment, as seen from Table 4.3.

4.4.5 Press Startup

Care should be taken to not load a new press to more than half its capacity on day 1. It is better to let all the worms and wear parts to run in for a day, before press is further loaded. On day 2, the loading may increase a bit, and the press may be loaded fully only by the third or the fourth day.

The first precaution to be taken for optimum performance of the press is continuous **adequate feed**. The first check, for the purpose, is to ensure that the flow of material from cooker is unhindered, and the horizontal feeder screw is fully loaded. The next step is to check for free flow through the vertical chute to the press. If vertical feeder screw is provided, its amperage should be checked to be below its 'overload' value.

At press **startup**, the press shaft speed is kept low at the beginning (with a VFD), to ensure well-formed cake from the outset. The choke may be kept at mid-opening, say 10–12 mm for medium size presses, up to cage dia 300 mm; and 12–15 mm for larger presses. Oil flow from all segments is observed closely. As oil flow stabilises, the oil-in-cake is checked. If it is within the desired range, then slowly the shaft speed is increased. The speed of the horizontal feeder should also be increased in tandem.

As the shaft speed is increased to take in more and more feed, oil flow starts from almost all segments. Once the oil flow is established over almost the entire length of the cage, the press should be operated at that speed for a few hours. Over time, the cages, and the press body heat up. The shaft cooling water, and water to cages if provision exists, should be started. The oil-in-cake should be checked. If within the desired range, choke may be opened in steps, to allow more throughput. Shaft speed may be increased slowly further until the OIC is at the top of the desired range. At this setting of shaft speed, the feeder speed may slowly be increased to fully load the press. As more feed is given, oil flow increases from segments close to the feed point. Care must be taken that the feed segment remains free of oil; oil slushing at feed segment would hinder air exit and interrupt free intake of the feed. In case of vertical feeder designed for forced feed, some oil flow at feed segment may be allowed, since all the air has been exited before material entry to the main cage.

4.4.6 Process Issues

4.4.6.1 Increase in Oil-in-Cake

During normal operation of the press, if oil-in-cake increases beyond the desired range, all parameters on preparation machines should be checked. If these are all ok, the flow to horizontal feeder should be checked as described above. Flow of oil from all segments to be checked and restored by increasing the feeder speed, if necessary.

If even after all the above steps the high OIC persists, cake thickness should be checked; if it has increased, the choke may be tightened to restore the thickness. If the cake thickness was ok, and if the higher OIC persists, the press shaft speed may be reduced a bit to restore the OIC.

4.4.6.2 Powdery Discharge

This problem occurs commonly with oilseeds that have low protein, say, sheanut and palm kernel. However, it may be encountered even with high-protein seeds. The reason in most cases would be low moisture in seed. Low moisture inhibits cooking action and does not allow proper coagulation of protein. This problem is more common long after harvest season, when moisture may have reduced during storage. A quick remedy is to add moisture in first stages of conditioner, in the form of live steam. For cases with very low moisture, hot water may be added up to 2%; this must be done in the upper stages of conditioner, to allow adequate time for moisture to be absorbed in the core of grits.

In another scenario, the press discharge may be powdery due to the problem of 'spurting', as explained below.

4.4.6.3 Spurting

At times, especially on 'full press' or 'second press' duties, a peculiar problem is encountered; the press starts 'spurting', making unusual sounds, and throwing out the cake with great force periodically. During this time, cake discharge is intermittent. This may happen for different reasons. One, the choke may be throttled too much. Second, the cake is too dry, so causes increase in pressure and may be pushed only with a great force. Or, sometimes a long collar is fitted on the shaft just prior to the choke, so as to develop high pressure for high oil recovery; such a collar does not allow smooth flow of cake especially at low moisture levels and can cause spurting.

If the problem persists, try increasing moisture at cooker outlet, by reducing air ventilation. Too much moisture is, of course, detrimental to press performance as seen from Table 4.1. If the last collar is very long, it may be replaced with a combination of a flighted worm, the same pitch and dia as the last worm, followed by a short collar. Spurting is detrimental to the life of press parts, especially the worm shaft and the thrust bearing, as it creates big 'shock' loads; it also creates too much **fines** at discharge, which is bad news for solvent extraction.

4.4.6.4 Foots Increase

Foots discharge with oil is a normal phenomenon. With a good press configuration, foots content of 6–8% on prepress of most seeds is common and acceptable. Should the foots content increase, however, it increases the load on filtration system downstream. Also, as the foots are recirculated to the cooker, these come back to the press and can affect the press performance; throughput may decrease due to the presence of too much 'slippery' foots. To arrest the problem, the following steps need to be taken.

The oil discharge from different segments of cage to be observed, and samples taken. If more foots are being discharged only from a few segments, maybe the shims at those segments are too thick and need to be replaced with thinner ones.

However, if the foots discharge is high from most of the segments, there is most likely a deficiency in seed preparation. Inadequate cooking may cause '**raw uncooked**' material to get in to press; proteins and starches within such materials are not adequately coagulated and aggregatedand can slip in with the oil easily. Such foots are also **sticky** and may create problems with filtration operation. In such a situation, the entire preparation process should be thoroughly checked.

4.4.6.5 Un-crushed Particles in Cake

Even cakes with low oil content may sometimes contain white or yellow particles that are clearly not crushed. This is, in fact, a common occurrence with 'soft' seeds such as rapeseed and groundnut. While the particles hardly reflect in higher oil content in the cake, they would most certainly affect the final ROC after solvent extraction. This may be checked by analysis of the milling defect.

In such a case, it is prudent to check for any seed, or grit, by-passed at flakers. However, most likely, the fault would lie with Knife Bars and Breaker Disks in the Press. Either these might be worn out, or were of wrong design in the first place. This matter is further discussed in the next chapter.

4.4.7 Press Load Control System

During normal operation, load on main motor, which drives the worm shaft, can vary to some extent. This may be due to various reasons, such as feed rate variation, moisture variation in the feed mass, fluctuations in upstream cooking operation, etc. The load fluctuation may even be due to internal causes, such as large clearances between worm flights and cage bars, worn-out knife bars, etc. which would cause material accumulations and its sudden intermittent movement. Large fluctuations in motor load cause abnormal mechanical stresses in pressure parts, as also in bearings and gearbox; these may even affect cake quality and oil recovery. It is a good idea, therefore, to regulate the motor load.

In small presses, just setting the speed of feeder screw, which must be well-machined to have uniform flight pitch and dia, offers a reasonable control of press load. However, in large presses, with motors rated higher than, say, 150kW, the potential load fluctuations may also be large and just a manual regulation is not adequate. For such cases, an auto-control system is preferred.

Auto-control system for press load is normally connected with the speed of feeder screw. It is a feedback mechanism which senses press load and regulates the feeder speed, so as to maintain the press load close to a pre-set value. It must be noted, however, that the response of the press load to feeder speed is not linear, hence a simple PID control is far from adequate. In fact, a simple PID control system may even lead to '**shunting**', i.e. continuous fluctuations in press load, thereby defeating the purpose of control system.

The response of press motor load to feeder speed is different in different load regions. Thus, the response curve at low loads would be quite different from that at medium loads, which would again be different from the response near full load. It is, therefore, beneficial to either incorporate different PID parameters for the different load regions, for a feed-back control system; or, write different response equations in the control algorithm, for a feed-forward control system. Either of these options may be accomplished with a **programmable logic controller (PLC)**. Many press manufacturers have designed such control systems; these not only ensure uniform press performance but also make the operation safe, thereby allowing remote press operation and control.

4.4.8 Oil Expression from Various Oilseeds

4.4.8.1 Soft Seeds - Rapeseed, Groundnut, Sesame, Castor

Rapeseed and groundnut may be categorised as 'soft' seeds, thanks to their soft kernel, with protein content of 20% and more and the absence of hard seed coat (hull). As these are crushed in a screw press, the pressures created are relatively low and so is the power consumption per ton of seed. Thanks to low pressures, and due to high viscosity of mass with high protein content (as seen in earlier Section 4.4.4), the press throughputs are high. Castor beans have a hard coat, but it is not thick, and the kernel is very soft, thanks to high oil content (50%) and high protein content (20%). Sesame seed also has a hard coat which is thin, and the kernel is soft, thanks to very high oil content (55%), although with medium protein (14%).

Thanks to the softness of the mass, the wear on worms and cage bars is also relatively low. Table 4.4 tabulates relative press throughputs, power requirement, and life of wear parts, for soft and hard seeds. This data was collected on a press with cage ID 305 mm (feed cage)/255 mm (discharge cage), cage length 2.5 m.

With proper seed preparation, foots content in oil can be kept below 7–8%; these are readily separated by filtration, and recirculated back to cooker.

Castor beans do have a hard seed coat, but the core is very soft, and when macerated, the mass is akin to that from other soft seeds as rapeseed and groundnut. If the seed has not been cooked well, the core can remain undercooked, which can cause too much sticky foots expressed with oil. These can cause major problems in oil filtration. However, with adequate cooking, the operation can be trouble-free.

TABLE 4.4
Comparison of Press Performance between Soft and Hard Seeds

Type of Seed	Power, kWh/T[a]	Throughputs for a Reference Press (Tpd)	Life of Wear Parts (hours)
Soft seeds – rapeseed, groundnut, castor, sesame	12–14	300	10,000
Soft seeds with hard hull (partially dehulled) – sunflower, cottonseed meats	20–25	200–240	6,000
Sunseed whole	30	175	4,000
Hard seeds – palm kernel, shea nut	30	125–150	4,500
Copra – first Press	30	125	4,500
Copra – second Press	60[b]	50	3,000

[a] Specific consumptions are on basis of feed tonnage.
[b] This figure may be seen as equivalent to just 24 kWh/T of copra.

A major issue with the 'soft' seeds is that these are compacted quickly under the pressure of worms. The quick compaction results in entrapment of uncrushed particles within the mass. Unless the mass is repeatedly cut and re-mixed, the chances of uncrushed particles escaping with the cake are high. This is the major reason why solvent extraction, as also desolventising, of rapeseed cake is at times difficult compared to soybean flakes. Thus, proper design, and maintenance, of mixing devices within a press, namely, knife bars and breaker disks, is critical for the pressing of soft seeds.

The problem is even more acute for groundnut, which has even higher protein content and is one of the softest seeds. The presence of uncrushed particles reflects in high Milling Defect of the cake; this can be analysed in laboratory. This factor will be discussed in detail in Part 2 of this book.

4.4.8.2 Sunflower Seed, Cotton Seed

Sunflower seed and cottonseed have soft kernel with hard seed coat or hull. These are mostly crushed after partial dehulling. The presence of even the partial hulls causes high friction with the worms and cage bars. This causes higher power consumption, lower press throughputs, and higher wear of worms and bars, compared to the figures for soft seeds (Table 4.4). The higher friction also causes higher heat generation. Shaft cooling is of critical importance to maintain the cake quality of these seeds. Cooling of cages, with water circulation through channels would also help in improving oil quality.

It is important that both these seeds be cracked (and flaked) well, so that the hulls are broken completely; otherwise the hard specs of hulls may carry through to the de-oiled meal, and make it unpalatable to chicken. When whole sunflower seed is pressed, it is always cracked, as described earlier. In all cases, the cracked grits or the flakes are cooked well prior to pressing.

When whole sunflower seed is pressed, the hard and abrasive hulls cause much faster wear of wear parts. The hard hulls also increase motor load and reduce the press throughput (Table 4.4).

If the seeds are full-pressed, the high pressures and high shear within the press ensure complete maceration of hulls; hence flaking may not necessarily be practised prior to full pressing.

It may be noted that when cracked grits are fed to the press without flaking, the initial portion of the press cage is utilised mainly for grits maceration, and oil flow starts only after the first few segments of the cage. Thus, the throughput capacity of the press reduces by 10–15%. This may still be acceptable, as the process saves the cost of, and power consumption in, flaker machines.

Foots production can be kept low with proper preparation, and these are readily circulated back to cooker and press, to be incorporated into the press cake.

4.4.8.3 Palm Kernel and Shea Nut

Palm kernel and shea nut are hard seeds, with low protein content @ less than 10%. These must be broken well with a hammer mill and further cracked or flaked prior to cooking. Even with these heavy preparatory operations, press throughputs are low, and the specific power consumption is high. These effects are quantified in Table 4.4.

Due to low protein content, the cakes are not strong and may crumble during conveying through the cake cooler to the solvent plant. This affects the percolation rates of solvent and reduces the extractor capacity.

Shea nut is a large seed. Unless the seed is carefully broken down into small pieces and properly flaked, the core remains under-cooked. This is the case with almost all sheanut mills, where the grits are not flaked. This results in generation of too much 'foots' with oil from the press cage. Also, due to under-cooking, foots are soft and sticky. These foots overload the oil filtration system. Also, foots stick to conveyor surfaces, necessitating manual handling. If these are recirculated to cooker, they start accumulating near the shell. Should these be sent directly to the press, the press operation is disturbed, throughput reduces, and the quantity of foots in oil increases further. Therefore, the common practice in large mills has been to make pellets from the foots, after mixing with deoiled meal and/or seed fibre collected from seed cleaner. The pellets are fed to the extractor along with the cake. However, the pellets crumble in contact with solvent in the extractor, causing fines to deposit on top of bed, and affecting the solvent percolation rate greatly. The extraction time, therefore, is required to be much longer than for other seeds.

Recently, a plant has been installed, incorporating a pre-heater and flakers, in between the grinders and Cookers. It is expected that the problem of under-cooking, and hence of sticky foots, will be minimised. Also, a separate 'foots press' has been incorporated, to form cake from the foots. This process should help overcome the problems of handling of foots and of poor solvent percolation in extractor.

It should be noted, however, that low protein content in sheanut and palm kernel will always be an issue for cake strength. Even 'good' cooking may not help achieve cake strength like for high-protein seeds. Novel preparation methods may be required for these seeds. This is further discussed towards the end of this chapter.

4.4.8.4 Copra

Copra is a soft material, thanks to very high oil content @ 65–67%. It is commonly pressed in two stages, with nearly 85% of the oil recovered in the first press, leaving 25% oil-in-cake; and only 10–12% of the oil recovered in the second press, final cake containing about 7–8% oil-in-cake.

The press performance on copra is peculiar and does not conform to those of 'soft' seeds. This is because, the copra is soft not because of high protein content (less than 10%), but because of very high oil content. So, as copra is pushed through the cage of the first press, the oil content progressively reduces, and copra mass becomes harder towards discharge end of the cage. Therefore the performance of the first press is also similar to that on 'hard' seeds, low throughput, high power requirement and high wear (Table 4.4). In fact, the throughputs are lower than even for hard seeds, thanks to the large quantity of oil to be expelled.

The copra cake fed to the second press is already hard, and it becomes even harder during passage through the press. Hence, the performance of copra second press is lower than on even the hard seeds. Press throughput is very low, power consumption is very high, and so is the wear on worms and bars. The cake from the first press must be broken into small pieces, to get a reasonable throughput of the second press. A hammer mill, with slow rotor speed may be used for the purpose. In fact, it further helps to add a small cooker on top of the second press. The press helps in two ways; one, the action of the rotating arm breaks down the cake pieces to sandy grain size; and second, it helps to complete the cooking of any uncooked particles in the cake. The second cooker may be sized for 10 minutes retention time. Only a small quantity of indirect steam may be given, just to maintain temperature, as the heat in the cake itself would be sufficient to cook the small quantity of uncrushed particles.

The cake emerging from the second press is hard and dry. Should care not be taken to have worms with flights until the very end of cage, close to 'choke', a lot of 'spurting' may occur. It helps to preserve moisture in cake by proper closing of conveying elements; whatever extra moisture is retained, it helps reduce spurting.

4.5 PRESS DESIGN ASPECTS

4.5.1 Methods of Design

Ever since the 'Expeller' was conceptualised and commissioned in the year 1900, screw presses have been designed on statistical and empirical models (Merrikin and Ward, 1981; Knott, 1991). Theoretical models have been attempted based on unidirectional pressing mechanism (Shirato et al., 1978; Mrema and McNulty, 1985), and efforts have been made to extend these to screw press mechanism.

A theoretical model was developed to predict the performance of a small screw press, fitted with a continuous flighted shaft, for pressing of oilseeds (Vadke, 1987; Vadke et al., 1988). The model calculations were mostly in agreement with experimental results, but it was not able to predict the extent of reduction in press throughput with throttling of choke. The effect of shaft speed and choke position on residual oil in cake was predicted reasonably well. The model predicted that cage cooling would improve the press performance, with higher throughputs and lower RO. It also predicted that the longer the cage length, the more will be the throughput capacity and the lower the RO; this matches well with the experience on screw presses in practice.

The major limitation of the above model was that it was basically developed for a press with continuous flighted screw shaft. Industrial scale presses have discrete worm pieces, distanced by collars. For such a shaft design, a suitable theoretical model is not available to date. Press manufacturers have developed their own statistical/empirical models for design of their presses. Let us consider the main features of an empirical model, which may conveniently be used to design the main features of a large press.

4.5.2 Cage and Worms Profile

4.5.2.1 Feed Worm Size

Say we want to design the cage and wormshaft for a specific capacity on rapeseed. We should know the bulk density of the feed material. For example, the bulk density of cooked flakes of rapeseed may be nearly 0.45 kg/L. For **prepressing** duty, we may consider shaft speed at say 35 rpm, to begin with. For the capacity requirement (T/h), and the bulk density and shaft speed, we can calculate the volume requirement of the first worm, the **feed worm**. The actual volume requirement (V_1) is obtained by dividing the theoretical volume by slippage factor, say 0.6.

Now estimate the power requirement for the given duty, from statistical data on earlier presses. For that power, and for 35 rpm, determine the shaft dia. It may be noted that allowable shear stress is just about 17% of the ultimate tensile stress for the material selected for shaft. As discussed earlier, it helps to select a material with high shear strength, so that shaft size is small as possible. Boss thickness of the feed worm may now be determined for a torque, which may be presumed to be just 5% of the total torque. Having known the shaft size, and the worm boss thickness, we know the ID of the feed worm. The next step is to assume the ID of the cage, which would give the depth of worm flight.

From the volume requirement, and the flight depth, the flight pitch is determined. If the flight pitch is much more than 1.35×Cage ID (assumed), then it is better to re-calculate using a bigger cage dia. Too big a ratio of Pitch : OD of the feed worm may cause much slippage of the flakes, and reduce the net volume displacement. Thus, now we have determined the cage ID, shaft dia, and feed worm sizes – boss thickness and pitch. The feed worm may be extended up to 1.25 pitch, to ensure positive feed into the compression zone of the cage.

4.5.2.2 The Last Worm

The next step may be to determine the volume requirement of the last worm, at the discharge end. This volume is quite different to that for the feed worm, for two reasons. One, after much of the oil has been expressed out, the mass flow rate has reduced considerably. And second, the bulk density of the material at the last worm is much different than at feed. All the air between flakes has been driven out, and the remaining material has been compressed repeatedly, to drive out even the air within oil cells. Thus, the density at discharge end is close to the true density of protein-rich mass, and may be considered as 1.3 kg/L. A slippage factor of 0.95 may also be considered for prepress.

Now, we have the overall volume ratio for the press:

$$VR_{overall} = \text{Mass ratio/density ratio/slippage ratio}$$

For example, if a prepress were to be designed for rapeseed, the overall VR may be computed as follows:

Mass flow ratio = 1.418 (for seed: oil 40%, moisture 5%; cake: oil 16%, moisture 6%) – see mass balance in Appendix 1.4.
Density ratio = Density at feed/Density at discharge = 0.45/1.3 = 0.346
Slippage factor ratio = slippage at feed/slippage at discharge = 0.6/0.95 = 0.63

Thus, $VR_{overall} = 1.418/0.346/0.63 = 6.51$

Once the required overall VR is determined, the volume of the last worm is known, from the following equation:

$$VR_{overall} = V_1/V_n \tag{4.2}$$

While the volume of the last worm is now known, we must await the determination of sizes of intermediate worms, to determine the pitch and depth of the last worm. This is so because the depth (and the boss dia) is determined in relation to the dia of the preceding worm.

4.5.2.3 Intermediate Worms

From the overall VR, intermediate VRs may be computed, for the determination of intermediate worm sizes. It is important to note that while designing the worm profile, the compression rate has to be maintained low. Sudden compression would lead to sudden increase in pressure, which would reduce the throughput capacity of the press to a significant extent. This relation is shown by the following equation (Shirato et al., 1983):

$$Q_x = G_3 * N - G_4 * 1/\mu * dP/dx \tag{4.3}$$

where, Q_x is the axial flow at a point 'x' in the cage, G_3 and G_4 are visco-geometrical terms (consisting of worm pitch and depth and the power-law index of viscosity of oilseed mass), dP/dx is a differential term, representing the rate of change of pressure over distance; other terms having the same meaning as given in Eq. (4.1).

Thus, it can be seen that axial flow is negatively affected by the rate of change of pressure over axial distance, dP/dx. Hence, sudden compression must be avoided as far as possible. This is one of the **most important principle**s in **design of worms** profile.

Two methods for the determination of sizes of intermediate worms have been discussed in Appendix 1.5. The two methods are 'constant compression ratio' and 'reducing compression ratio'.

The volumes of successive worms, with the first method, with constant compression ratio of 0.95, taking $V_1 = 10$ units, are plotted in (Figure 4.12). The successive volumes are 6.66, 4.6, 3.33, 2.54, 2.05, 1.73, 1.54 and 1.45 units, respectively. It may be noted that the plot is only indicative since the successive worm positions are marked at equal distance on the axis; this, in fact, cannot be, since the length of successive worms will be shorter. Anyway, it is clearly seen that the compression is milder towards the discharge end. This is how it should be, as even small compression can increase the pressure considerably towards discharge end. This is so because, towards the discharge end, as the oil content in the mass reduces, its viscosity increases, and so does the pressure (Vadke and Sosulski, 1988). The plot also shows that the progression in volume reduction is smooth, without any sudden changes; this is important to prevent pressure peaks. The profile would be even better with the second method, of reducing compression ratios.

With the above computation method, we have determined to have nine worms for the duty. The calculated VRs are to guide the determination of worm sizes. As the worm sizes are determined, the actual VRs may vary slightly from the calculated ones, since the worm pitch and boss dia are selected to round figures, for ease of machining, and for the sake of standardisation. In any case, the size of the last worm must be maintained close to the calculated value per Eq. (4.2).

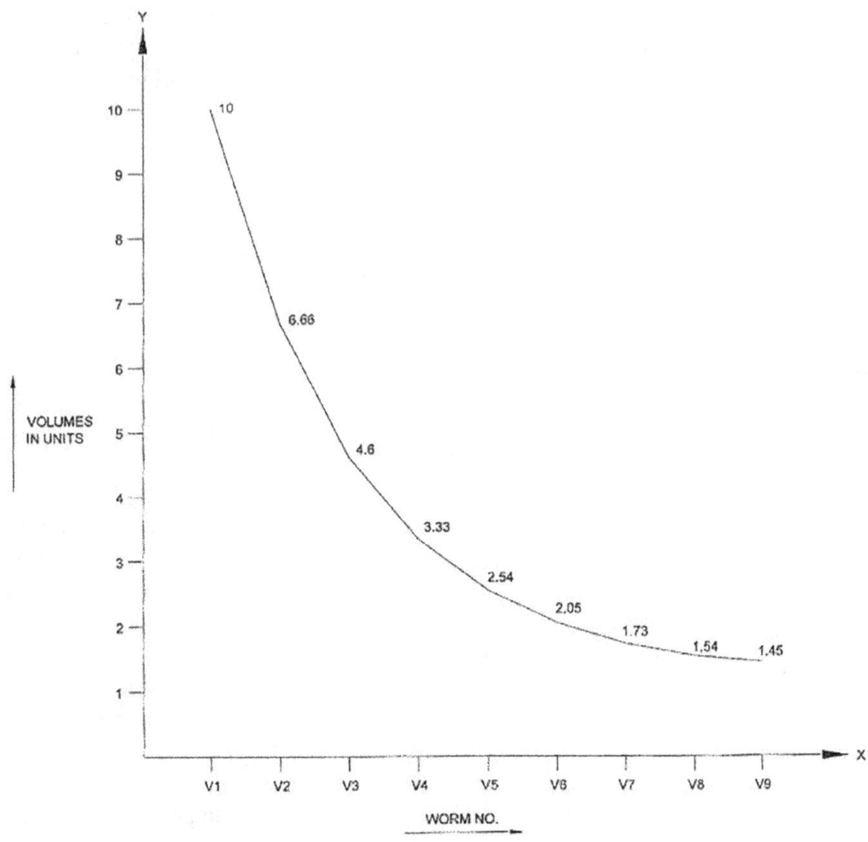

FIGURE 4.12 Reducing Worm Volume Profile.

New machines coming from major press manufacturers are now seen to have fewer number of worms, just five or six, with more intermediate humps and breaker disks. The procedure to compute the volumes, for less number of worms, while still maintaining the overall VR, could be as follows.

Consider the first VR as 1.6, in place of 1.5, and follow the same method, of a constant compression ratio of 0.95. We may then achieve the overall VR with just six worms. A similar procedure may be followed even with the 'reducing CR' method.

4.5.2.4 Intermediate Collars

As the worm boss dia is increased progressively, to cater to the reducing volume requirement, there would be mismatch between the boss dia of adjacent worms. A taper collar is fitted between the two to ensure smooth flow of material. Care must be taken not to have big difference between diameters of adjacent worms, so as to avoid high rate of pressure increase. High rate would impede the throughput severely, as explained with Eq. (4.3). Even where boss dia of two adjacent worms is the same, a short collar may be provided in between. Such collars serve two purposes, one to

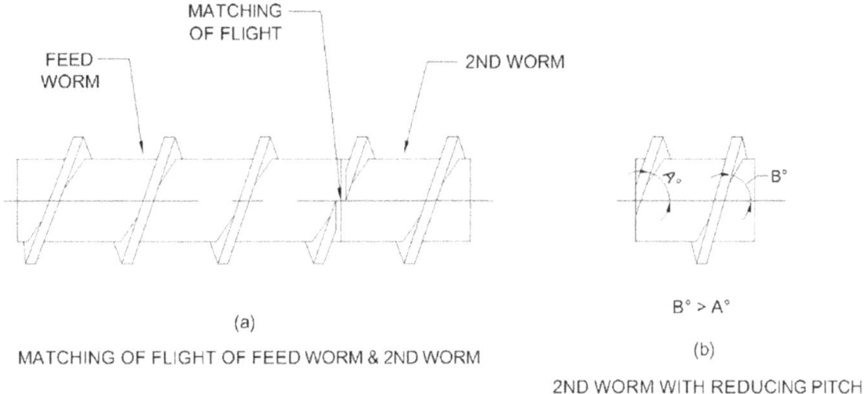

FIGURE 4.13 Feed Worm and Second Worm Details.

create extra pressure on the mass, and two, enable the fitment of a 'knife' between the two worms, for cutting and re-mixing action, as explained in the following section.

No collar should be placed between the feed worm and the second worm because such a collar would interfere with the intake capacity of the feed worm. Also, the starting flight of second worm should be matched with the flight end of the feed worm (Figure 4.13a). In fact, if the second worm could be machined for varying pitch, so that the starting pitch may match that of feed worm, and end-pitch would be as calculated by VR_1, it would prevent sudden contraction at the intersection of the two worms, and help improve the throughput capacity of the press (Figure 4.13b).

4.5.2.5 Wormshaft Assembly and Cage Length

Additional pieces such as pressure cones, or humps, followed by cake breaker disks, maybe added on the shaft. In a long press, typically two intermediate humps may be utilised. Once the lengths of all the pieces are added, the cage length is calculated. Since the cage length is a sum of individual segments, each equal to length of lining bars, some adjustments are usually required on the length of shaft assembly. These may be accomplished by varying the length of collars. Should this mean that the last collar, just before the choke, needs to be long (more than say 0.4 times cage ID), an extra worm, the same size as the last worm, be fitted just after the last worm, followed by a short collar.

Collars do not have flights, hence do not push the material forward. Therefore, material gets compacted on the collars and is pushed by the previous worm. A long collar would cause high pressure peaks and is detrimental to throughput capacity. Collars, therefore need to be short, say just about 0.1–0.15 times the cage ID. Only when the collar taper is high, to match the boss dia of the next worm, should the collar be longer, say up to 0.2 times the ID.

The assembled worm shaft in the example cited above would appear as shown in Figure 4.8. The location and lengths of intermediate humps are decided as per the guidelines discussed in the section below.

Matching of the feed worm and the second worm is important, as explained above. There is to be no intermediate collar, and the boss dia of second worm should be the same as that of the feed worm, to ensure smooth material flow into the cage (Figure 4.13).

4.5.2.6 Full Press Design

For the design of a press for full pressing application, shaft speed of 20 rpm may be considered. The required compression ratio would be different from that for prepress. For example, for full pressing of rapeseed, the mass flow ratio would be 1.58, against 1.42 for prepress, for the condition of cake with 8% oil and 5% moisture (see Appendix 1.4). Also, the density at discharge worm would be higher, due to further compaction (release of more air from cells), say 1.4 as against 1.3 for prepress. Slippage factor ratio would also be different because of more churning at discharge due to higher exit pressures. Thus, the required overall compression ratio would be as follows:

Mass flowrate ratio = 1.58
Density ratio = 0.45/1.4 = 0.321
Slippage ratio = 0.6/0.9 = 0.667

Hence, required $VR_{overall} = 1.58/0.321/0.667 = 7.38$.

Other calculations maybe conducted on similar lines as explained for prepress application.

4.5.3 Intermediate Humps

Large presses are designed with multiple squeezing zones, each with a high-pressure collar, or hump, at the end. For a press with two cages together nearly 2.5 m long, with total of 8 segments each say 300 long, and a transition piece 100 mm between two cages of different ID, it is beneficial to have at least two such humps. These are in addition to the final cone which is a part of the exit choke mechanism. For a full press machine, an additional third hump might help recover the final oil.

Successive squeezing, followed by re-mixing, of the oilseed mass helps improving the oil recovery without impacting the press throughput. This phenomenon may be explained by the **Theory of Consolidation** (Terzaghi, 1943; Mrema and McNulty, 1985).

Consider a case of **unidirectional pressing**, oilseed mass being pressed in a vertical cylinder, fitted with a perforated bottom, with a piston acting on the mass vertically down (Figure 4.14a). Initially, the mass is evenly distributed thru the cylinder. As the piston starts to press down, air goes out of the perforated bottom. As it presses more, some oil starts to come out. At this stage, a clear consolidation of mass can be observed in the layer just above the bottom screen (Figure 4.14b). As the piston presses further down, as some more oil is expelled, the consolidation layer becomes denser and deeper.

It is through this thick layer that any further oil has to be expelled, hence, too much power is required to expel incremental quantity of oil.

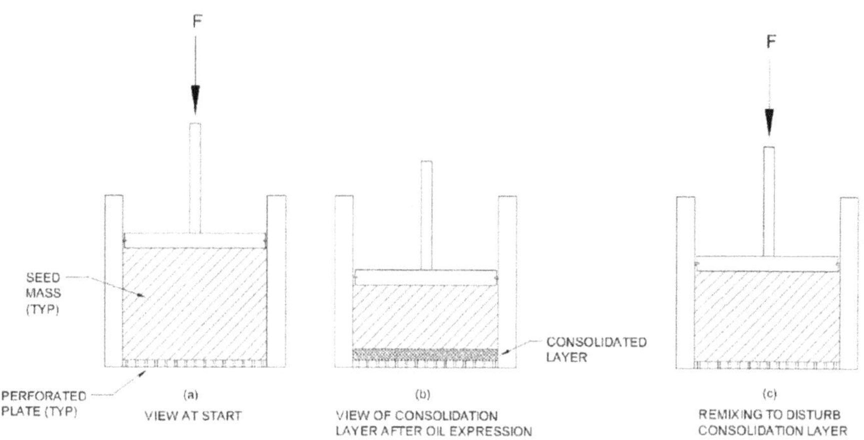

FIGURE 4.14 Uni-directional Pressing.

However, if the piston is taken out, and the oilseed mass is mixed thoroughly, the consolidated layer is disturbed, and the mass becomes evenly distributed (Figure 4.14c). If now the piston is pressed down, oil comes out readily without much power. With more pressing, a consolidation layer forms again, which should be re-mixed if more oil is to be expressed without using too much power.

This theory may be extended to a continuous screw press. Even here, the pressure is acting on oilseed mass from wormshaft radially. As the oil is expelled from the slots between cage bars, the layer near the wall tends to consolidate. This layer must be disturbed to maintain the oil flow rate without the need for high pressures. This job is accomplished by knife bars placed at the cage wall.

As the mass is subjected to a good squeeze at the hump, a strong consolidated layer would form. For quick thorough mixing of the layer, a specially designed breaker disk is fitted right after the hump. Both these devices, knife bars and breaker disks are detailed in the next section.

To benefit most from intermediate squeeze and re-mix, the first hump may be placed where nearly 40–45% of the overall VR has been achieved, and the next at 75–80%. For full press design, an additional hump might help recover the last oil; thus, humps at 40–45%, 65–70% and 80–85% of the overall VR would be beneficial in the latter case. Thus, in the example above (prepress for rapeseed), the first hump may be placed after the fourth worm, and the second hump after the seventh worm.

The size of hump should be decided to meet two objectives; one, to give sufficient squeeze for significant oil flow, and two, to not create too high a pressure peak. To meet both objectives, the discharge diameter may be determined such that the ratio of open cross-section area at the discharge end may not be less than 0.5 times the area at feed end (Figure 4.15). The length of the hump may be computed to keep the taper around 10–15° for the first hump, and around 5–8° for the second. A hump is usually built with a taper portion, followed by a short flat portion, which helps to create sustained pressure on the oilseed mass.

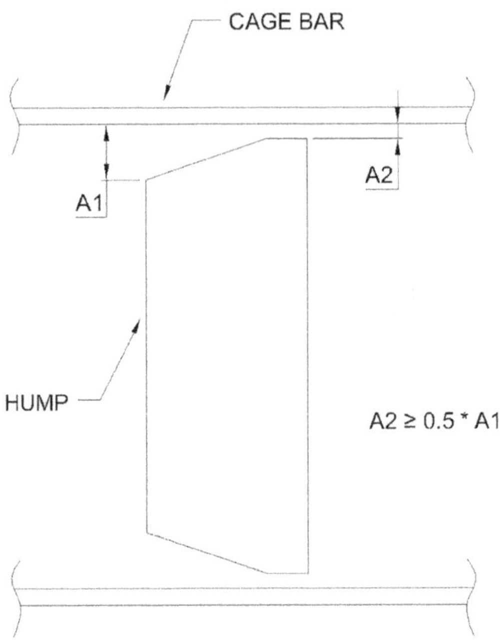

FIGURE 4.15 Pressure Hump.

4.5.4 Knife Bars and Breaker Disks

Knife bars and breaker disks are the cutting and re-mixing devices that disturb the consolidated layers and restore porosity of the oilseed mass within the cage. Generally a lot of attention is paid to the configuration of worms, as also to their maintenance; while knife bars and breaker disks are not paid the same attention. That is a mistake since these devices help not only achieve low OIC at low energy consumption, these are chiefly responsible for ensuring that no particles remain uncrushed, thereby ensuring a low milling defect.

Knife bars are fixed on each segment of the cage, and help mix the mass after each pressure worm. These are designed such that the individual knives sit above the intervening collar between two adjacent worms. The length of the knife may be determined, so as to leave 10mm gap from adjacent worm flight at each end (Figure 4.16). The depth of each knife may be decided to leave at least 6–8mm clearance from the collar. In fact, for the initial part of the cage, where worm fights may be quite deep, say more than 30mm, the knife may project only up to half the depth.

A breaker disk is fitted immediately next to a hump, to achieve complete re-mixing of the mass after high compression at the hump. A breaker disk may be designed to have 6–8 'teeth', tapered towards the feed face (Figure 4.17). The width of each tooth may be such as to leave at least half the cross-section area open for discharge. The clearance between the teeth and the cage bars should be minimum, say 2–3 mm, to prevent material by-pass. The depth of the teeth should be such that the cross-section area at the exit of breaker should not be less than the exit area of the hump (B, in Figure 4.17).

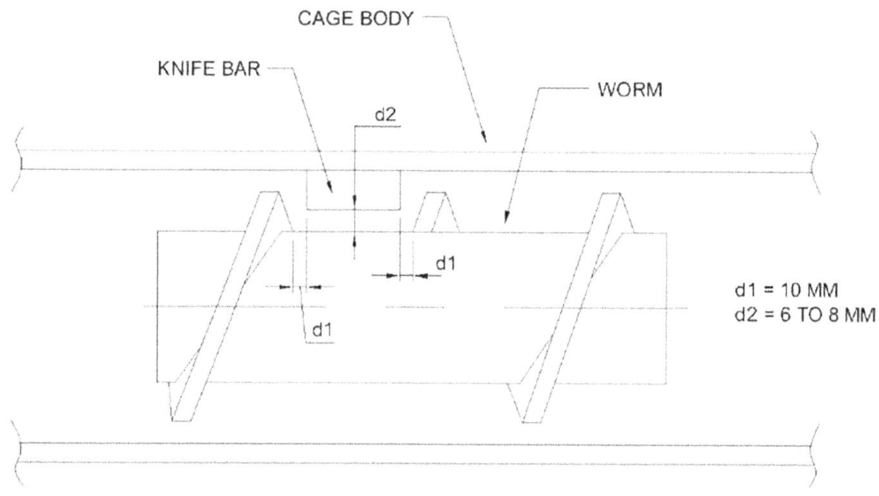

FIGURE 4.16 Clearance between Knife Bar and Worm.

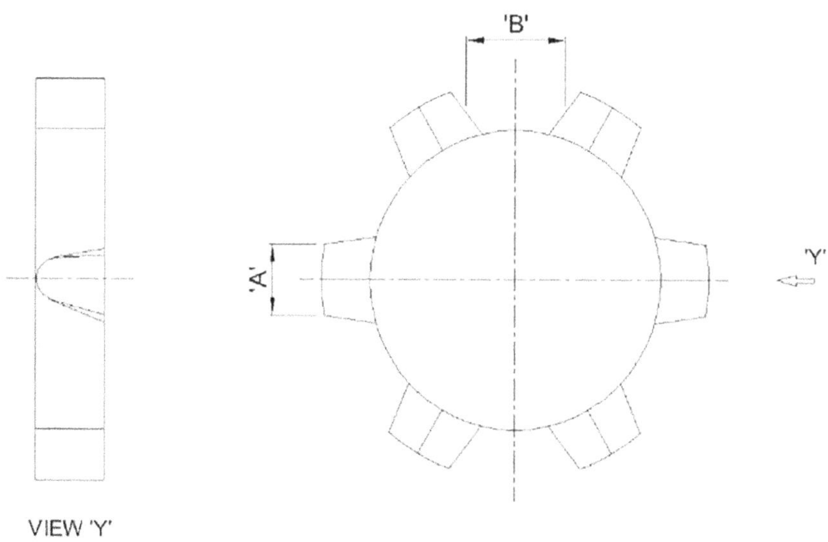

FIGURE 4.17 Breaker Disc.

In some presses, it is seen that the breaker disks are designed wrongly. The teeth have just a slight taper and are almost square-shaped. While there is still some mixing action, thanks to the combination of teeth and open passages, the disk acts largely as a pressing component. This effect may be observed from large oil flow from the cage segment at the location of breaker disk.

4.5.5 Choke

The choke consists of a cone, like a taper collar, moving inside a fixed ring (Figure 4.18). Both the cone and the ring are tapered to the same degree. As the cone is moved inside, towards the ring, the gap between the two reduces, and so does the cake thickness. For prepress machines, the gap may be adjusted between 10 and 15 mm for medium size machines up to 300 mm cage dia, and 14–20 mm for larger presses, so as to maintain cake thickness between 12 and 17 mm and 16–22 mm, respectively. For full press machines, the cake could be thinner, say 8–12 mm thick.

The length of the choke should be such as to create sufficient pressure on the exiting cake. The pressure may be calculated by equations governing the flow of pseudo-plastic fluid through an annulus. In large presses, typically, the length of taper choke may be nearly 0.25 times the cage ID. The choke is usually tapered with increasing dimension towards discharge. The taper of the cone, and the ring, may be around 20–30°.

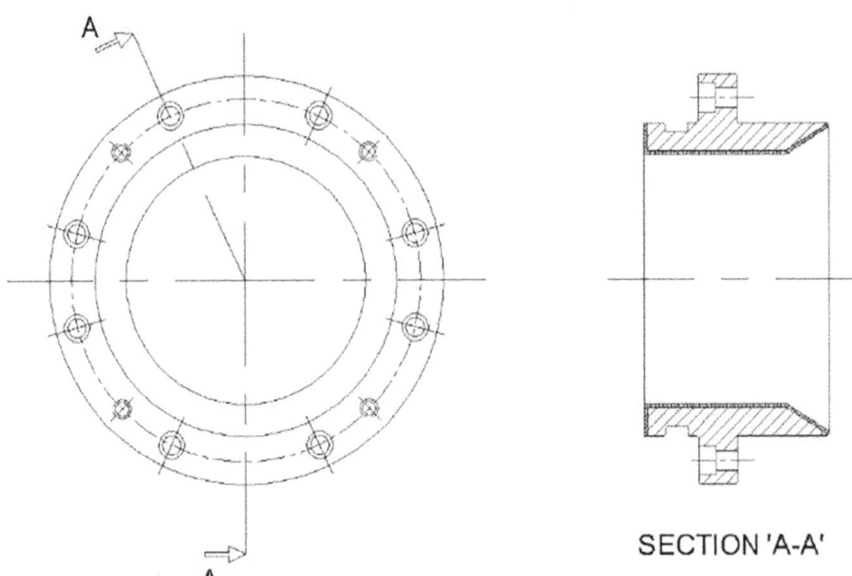

FIGURE 4.18 Choke.

4.5.6 Drive Power

As discussed before, the drive rating of a press is mostly determined from previous statistical data. Theoretical equations have been attempted for food extruders and may simply be represented as:

$$P_t = P_s + Q * P \tag{4.4}$$

where, P_t is the total power, P_s is the power utilised for shearing of mass inside the barrel, Q is the net volumetric flow rate and P is the exit pressure.

The shear applied on the mass is dissipated in form of heat, which raises the temperature of the mass. It is known that the major part of energy applied to shaft is dissipated as heat. In fact, the portion of the mechanical power used for building pressure and for pushing the mass through the cone represents less than a third of the total energy input (Janssen, 1989).

However, these relations usually give lower values than required for oilseed presses. The main reason is that the power required for breaking the cell structure of the oilseed mass is missing from the equation, so also is the power for the expression of oil through the mass and thru the slots in bars. Another factor missing is the energy wasted in the form of vibrations, both in sound range and otherwise.

Thus, press manufacturers select the motor size mostly with empirical correlations, based on their statistical data.

Once the drive is installed, and press started, it is educative to check if the drive selection is appropriate for the crushing duty. Press may be loaded fully, and motor amperage checked and compared to the rated amperage of the motor. The method to load the press fully is described in the earlier section, on press operation practices. Feeder speed is slowly increased until oil flow is observed from cage segments close to the feed chute. When the oil flow from the segment adjacent to feed chute is stabilised, the motor amperage is noted. If the consumed amperage is far too lower than (less than 80%) the rated amps of the motor, it indicates the motor is oversized. On the other hand, if even at full load of motor (say 95% of the rated amps), the oil flow is not observed from segment next to feed chute, it indicates the press is not fully loaded, and the motor is undersized.

4.6 OIL MILL LAYOUT – PRESS LOCATION

In traditional oil mills of yore, presses used to be placed on ground. The oil would exit from an opening in base box and would flow by gravity to an underground tank. Oil from this tank would be pumped out to filters by a positive displacement pump, typically a piston pump. Some foots would settle at the bottom of the tank and would be scavenged manually every few days. Also, any leakage in the tank would go unnoticed for many days or weeks, until significant quantity of oil was lost to the ground below.

In modern mills, manual scavenging of foots must be avoided, as it is always a risky operation. Thus, the oil tank must be placed over-ground. This requirement automatically requires that the presses be placed on the first floor. Such placement also allows to place downstream elevator, which receives foots from the screening conveyor (see section on oil filtration), on the ground, in line with other elevators of the seed preparation section.

Thus, the presses are conveniently located on the first floor, along with other heavy machinery such as crackers and flakers. This floor becomes the main operation area of the mill. The floor, of course, has to be constructed in RCC, not only for the static load, but more so to absorb vibrations. Sufficient space is kept vacant in front of the presses, to allow easy removal of the long press shaft. Cookers are usually placed just above presses, with the feeder of the press serving as the discharge screw of cooker.

A typical material flow sheet, indicating various levels of the layout, is shown in Figure 4.19.

Oil filters are conveniently located just above the main operation floor, and the valves for filter operation are located on the main floor. Ground floor houses, apart from the oil tanks, the Cake Cooler. This allows the press cake to be carried gently, with a drag chain conveyor, to the cooler. The cooled cake is then conveyed, again by a chain conveyor, to the solvent extraction plant located next door.

Control panel of oil mill should be placed on the main operation floor, while the motor starters panel, which tends to be much bigger than the control panel, may be located on the ground floor, which has much free space.

Two staircases should be provided on two sides of the oil mill, as a matter of safety. An overhead monorail should be fitted to lift and lower machine components down the building from a side, or through openings in floors within the building.

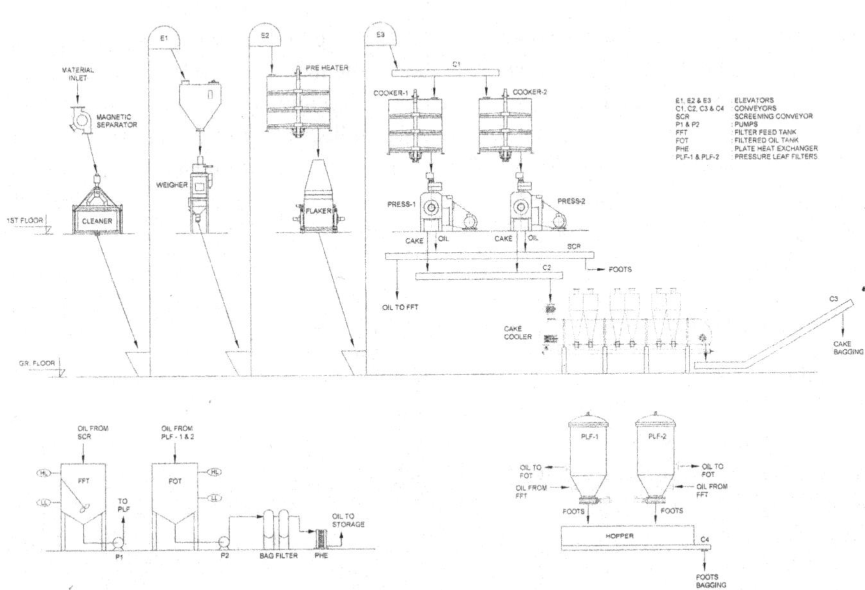

FIGURE 4.19 Oil Mill (Rapeseed) Flow Sheet with Level wise placement of Machineries.

4.7 FUTURE POSSIBILITIES

4.7.1 Large Press Capacity

Press manufacturers have been designing bigger and bigger machines to achieve more and more crush capacity per machine. This trend will continue, and we will see even larger presses in the coming decades. As prepress capacities have reached 800–900 Tpd with a single machine, medium-sized oil mills are already seen with a single press, as against a battery of presses of yore. As press capacities increase further, we will see even large oil mills with a single, or two, presses.

4.7.2 Energy Reduction

Efforts will also surely be made to reduce specific energy consumption (kWh/T) further. For this purpose, two directions appear plausible. One, to improve seed preparation, so that less work would be required within the press to macerate the cell wall structure. Second, bigger dia presses, with deeper channels of worms, would help reduce the heat dissipated due to friction at boundaries, namely the surfaces of worms and bars.

As discussed in the earlier section, the portion of the mechanical power used for building pressure and for pushing the mass through the cone represents less than a third of the total energy input, and the major part of energy applied to shaft is dissipated as heat. Thus we should see more and more efforts directed towards the reduction in heat dissipation from presses.

The heat dissipation is even more pronounced in case of full pressing operation. This is a major reason why the specific energy input is very high for full pressing. The other major reason, of course, is the higher pressure required for expression of the later part of oil from mass which is already hard, due partly to reduction in oil content, and partly to the consolidation effect. The consolidation theory has been explained in section on design of humps, knife bars and breaker disks. It would help to design devices which would cut and re-mix the mass better within the press.

4.7.3 Lower RO in Full Pressing

Press manufacturers have already declared their intent to design presses that would recover maximum oil, leaving only 3% oil in cake (Vavpot et al., 2012). This has to be done without the generation of excessive heat, which would otherwise cause charring of the cake and darkening of oil. More effective cutting and re-mixing devices would certainly help in this endeavour. Deeper channels, achieved with larger cage dia, would also help reduce the heat generation as explained earlier. Thus, steps taken to reduce energy requirement would also help reach lower RO, without affecting the quality of oil and cake.

Such presses would also support the movement towards ‘green technology’, by offering a practical alternative to the solvent extraction process. A low RO of 3% would mean most of the oil has been recovered. There would not be much incentive to reduce the oil-in-cake further, since the cake would be directly useful for animal

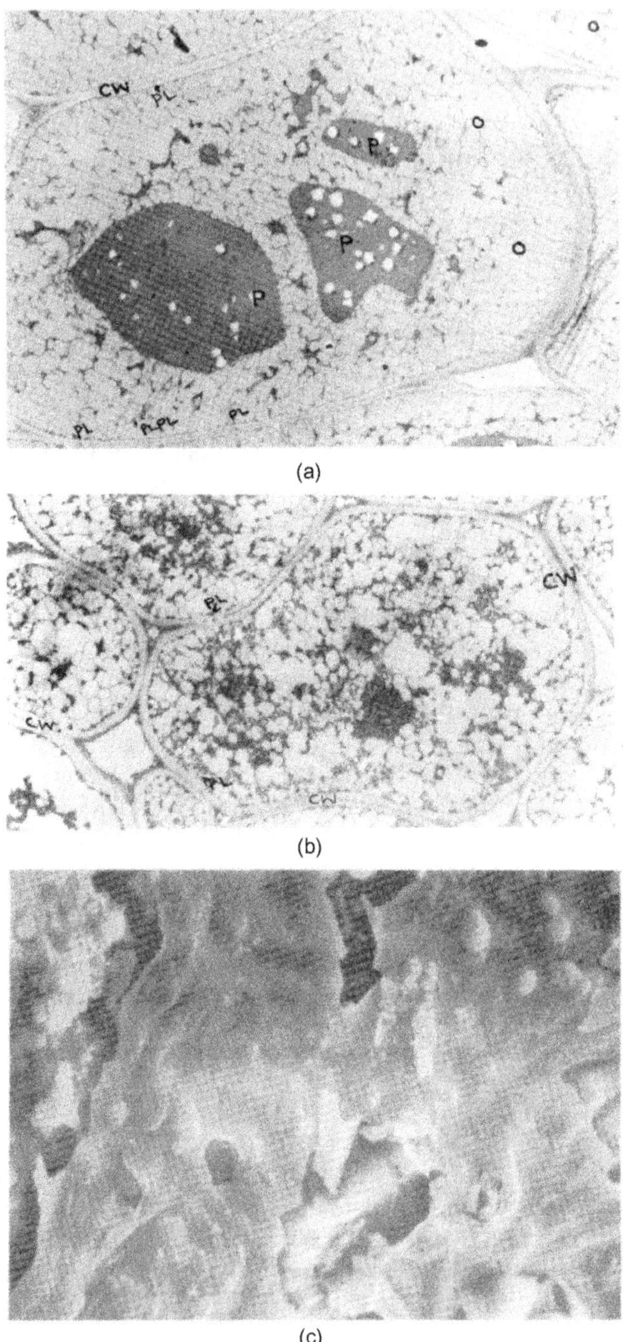

FIGURE 4.20 Transmission Electron Micrograph of typical sections of Oil seed. (a) Rapeseed Kernel prior to oil expression, (b) Rapeseed Kernel after oil expression (drained), (c) Cashew Kernel after undrained loading.

feed which requires between 2% and 4% oil anyway. Apart from the lower RO, less specific energy consumption would also be necessary, if the process is to replace solvent extraction.

4.7.4 Cell Wall Rupture by Hydraulic Pressure

Mrema and McNulty (1984) showed that it should be possible to demolish the structure of cell walls by means of 'undrained' compression (Figure 4.20). Thus, this could be an important pretreatment of the seed, preceded by flaking and cooking, prior to pressing.

If complete cell wall rupture were achieved in the mass, a press could then be designed to create radial pressure, without too much shear. Low shear would lead to low heat dissipation and less energy input to the press. Low shear would also ensure low temperatures, and better quality of oil and meal. Furthermore, if all the oil had been released from the cells in the undrained compression step, a very low RO could be targeted.

So, pretreatment of the kind mentioned above could achieve low RO without using too much energy, and help produce better quality oil and meal. Thus, this pretreatment step could meet all the requirements of a 'green' technology that the world awaits.

5 Cake Cooling, Storage and Utilisation

5.1 PROCESS STEPS

The cake discharged from the screw press is hot and moist. The large pieces discharged from the press (in the case of well-formed cakes of high-protein seeds) break down to medium size, say 50–150 mm wide during conveying to downstream process machines. Cake must be cooled, and surface moisture removed, either for safe storage or for feed to solvent extraction. Breaking the cake further, down to 15–30 mm size, also helps in solvent extraction to a significant extent. This aspect is discussed further in Part 2 of this book.

5.2 CAKE BREAKER

Size reduction may either be practised before or after cake cooling. The breaker operation is easier if done on a cold cake. Warm cakes generally emit water vapours which condense inside the breaker and may lead to the accumulation of cake fines. The wet fines deposition leads to corrosion and may also lead to process interruption. Warm cakes are also somewhat plastic, and may clog the breaker rolls. However, one major advantage in performing the size reduction upstream of cooling is that the small size of cake lends itself better to uniform cooling. Hence, breaking prior to cooling is the preferable sequence; vapour condensation, and fines deposition, may be avoided by ventilating the breaker. To overcome the issue of plasticity, the conveyor carrying the cake to the breaker may be ventilated; air contact over a minute, as with a slow-moving conveyor, can make these cakes sufficiently firm for effective breaking.

In a cake breaker, typically, the cake is passed between two rolls, fitted with spike disks (Figure 5.1), which give a gentle poke to the cake and break it into smaller pieces, without creating much fines. The disks are cast from alloy steels.

For large size cakes, as obtained from rapeseed or groundnut, a breaker with two pairs of rolls is preferred, so as to minimise fines. In double pair machines, the disks on upper pair rolls are configured with deeper spikes, spaced on wider pitch. Mid-sized pieces are obtained from the upper rolls. These then fall through the lower pair of rolls, fitted with shallow spikes spaced closer, to obtain small pieces of cake. Care must be taken to reduce the pitch in both directions, between spikes on a single disk, as also the pitch between two adjacent disks. Otherwise, the size spread can be large, and the product is a mixture of small and medium size pieces. Lower rolls must be rotated at higher speed than the upper rolls, to ensure the free flow of broken cake.

Unlike seed crackers, cake breakers are not heavy-duty machines. The specific power consumption is only between 0.5 and 0.8 kWh/T.

DOI: 10.1201/9781003309475-6

FIGURE 5.1 Cake Breaker. Courtesy – Jo Jon Engineering Pvt. Ltd., Mumbai.

If the cake is to be sent for second press, a **hammer mill** may be used for cake breaking. Here, the fines created during breaking action do not pose a problem, since these would be agglomerated within the second press. This is usual practice in copra crushing and is highly recommended for multi-stage crushing operations as practised on various seeds in countries in Asia and Africa. For this application, the speed of rotation of the rotor should be low, say 300–400 rpm.

5.3 CAKE COOLER

Cake should be cooled to 50–55 C if it is conveyed to the solvent extraction plant. On the other hand, lower temperature, close to 40 C, is preferred for storage. Should simultaneous moisture reduction be desired, slow cooling is preferable to quick cooling. Slow cooling allows for moisture from the core of the cake to rise to surface and evaporate. Quick cooling may produce cake which would appear dry on surface, but with more moisture in the core which could create problems later. During solvent extraction, the higher moisture in the core would interfere with solvent flow in and out of cells; while in storage the core moisture would slowly rise to the surface and might cause lumps formation. For cases where cake cooling is the main objective, together with moisture removal only from the surface, quick cooling processes are preferred for their simplicity.

A vertical zig-zag channel is a simple device for quick cooling of cakes (Figure 5.2). As cake falls through a series of inclined diverters fixed in a wide chute, it is contacted by air sucked from the bottom through the falling cake. Multiple counter-current

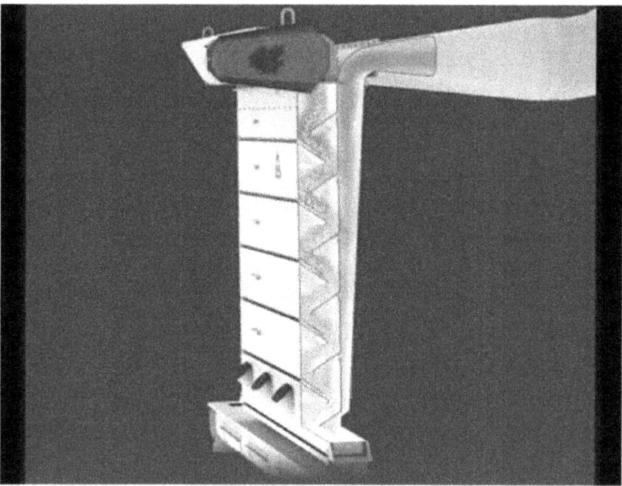

FIGURE 5.2 Multi Aspirator. Courtesy – Kice Industries.

contacts with air ensure quick evaporation of moisture from the cake surface. Should warm air be used, in place of ambient air, moisture from the core would rise to the surface faster, thanks to the lower relative humidity of warm air. Discharged cake would be a bit warm, say at 45–60C, depending on warm air temperature and the moisture level. This cake temperature would be appropriate for solvent extraction. If cake is to be stored, however, the warm air section would be followed by a cold air section at the bottom, to cool down the cake to near ambient condition. Some designs utilise air suction from multiple points, in place of only bottom suction, making it a cross-current operation. These devices are often referred to as **Multi-Aspirators** (this is the brand name registered by Kice Ind.). The multiple tumbling action on the cake often leads to cake breakage and the creation of fines, which is a major disadvantage of this design.

The most common design of coolers used for cake cooling is the horizontal, slow moving **Screen Belt Cooler** (Figure 5.3). The screen is built in a series of segments, each 150–250mm long, fixed to two chains at either end. The screen and the chains are mounted in an enclosure. As the cake moves along with the screen, the air is sucked from the windows below at multiple locations, in a cross-current configuration. The advantage of this design is longer cooling time, as also gentle handling of cake which prevents cake breakage and fines production.

Cake is carried normally in a shallow bed, depth 200–300mm, depending on the percentage of small particles in the cake. If small particles content is high, a low bed depth is preferred to avoid too much pressure drop on air stream. The screen belt should be provided with a variable speed drive, to allow regulation of bed height, and also to vary the cooling time. For most cakes, a cooling time of 6 minutes is sufficient to achieve effective cooling, Should significant drying be desirable, warm air may be forced through the cake just after the feed, followed by cold air draft up to the discharge end.

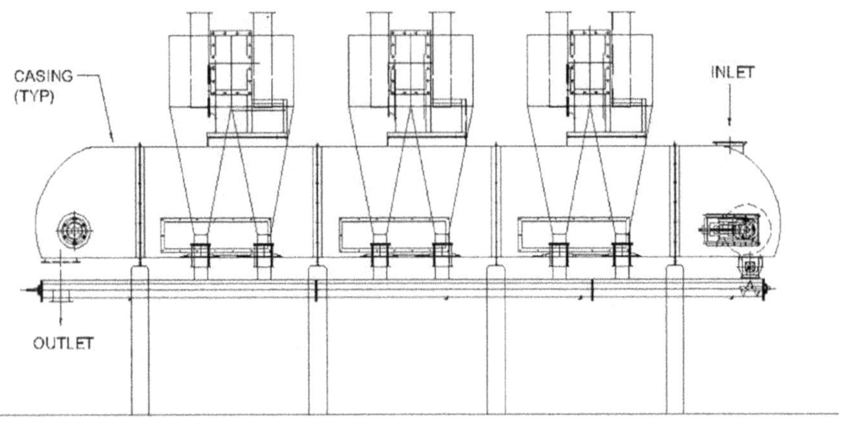

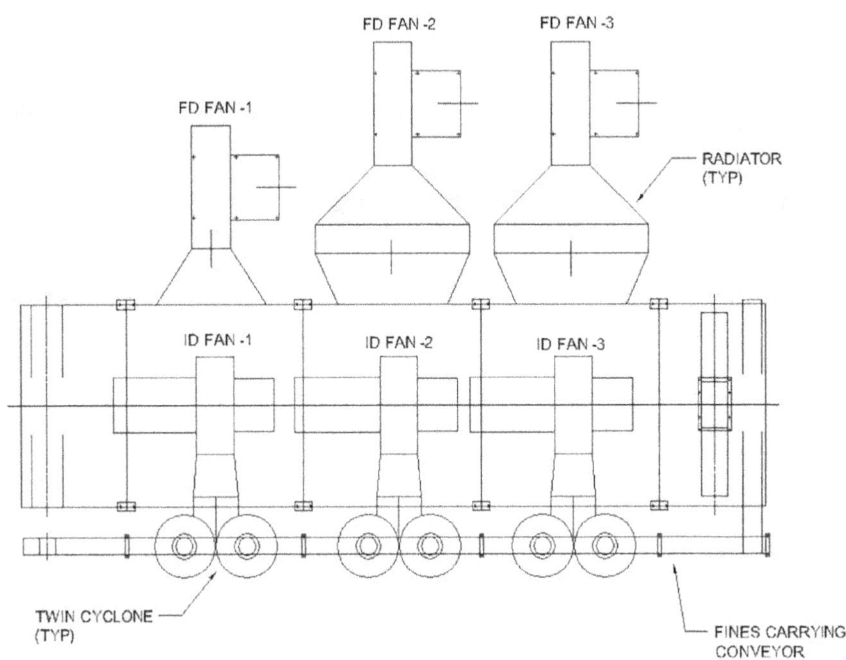

FIGURE 5.3 Screen Belt Cooler.

A critical factor in the operation of a horizontal moving bed cooler is the uniform spread of cake over the entire width of the screen belt. If part of the screen is not covered, air would bypass from the open area, drastically reducing the effectiveness of the cooler. One way to ensure uniform spread is to install a full-width rotary valve (RV) at the inlet and maintain a small buffer above. This method, however, leads

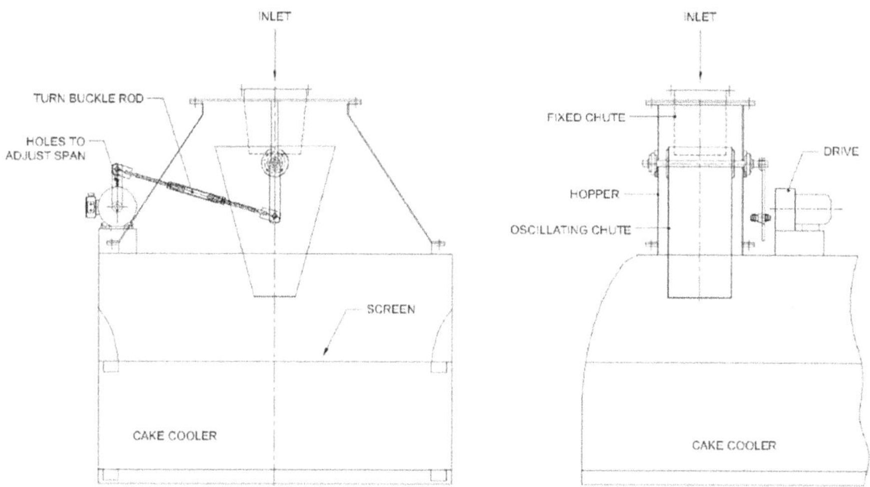

FIGURE 5.4 Oscillating Feeder.

to cake breakage in the vanes of the RV and is usually not preferred for cooling of cakes, especially if the cake is to be conveyed to solvent extraction.

A more suitable method to ensure uniform spread is to utilise an oscillating feeder (Figure 5.4). Here a feed chute is oscillated to spread the cake over the width of screen. The device should come with two adjustments, one to vary the span of oscillation, thereby adjusting the spread span; and two, a centering adjustment to adjust the spread in one direction or the other.

As the cake may usually contain some fine particles, the outlet air would carry some of the fines. Air cyclones should be used to recover the fines and make the air clean. The fans and the cyclones may conveniently be fitted on the cooler body, to make the unit compact, reduce its footprint, as also to make it easy to erect.

The presence of moist air can cause corrosion on the internal surfaces. One solution would be to use stainless steels for the enclosure walls. A cost-effective solution, however, is to spray- galvanise the internal surfaces. The screen belt is preferably made of stainless steel, as it is also in contact with the cake. The twin chains used to move the screen are mostly made out of high-tensile steel, with the rollers being made of hardened steel. As this is a slow-moving screen, carrying shallow bed of cake, the power requirement is minimal, much less than 0.1 kWh/T. Most power is used for the air drafts, in the range of 0.8–1.5 kWh/T.

Air requirement for cake cooling may be determined with heat balance, assuming the outlet temperature of the air. Should moisture reduction be desired along with cooling, the air quantity may be determined using the air-water psychrometric chart. For higher moisture reductions, say more than 1%, warm air should be used to avoid having to pump in large air quantities. The warm air requirement may also be determined using the psychrometric chart, by assuming air outlet humidity.

5.4 CONVEYING TO SOLVENT EXTRACTION PLANT

The broken cooled cake maybe conveyed to the solvent extraction plant, which is housed in adjacent building. Care must be taken not to break the cake further during conveying. For this purpose, either a slow-moving drag-chain conveyor, en-masse type, or a belt conveyor is preferred. Drag chain conveyors are compact compared to belt conveyors, but consume marginally more electricity. Chain speeds may be restricted to 20–24 m/min to avoid cake breakage. As the conveyor would typically have a bend, for upward lift towards the extractor inlet, care must be taken to avoid cake accumulation at the bend; any such accumulation is a sure recipe for the production of fines. The accumulation may be avoided by using long-curve bends, fitted with a thin pusher wheel to keep the chain close to the bottom plate. It must be noted that the presence of fines affects the solvent percolation rates in the extractor; hence, fines content must be minimised.

The conveyor should be continuously ventilated at the discharge point, to eliminate the risk of passage of solvent vapours, escaping accidentally from Extractor, towards the feed point of the conveyor.

5.5 BAGGING FOR STORAGE

If the cake is not to be solvent-extracted, it may be bagged and sent for storage. This may be cake from full press operation, which is used in animal feed formulations. Alternatively, in small oil mills in Asia and Africa, which are not attached with a solvent extraction plant, the bagged cake may be sent to a solvent extraction facility nearby.

Bagging of cake may be done manually in small oil mills. Typically, bags would be held below a Y-chute from bagging conveyor, placed on a weigh platform. As the bag fills up to the desired weight, the cake is directed to the other arm of the chute by a manual flap, to fill another bag placed below. Bags may be stitched manually with a hand-held stitching machine. Polypropylene woven bags are mostly used for the purpose, filled up to 50–80 kg.

In large mills, engaged in full press operations, auto-bagging machines may be used. These machines are typically designed for 6–8 bags/minute. Since bags have to be attached to the machine manually, the average practical bagging rate may be 5–6 bags/min. Thus, with 50 kg bags, bagging rate per machine may be 15–18 Tph.

Each bagging machine may be provided with a buffer hopper above, sized for 10–30 bags, to ensure continuous flow to the machine. Dusty atmosphere should be avoided for trouble-free operation of the machines; a central dust collection system for the bagging house is desirable for the purpose.

These bagging systems are used extensively for deoiled meals after solvent extraction. The bagging and storage systems are explained in some details in the second part of this book, on solvent extraction technology.

5.6 CAKE UTILISATION IN ANIMAL FEED

The cake, used in animal feed, is ground and mixed with other ingredients required for the formulation. In most cases, the mix may be pelletised and bagged.

Some cakes may need special treatment prior to being fed to animals. As mentioned in the section on seed preparation and crushing technology, in India, rapeseed is crushed for the production of 'pungent' oil. The cake, containing 8–9% oil, with polyphenols and phytates which give it bitter taste, is mostly used in cattle feed formulations. Before being fed to cattle, however, it is soaked in water for a couple of hours, until it disperses evenly to form a paste; the paste is then mixed with other feed components and fed to cattle. The hydration reduces the bitter taste and makes it palatable to cattle.

6 Oil Filtration, Cooling and Storage

Oil expressed at the screw press contains foots, particles from seed mass. The foots must be separated, and the oil cooled, prior to storage.

6.1 PRE-FILTRATION

The foots may be a mixture of coarse and fine particles. It helps to screen out the coarse particles prior to filtration, to reduce load on filtration systems. It must be noted that the removal of all coarse particles at this stage, may cause operational issue at the downstream filtration stage. If only fines were sent to a filter, these would clog the filter screen and reduce filtration rates. Hence screen opening in pre-filtration operation should be selected carefully. Three types of equipment are commonly in practice.

6.1.1 Screening Tank

The most common equipment used for screening out the coarse foots is the 'Screen Tank'. This is a tank sized to hold oil for nearly 1–1.5 hour, to allow settlement of the coarse foots, which are then scraped off the tank bottom by a slow-moving slat conveyor, that brings these up onto a screen fitted at the top of tank (Figure 6.1). As the oily foots are moved slowly on the screen, oil drips down, the foots are carried forward, and discharged into a screw conveyor at the end of the screen. Pre-filtered oil is let out at a height from the tank bottom, so as not to disturb the solids settled at the bottom, and pumped to oil filters. It may be noted that the over-an-hour oil hold up also helps in absorbing the variation in pumping rate to filter, and the oil recirculation at the filter startup.

Until about 30 years back, the screen used to be typically made of wire mesh, with a support frame below to maintain straightness of the screen. However, no design of support frame would prevent sagging of the screen between frame elements over a period of time. As portions of screen sagged, foots would accumulate on the portions, interfering with screening process. Fouling of screens was also a recurring problem. Over the past two decades, wire mesh screens have been replaced with **wedge-wire screens**, which are not only strong enough to eliminate sagging but are also self-cleaning, thanks to the tapering profile of individual wires (Figure 6.2). Wedge wire screens are usually made of stainless steel, while the tank, and the slat conveyor, may be of carbon steel. The screen area is determined to offer dwell time of foots, on screen, of nearly 4–6 min. The foots discharged from a screening tank may contain 50–65% oil, depending mainly on the degree of cooking that the proteins-in-foots have undergone. Well-cooked foots may be drained to lower oil-in-foots.

DOI: 10.1201/9781003309475-7

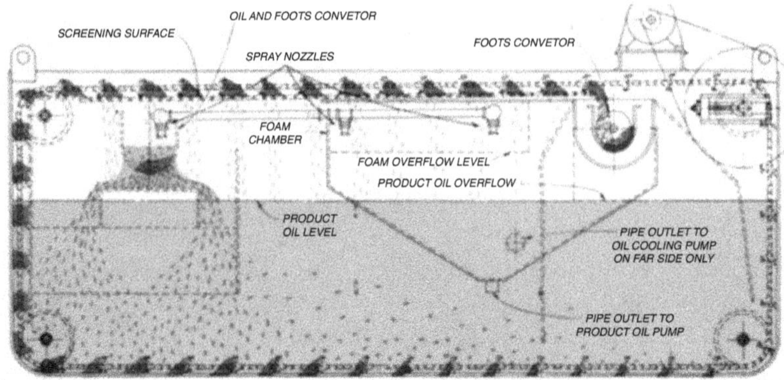

FIGURE 6.1 Oil Screening Tank. Courtesy – Anderson International.

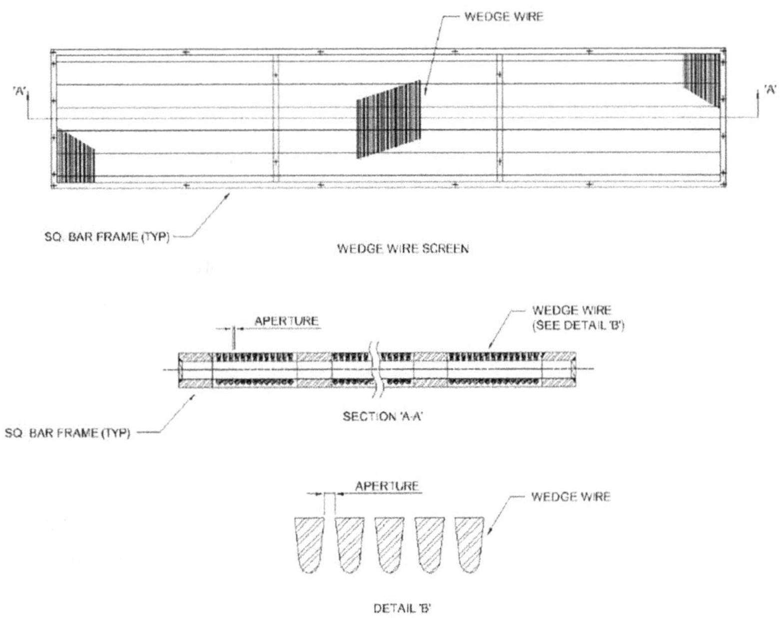

FIGURE 6.2 Wedge Wire Screen.

The opening between adjacent wires is usually between 0.4 and 0.6 mm. For a wire width of nearly 1.5 mm, the open area is nearly 20–25%. Opening less than 0.4 mm is not recommended, as these would remove even the mid-size particles (0.3–0.4 mm), which are actually necessary to form the first layer on filter screens in downstream filtration operation.

Sticky foots, produced mainly due to inadequate seed preparation, would tend to form a thin layer on the screen, unmoved by the slats which travel on the screen at a small clearance. Such a layer would stop further screening of oil and lead to very oily foots from the screen tank. To prevent such an occurrence, the slats may be fitted with a flexible flap touching and scraping the surface of screen.

6.1.2 Screening Conveyor

A recent alternative to screening tank is a Screening Conveyor (Figure 6.3). Here the screening action is similar to that in screening tank. However, there is no oil hold up for the settlement of foots. The oil recovered at presses is sprayed directly onto the screen. Foots are pushed by slow-moving slats and are discharged into a conveyor at the end. The dripped down oil, containing some fines, is pushed in reverse direction by the return slats and is discharged in to a tank, which feeds the filters downstream. The body of a screen conveyor should be of welded construction, and segments should be properly sealed with gaskets which withstand action of oil and heat.

The screening conveyor offers significant advantages over the traditional screening tank. One, since there is no oil hold up, the duration of contact between oil and the foots is minimised. It may be noted that the foots, being small particles, are of a darker colour than the general mass of cake. During the long contact, foots impart some of the red colour to oil. Second, in case of multiple presses, a conveyor is required to carry the oil and foots to the screening tank. The screen conveyor simply replaces the oil-n-foots conveyor, hence saves the additional footprint of a tank. Finally, of course, the conveyor is much cost-effective compared to the tank.

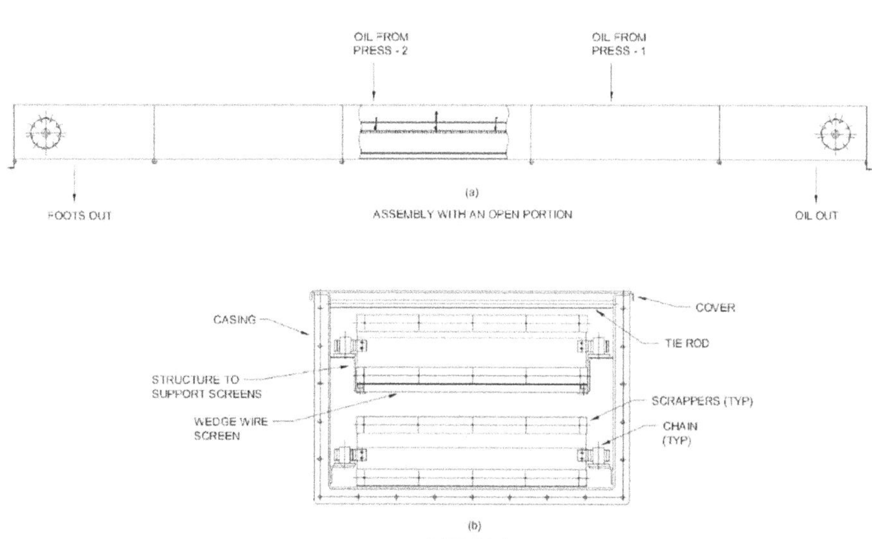

FIGURE 6.3 Screening Conveyor.

It would be pertinent to note some operational issues related to screen conveyor. One, sticky foots would also overflow to chains on either side of slats and hinder the chain movement. In case of fats which solidify at ambient temperature, even a short stoppage of the plant, would cause solid fats accumulation on the chains. Such a problem is faced especially with shea butter. For such application, the entire length of conveyor may be steam-jacketed on either side.

Second issue with screen conveyor may be the difficulty in spreading of foots over the entire width of screen, especially with screens wider than 500 mm. Typically the feed to a screen conveyor is via a chute connected to the bottom conveyor of a press. If a chute of say dia 150 mm feeds the screen at the center, it is likely that the foots may spread only around the center, leaving the edges empty. A diverter, placed within the chute may help spread the foots wider. Alternatively, the chute may be shaped elliptical, at discharge point, to cover most of the screen width.

The screen opening, and the oil-in-foots, is nearly the same as in a screening tank. The dwell time on screen is somewhat shorter, say 3–4 min, due to the narrow screen compared to that on screening tank; this time is, however, adequate in most cases.

6.1.3 Vibro-Sieve

Vibro-sieves may be circular (Figure 6.4) or rectangular in shape. The sieves are vibrated, in high frequency but low wavelength mode, with the help of eccentric motors, in such a way that the foots move forward in one direction. In circular vibro-sieves, oil and foots are fed at one point on the sieve, and as the mixture moves round through nearly 300–320 degrees, oil drips down, and the foots are discharged from an opening in the peripheral shell. In rectangular sieves, oil foots mixture is fed at one end, and the foots are discharged from the other end.

Because of high-frequency movement of the sieves, the separation of oil from foots is much better than with the case of gravity separation on stationary screens (tank or conveyor), and the foots typically carry only about 40–50% oil. Wire mesh, of mesh size 30×30, having opening 0.55 mm, or 40×40 (0.37 mm), is commonly used for the sieve. Standard circular vibro-sieves are available in sizes up to 1,800 mm dia, which can handle foots up to 300 kg/h.

Vibro-sieves should be mounted on strong anti-vibration pads and preferably placed on RCC floors. In spite of the better separation efficiency of vibro-sieves over stationary screens, the former are not preferred widely, because of maintenance issues, and noise, associated with high vibrations.

6.2 OIL FILTRATION

6.2.1 Filter Feed Tank

An oil buffer is required to ensure continuous pumping of oil to filters. In traditional mills, the Screening Tank provides the buffer. Should the Screening Conveyor or vibro-sieve be used instead, a separate tank must be provided upstream of the filter, for the purpose. Such a tank is commonly known as the filter feed tank (FFT).

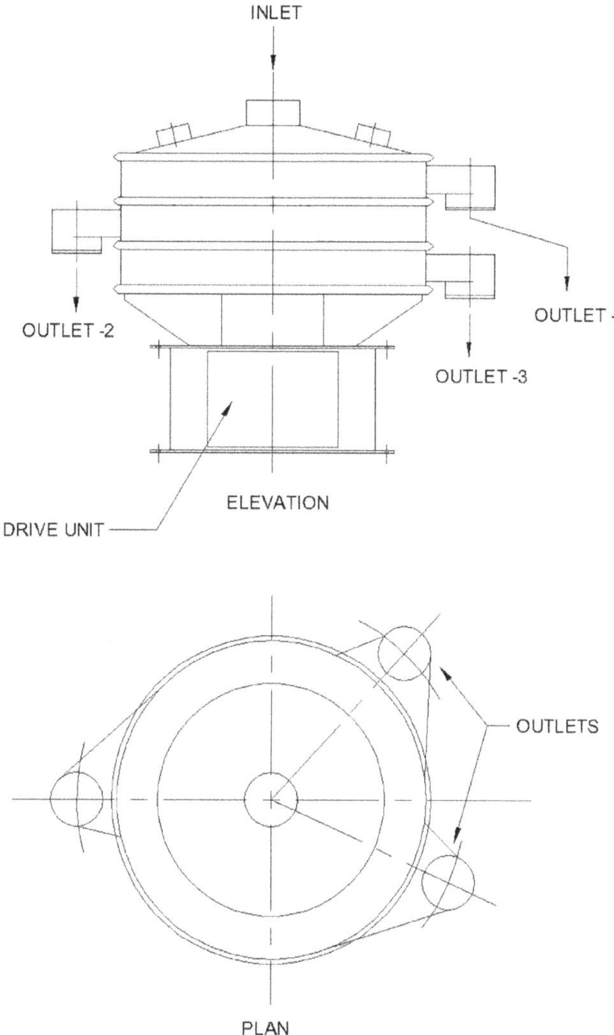

FIGURE 6.4 Vibro Sieve. Courtesy – Sharplex Filters (India) Pvt. Ltd.

FFT should be fitted with an agitator designed to keep all foots in suspension and to prevent settling. Settling of foots would clog the inlet of pump. If foots were settled at the bottom of tank, and connection to pump was given from a height, the oil to filter would be devoid of mid-sized particles necessary for proper filtration; also, the settled foots would have to be drained, and the slurry handled separately.

As discussed in the chapter on press layout, the Screw Press should be located on the first floor. Oil should flow by gravity to the FFT. In old oil mills, where presses were located on the ground floor, oil tanks were placed underground. Oil was pumped out by means of positive displacement pump placed on top of the tank. These tanks were typically flat-bottomed, and part of foots settled at the bottom. These had to be excavated manually, which was a risky operation. Also, the oil mill had to be stopped, say once a week, for this manual cleaning of foots. Leaks in the tank could go undetected for days or weeks, and oil lost to the ground below. In modern oil mills, FFT must be placed over ground, and pump, connected to bottom of tank, to have positive suction always.

The FFT should be adequately sized to cater to accommodate the recirculated oil from filter at the start of the filtration cycle. Typically, an FFT, sized to hold oil for 60 minutes, and operated normally at mid-level, is sufficient for the purpose. FFT should be fully covered, provided with a vent pipe, and with sight glasses provided to observe the oil level. An oil level indicator, mounted on the side, helps plant operator, to ensure continuous oil flow to filter, as also avoid the risk of overflow. At the least, low-level and high-level switches should be fitted on the tank to ensure (a) pump not running dry and (b) no overflow from tank, respectively. In automated mill operation, level transducers work to show the level on remote computer, as also indicate it locally.

Centrifugal pump, with open impeller, may be used to pump oil from FFT to the filter. It should be sized for dual set of parameters; one, for nominal filtration flow at high pressure, and the other, for high flow at low pressure at the beginning of the filtration cycle.

6.2.2 Filter

Amount of fines in the screened oil is generally 3–4%. These are to be removed in main filters. Different types of filters are used for the purpose.

6.2.2.1 Plate-n-Frame Filters (Plate Filters)

In small oil mills, plate-n-frame (or, recessed plate) filters are used for oil filtration (Figure 6.5). These are sturdy and versatile equipment and have been used for decades all over the world. The filter element is a woven cloth with apertures of 5–10 microns (air permeability 10–20 L/m^2/s). After every filtration cycle, the filter plates

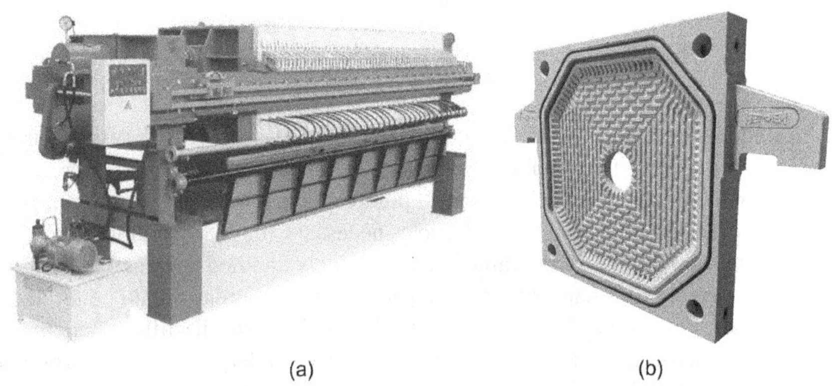

(a) (b)

FIGURE 6.5 (a) Plate Filter. (b) Polypropylene Filter Plate. Courtesy – Metchem.

have to be opened, and the filter cake scraped off the cloth pieces manually. Thus, it is a labour-intensive operation. After a few filtration cycles, if the cloth is fouled by fines, the filtration rate reduces, and it is time for washing the cloths.

In recent times, cast steel plates have been replaced with those of polypropylene or high-density polyethylene (HDPE). Also, the open discharge of filtered oil from each plate (or frame), has been replaced with closed piped discharge. Due to the manual labour involved, these filters are not preferred in large mills. However, in cases where the filter cake tends to be sticky, as for pungent mustard oil or shea butter, plate filters may still be preferred over pressure leaf filters (PLF).

6.2.2.2 Pressure Leaf Filters (PLF)

These are the filters of choice in modern mills. Even among small oil mills, there is a trend to replace old plate-n-frame filters with these. Two types are available: vertical VPLF and horizontal HPLF. The vertical types are easy to operate and involve practically no manual work. These are also cost-effective compared to the HPLF and also require less area for operation. The HPLF are used for special cases, where filter cake may be somewhat sticky and may need personal attention.

6.2.2.2.1 Vertical PLF

Construction: VPLFs consist of a series of filter screens (leaves) assembled vertically on top of a common horizontal header, enclosed in a vertical cylindrical body with conical bottom (Figure 6.6). The leaves typically have a 60 × 60 mesh wire mesh supported by sturdy grills inside. The leaves are held on a tie-rod at top, with distance pieces to maintain the exact distance between leaves. A vibrator is mounted on the tie-rod extended outside the shell, and is used to release the cake off the leaves at the end of the filtration cycle. The cake falls down in the conical portion and is discharged below by opening the bottom valve.

Startup: At least two filters are used for the continuous operation of the filtration section. When a filter is ready, and boxed up, for filtration, oil feed pump is started. As oil enters the filter body from bottom up, it displaces the air within. The air goes out from the overflow pipe. As oil fills up, and slight overflow is observed in sight-glass provided on overflow pipe, overflow valve is closed. As more oil is pumped, pressure builds up within the filter body. The pressure forces oil through the filter leaves, and oil flow in outlet line is observed. Initially, some fines may be seen in filtered oil. This oil should be circulated back to FFT. As oil flow continues through the leaves, mid-sized particles in oil form a thin layer on the leaves. This layer then acts as the filter medium for subsequent oil flow, to trap all particles, mid-sized as also fines. Once that happens, the outlet oil is seen clear of fines. This oil may then be sent to filtered oil tank downstream. The oil re-circulation may be necessary for 20–30 min, depending on the size distribution of foots in oil. This is the reason why the FFT must be sized for at least 60 min hold up.

Operation: As filtration progresses, and the thickness of filter cake increases, pressure rises slowly, and filtration rate reduces. If the leaves are already fouled, the pressure can rise high. It the pressure rises beyond the rated pressure of filter, normally 4.5 Barg, pumping must be stopped, and be diverted to second filter. Should the leaves be clean, pressure may not rise high even when the filter may be fully

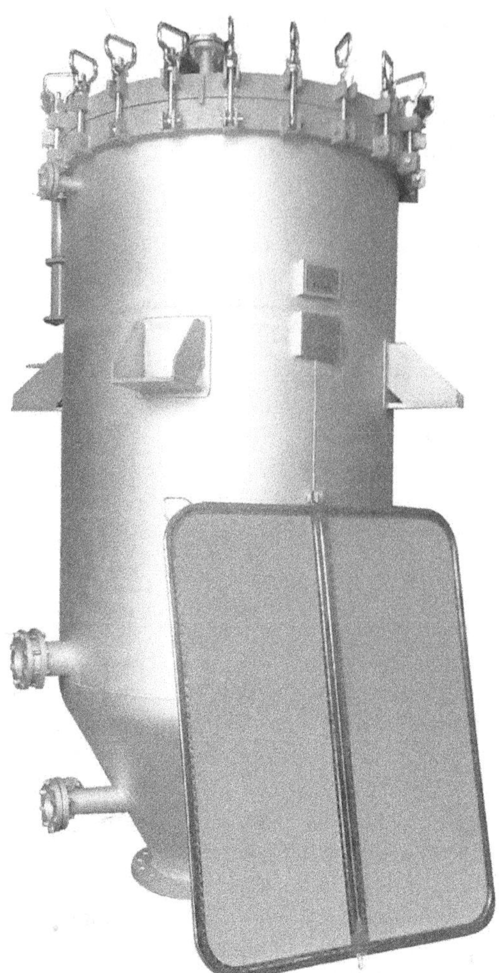

FIGURE 6.6 Vertical PLF. Courtesy – Sharplex Filters (India) Pvt. Ltd.

loaded of filter cake. If high pressure is viewed as an indication for the completion of the filtration cycle, and filtration is continued further, excessive loading may result in the increase of cake thickness to the extent of closing the space between leaves. At this stage, excessive forces act on leaves frames and may cause leaf deformation. Such a situation must be avoided. Hence, a filter should be stopped after a pre-determined time, calculated to achieve about 80% loading.

Full loading means cake thickness equal to one-third of the space between adjacent screens. At full loading, the empty space between screens is just enough for the ready discharge of cake. Hence, a safe operation is to load filters to 80–85% of full loading. A timer may be provided on the control panel, to set time calculated for 80% loading. As alarm goes off, it is time to change the filter. Oil pumping is then diverted to the second filter.

Filter Changeover: The first filter is to be emptied, and made ready for the next filtration cycle. Oil outlet line valve is now closed. Compressed air is let in to force the unfiltered oil out through the inlet pipe, which is then diverted to another tank, called Heel Tank. As oil flow from filter inlet line subsides (as seen in sight glass on the line), inlet valve is closed and outlet valve is opened again. Compressed air now forces oil out of the cake layer. This oil is also diverted to the Heel Tank. As the oil flow in outlet pipe stops, compressed air is stopped and outlet valve is closed. It is now time for discharge of cake.

The bottom valve is opened. Vibrator is turned on, and vibrations transmitted to the leaves. The high-frequency, low-amplitude, vibrations cause the cake to fall off the leaves, and fall through the bottom valve, and is collected into a hopper below. As all the cake is discharged, vibration is stopped. The bottom valve is closed. All valves are set to filtration mode. The filter is now ready for the next filtration cycle.

Filter blowing, with compressed air, should reduce the oil content in filter cake to nearly 20–25%, and may take around 20–40 min for the purpose.

Compressed air requirement for filter-blowing is quite significant. Large air compressors are required for the purpose, and these are a major part of utility services for the oil mill. The air requirement is further discussed later, in the section on utilities.

The entire filtration cycle may be automated with fitting of flow switches, and a set of auto ON/OFF valves, along with a small PLC to control the timings and start/stop sequences.

Maintenance: After about ten filtration cycles, the top cover of a VPLF should be opened, and condition of leaves should be observed, to check if all the cake has fallen, and if the leaves are clean. If the leaves appear dirty, fouled by fines, these should be removed for cleaning. A cleaning tank should be provided nearby for the purpose. Manual handling of leaves should be minimised, to avoid damage to the screen. An overhead monorail, to carry the assembly of leaves together with the tie-rod, may be provided for the purpose.

Filtration area and Leaves Spacing: It may be noted that all the leaves of a VPLF are not the same size. Height being the same, width varies, to fit into the circular vessel. For most oilseeds, filter cake thickness up to 20 mm may be allowed. Thus, the clear space between leaves should be 60 mm minimum. This is achieved when adjacent leaves are set at 75 mm pitch. For coconut oil, the foots are fibrous and a thicker filter cake may be possible without affecting the filtration rate. To take benefit of these characteristics, the leaves spacing may be increased to 90 mm. Filtration area requirement may be computed from the cake thickness and the quantity of foots expected in 3 hours. A filter may be emptied, cleaned, and boxed up, within a span of 2 hours; hence, 3 hours is a safe cycle (filtration) time.

Filter Cake Hopper: Cake hopper must be sized to receive the entire load of filter cake from a filter. For two, or more, filters, a common long hopper may be provided. A screw conveyor is fitted at bottom of cake hopper, to take the filter cake back to Cooker. Variable speed drive is provided to regulate the cake flow, so as not to overload the cooker with filter cake. The speed is so adjusted that the hopper is emptied just before the time for the next cake discharge.

Foots may not be recirculated to rotary cooker, where heating tubes are prone to fouling. In such a case, foots may be fed directly to the Feeder Screw of the Press.

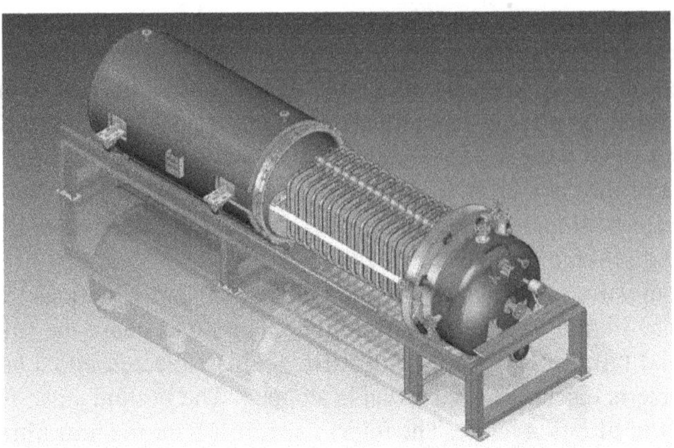

FIGURE 6.7 Horizontal PLF. Courtesy – Sharplex Filters (India) Pvt. Ltd.

6.2.2.2.2 Horizontal PLF

HPLF consists of vertical leaves mounted on a common header inside a horizontal cylindrical body (Figure 6.7). The entire shell may be de-coupled from an end-flange and retracted, on a rail, to help discharge cake into hopper below. This allows a clear view and access to all the leaves. Should any cake be sticking to a leaf, even after sufficient vibration time, it can be cleared manually. Filtration procedure and parameters are similar to those for VPLF. However, filtration area requirement may be computed using a 3.5–4 hours cycle, in place of 3 hours, since the opening and box-up time is a bit longer than for VPLF.

HPLFs are more expensive than VPLFs; however, the cost of spares is less, since all the leaves are the same size.

6.2.2.2.3 Screen Selection

As seen earlier, in section on VPLF, the standard screen used in PLFs for filtration of press oils is wire mesh 60×60 (opening 0.25 mm). A question is often asked as to why a finer mesh is not selected. For example, in oil refineries, bleaching filters commonly use a finer mesh 110×24 (opening 0.13 mm). The use of such a mesh would then allow a finer opening to be used in pre-filtration screening operation, which would help the removal of more foots at that stage, and reduce drastically the load on filtration?. A reduced load of foots could, in theory, allow smaller filters to be used and save significant equipment cost?. However, it should be noted that finer mesh cannot be used in the filtration of press oils, as the proteinaceous foots are prone to compaction and can clog the screen. That is why relatively bigger particles, the mid-size 0.3–0.4 mm, are required to form the first layer on the screen and act as the filter medium. Thus, pre-filtration must allow passage of these mid-size particles, and then the filter screen is selected to have opening just smaller than the mid-size. Hence, the idea of using finer screens, however tempting it might be, must be avoided for the filtration of press oils.

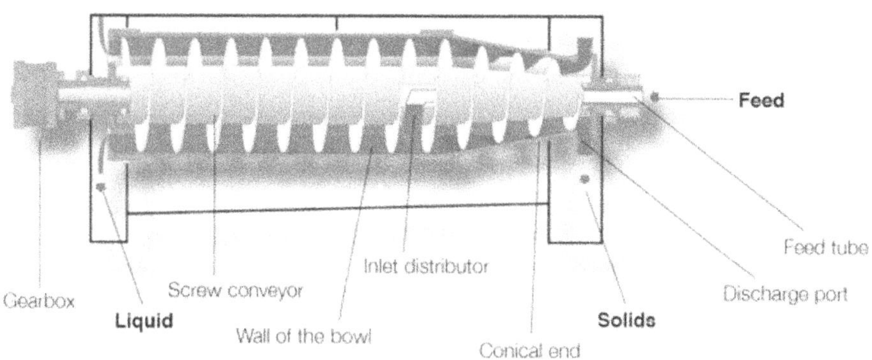

FIGURE 6.8 Decanter. Courtesy – Alfa Laval (India) Pvt. Ltd.

6.2.3 Decanter

For oils, whose foots are sticky and do not lend themselves to filtration, separation by using high-speed decantation may be utilised. High-speed decanters consist of a horizontal cylindrical bowl rotating at high speed, pushing solids to periphery by centrifugal action, and a screw rotating at slow speed to discharge the solids out of the bowl (Figure 6.8). The bowl is encased in a semi-circular cover.

Such decanters are in use in mills which crush shea nut and castor beans. Decanters have also been employed to remove foots and gums together from castor press oil (Keshre, 2019).

Typically, the separated solids contain up to 60–70% oil, due to the 'sticky' texture.

Decanters are compact machines, which use much smaller operation area compared to filters. The energy consumption, however, is much higher than for filters which only use the pump power and compressed air. Also, the fines in outlet oil tend to be higher @ up to 0.15%, compared to just about 0.01% after filtration.

6.2.4 Heel Tank and Filtered Oil Tank

Heel tank is used to receive oil during emptying of filter. Thus, it may simply be sized to have useful volume equal to 1.5×volume of filter. A small scrubber cyclone may be fitted at the vent, to recover oil droplets from the air, which is used to blow the filter cake. The oil is circulated back to FFT for filtration. It is better to fit an agitator in heel tank to prevent settling of foots. Hi-level and low-level switches may be fitted for safety.

The filtered oil tank (FOT) receives filtered oil from the filter. Since the oil is devoid of foots, no agitation is necessary. High-level switch may be fitted to prevent overflow, and low-level switch to prevent the pump from running dry. An FOT may be sized to hold oil for just up to 15 minutes, since the oil is continuously transferred either for drying, or cooling, prior to storage.

Both the tanks are usually made of mild steel, and so is the filter feed tank.

6.2.5 Safety Filter

Filtered oil must be cooled prior to storage. Usually, it is passed through a safety filter first, to ensure the total absence of fines, which could foul the surfaces of a cooler, or that of a heater in case of oil drying. In case of any damage to any of the screens in the main filter, some fines may pass to the filtered oil. Such accidentally escaped fines must be trapped in the safety filter. Thus, safety filters are usually sized at twice the area required for normal fines content in filtered oil.

A safety filter, typically, is a bag filter, wherein a cloth bag is supported on a strainer basket. Cloth may be woven or non-woven polypropylene, with openings of 5–10 micron. A duplex-filter unit is preferred, so that when one is fouled with fines, the other may be used just by changing the position of inlet and outlet valves, for continuous operation. Pressure gauges before and after the filter indicate the extent of fines trapped. A pressure drop up to 0.5 bar is normally allowed before the change-over. If the pressure drop is more than 0.2 barg at the start of a cycle, it indicates the bags are dirty and need washing before re-use.

In large oil mills, with oil flow rates of 300 kL per day, or more, where main filters may be sized at around 100 m^2 each, or bigger, the size of safety filters should also be increased in proportion. In such cases, a plate-type filter may be preferred, over bag filters, for its larger fines holding capacity.

6.3 OIL DRYING AND COOLING

6.3.1 Oil Drying

Most oils, obtained by mechanical expression, are filtered, cooled and sent to storage. However, some oils, e.g. shea butter, may be required to be dried, to reduce moisture content to below 0.1% to prevent deterioration due to oxidation during storage. Typically, press oils would contain 0.2–0.3% moisture. Process employed for moisture reduction is vacuum flashing of warm oil. In case moisture in oil is higher than 0.3%, filters may be prone to clogging; in such cases, oil may be dried prior to filtration.

Oil is heated to around 90–95 C, and flashed in a vessel under vacuum of 690–700 Torr (absolute pressure 50–70 Torr). This level of vacuum may be obtained by means of either a vacuum pump (water ring) or a steam ejector. Should the starting moisture content be higher than 0.3%, a water-cooled condensor, placed between the flasher and vacuum pump, would help by condensing the water vapour, to reduce load on the vacuum pump.

The vacuum created by a water-ring pump depends on the cooling water temperature, which in turn depends on the wet-bulb temperature of the air. In cold atmospheres, with ambient air below 20 C, vacuum up to 720 Torr may be obtained. In such a case, oil temperature may be reduced.

Temperature control is necessary to prevent overheating, which could cause darkening of oil. A simple feedback control scheme, with an RTD, a PID controller, and a proportional steam control valve, would enable a fair regulation.

The dried oil must be cooled prior to storage.

6.3.2 Oil Cooling

Oil is cooled to below 45 C, prior to storage, to avoid oxidation reactions due to long contact with air. If cooling water is available at 32 C or lower, cold oil temperature of 40 C may be achieved. Since the oil is free of all solids, plate heat exchangers (PHE) offer a simple, compact, and cost-effective solution. A temperature control scheme, with a control valve for water flow, may be used. However, this is not always practised, and water flow kept at a bit excess, to maintain oil temperature at 40 C or lower.

Pressure gauges installed on oil lines before and after the PHE indicate the state of fouling of plates. When the PHE is clean, the pressure drop would be in the range of 0.6–1.0 Bar, as per the design. If it increases more than 0.2 bar beyond the design value, it is time to open the PHE and clean the plates.

After cooling, oil is sent to storage. A mass flow meter may be installed on the line, to record the quantity of oil sent to storage. At the very least, a volumetric flow meter, with a totalizer, must be kept on line.

6.4 OIL STORAGE

Press oil may be stored separately, or mixed with solvent-extracted oil and stored. Since the oils are to be refined in most cases, except where press oils are used for direct human consumption, storage may be designed for a few days hold up only. Where the refinery is operated continuously, storage for 3–5 days would suffice. In cases where disruptions to refining operations may be expected, longer storages, up to 7–10 days may be foreseen.

Storage tanks should be fitted with level indicators, to help keep the oil stock. For a close tab on the stock, tanks should be calibrated, by filling water or weighed quantity of oil, every 3–4 years. Such a measure helps to counter-check the quantity of oil released from the oil mill, as measured in the outlet flow meter.

Tanks may be constructed of mild steel. Outer surfaces should be duly painted. Inner surfaces may be protected from rust by the application of coat of a drying vegetable oil such as linseed oil.

6.5 OIL FILLING

Press oil is sent to bottling plant only where it is used for direct human consumption. Packing sizes may vary from 1 to 20 L. The usage of pouches for 1 L filling, is increasing the world over. Even smaller size pouches, or bottles, may be utilised for oils with high value, such as coconut oil, olive oil, or mustard-pungent oil.

A polishing filter is preferred prior to filling. A polishing filter must remove any suspended impurities, such as gums along with micro-fines. Typically, a paper cartridge, or papers on stacked horizontal leaves, may be used for the purpose.

A polishing filter is, as a rule, made of SS, and so are the buffer tanks in the filling section.

7 Process Control and Automation

7.1 PROCESS CONTROL PARAMETERS

While it is important that individual machines be designed optimally to suit the specific requirement of processing the given oilseed, it is equally critical that the process parameters be controlled, so that the oil mill performs with consistency and high efficiency.

7.1.1 Feed Control

The very first parameter which must be regulated is the flow of seeds from the Day Silo to the Oil Mill. Uniform material flow allows optimal utilisation of various machines and helps achieve consistent quality. An easy method to accomplish this is to install a Rotary Valve, with a variable speed drive, at the bottom of the Day Silo (Figure 7.1). The speed may then be set for the production level desired for the day. Care must be taken to pre-clean seeds prior to storage in silo, as impurities, especially farm chaff or twigs, may block the RV and disturb the material flow.

For materials such as copra, which tend to bridge over conical bottoms, silo openings are designed to be wide where a screw conveyor may be fitted. Even if the speed of screw is set as desired, the flow would fluctuate to some extent, due to the uneven discharge of the large copra cups. Hence, it is desirable to have a buffer bin, after the first Hammer mill inside the Oil Mill, which may be fitted with an accurately-machined, varied pitch, screw discharger, with variable speed (Figure 7.2). Copra flow to downstream machines may be regulated at the Bin, so that load on the second Hammer Mill is uniform, and so is the size reduction.

7.1.2 Overflow Bins

When a production line employs multiple crackers, or grinders, an Overflow Bin should be installed. As feed to individual machines is regulated, with their own feed mechanisms, excess seed coming from upstream equipment, namely cleaning section, can overflow to the bin. The material in the bin is evacuated by means of a screw discharger with variable speed drive and is recirculated to the feed point. Should this bin be not provided, the last cracker, or grinder, in line would be operated with the fluctuating level in its feed hopper. Such a fluctuation might cause the last machine to produce inconsistent quality product and to even run empty at times. This principle also applies to lines with multiple dehulling machines for cottonseed or sunflower seed.

DOI: 10.1201/9781003309475-8

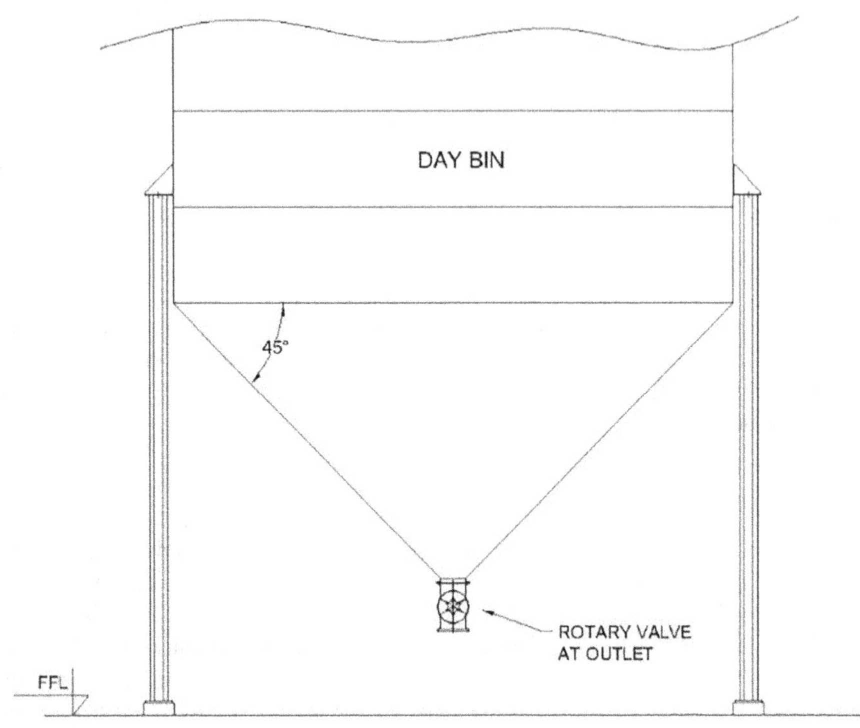

FIGURE 7.1 Rotary Valve Below Day Silo.

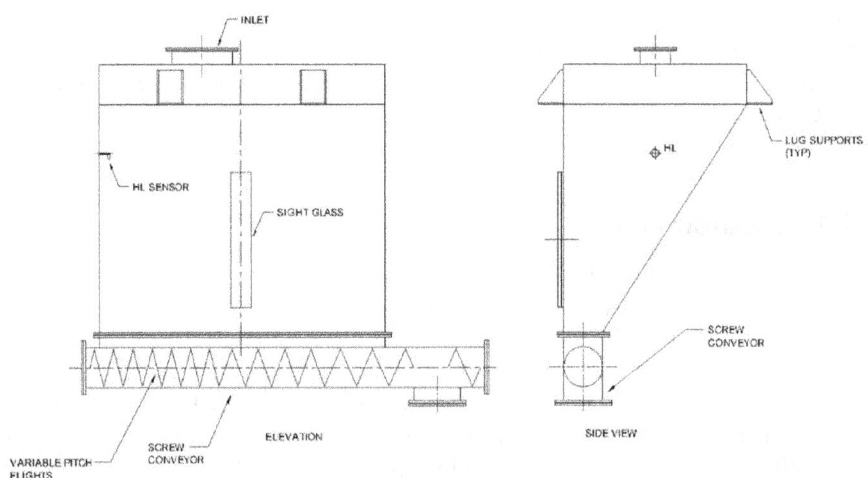

FIGURE 7.2 Screw Discharger below Metering Bin.

Such an overflow bin becomes even more important in case of multiple flakers, because inconsistent feed to the last flaker could cause major deficiencies in flakes quality. Also, frequent empty run is not desirable, as the short delay in re-engagement of the rolls, after each empty run, yields a load of thick flakes. With the overflow bin in place, all the flakers can be made to run at full load consistently.

Overflow bin is also beneficially employed in line with multiple cookers, installed on top of multiple presses.

7.1.3 Cooker Level Control

Pre-heaters and Cookers are vertical multi-stage vessels. Level regulation on all stages allows control of cooking time. Various types of level controls are detailed in Chapter 3.

7.1.4 Cooker Temperature Control

Temperature control, together with level control, on pre-heaters and cookers helps achieve the uniform quality of cooked grits or flakes, in terms of softening of oil cells, as also the moisture content required for effective pressing. The control mechanisms for grits/flakes temperature, as also for injection of open steam, are described in Chapter 3.

7.1.5 Feed Control to Presses

Material flow to presses is regulated by varying the speed of the Feed Conveyor, which is fitted just below the Cooker. The speed may either be set manually or in auto-control mode to maintain the load on the press motor. This is detailed in the chapter on Screw Pressing.

7.1.6 Foots Recirculation to Cooker

The seed fines, which are squeezed out along with oil, are known as 'foots'. These foots are separated from the oil in filtration equipment and circulated back to Cooker, except for Shea. The foots recirculation has to be controlled, so as to not overload the Cooker with too much foots at any time. Too much foots would also affect the performance of the Press. Hence, a machined screw conveyor is fitted below the hopper which holds the foots dumped from Oil Filter. Speed of the screw is set low, so that the hopper is emptied over a long time, close to the cycle time of Filter, so that the hopper is emptied just in time to receive another load from the Filter. This system ensures near-uniform load of foots to the Cooker, and to Press, at all times.

7.1.7 Cake Cooling Time

The Cooler screen belt is provided with a variable speed drive. Setting the speed allows regulation of the cooling time. Optimum cooling time ensures sufficient hardening of the cake. Inadequate cooling, and hardening, could result in compaction inside the Extractor in the solvent extraction plant, which would affect the solvent percolation rates and result in incomplete oil extraction. Such a cake would also be difficult to desolventise. In cases when the cake is sent to storage, inadequate cooling may result in lumps formation during storage and possible fungus development.

7.1.8 Oil Level in FFT

Continuous flow to filters has to be ensured to prevent falling of the cake from the screen surface. Disruption in flow also makes the pump run dry. This can be avoided with a level control mechanism mounted on FFT. The mechanism can also incorporate low-level switching action to prevent pump running dry at any time. Another switching action, for high-level, prevents overflow from the tank. In absence of level control mechanism, simple switches may be fitted. The types of switches are discussed in the section on filtration.

7.1.9 Oil Filtration Time

It is important to avoid overloading of filter screens. Any overloading would cause deformation of the screens and reduce the filtration capacity of the filter. From the known maximum loading allowed, and the foots content in screened oil, the time limit of filtration may be determined. A timer may be set on the control panel which would give an alarm at, say 80–85% loading, to indicate the time for change of filter. The change-over could be automated, if auto on/off valves were provided on piping to filters.

7.1.10 Oil Level in FOT and in Heel Tank

High-level switches in both tanks ensure no overflow of oil; and low-level switches ensure that the discharge pumps do not run dry. If HL were reached in the FOT, one remedy would be to stop oil pumping to the filter. However, start-stop of filtration is not recommended, as loss of pressure inside the filter may cause disturbance of the filter cake. A better option would be to install a level control system on the FOT, with the variation of discharge pump speed (preferably with a positive displacement pump), to ensure the HL is not reached. In any case, the latter is a common feature when the plant is fully automated – see next section.

7.1.11 Oil Drying Temperature

When oil is to be dried before storage, it is heated and flashed under vacuum to achieve the desired moisture reduction. It is important to control the process temperature to achieve the required moisture reduction, while not overheating the oil.

7.2 OIL MILL AUTOMATION

As press operation, as also operation of other machinery in the oil mill, allows trouble-free operation, without the need for manual intervention, there is a growing tendency to opt for automation of the mill, to allow remote operation. It should be noted that plant automation not only reduces dependence on manpower, but also has the potential to achieve steady-state high-efficiency performance.

7.2.1 Level 1

Oil mill automation can be done at different levels. At a primary level, the following features may be automated:

- All process control schemes are included in a central controller (Programmable logic controller, or PLC).
- All motor interlocks, designed for the safety of motors and machinery, are to be incorporated in the PLC.
- All the sensed parameters to be displayed on a computer screen, recorded, and charts of behaviour of control parameters may be retrieved from the computer.

Such a level of automation allows for better control of critical plant parameters, by the plant engineer.

7.2.2 Level 2

The next level of automation may include more parameters as also some operation features:

- All process parameters which were otherwise monitored with field-mounted gauges, including pressure, vacuum, temperatures, motor amperes, etc.
- All motors ON/OFF function

This level of automation allows remote-controlled monitoring and operation of the plant. However, some manual presence is still required, especially for changing/adjustment of valves on oil and steam lines.

7.2.3 Level 3

The next level goes towards full plant automation. Additional features may be:

- Installation of auto on/off valves, which may be operated from the remote station.
- Installation of level sensors on all tanks, as also possibly on overflow bins, to allow monitoring of all levels.
- Installation of online weighers for cake at the discharge of cake cooler, and mass flow metres for oil at the outlet of oil cooler. Such flow readings, together with the online seed weighment, and the liquid levels, enable real-time material balance for the plant.

This level of automation requires significant additional investment, and may be cost-effective mainly for large oil mills. Some oil mills have already attained all three levels of automation.

7.2.4 Level 4 and Future

There can still be a higher level of automation, with 'supervisory controls', which would allow intelligent diagnosis of plant disturbance and take auto-corrective actions. With the advent of artificial intelligence technologies, supervisory control system might see the day in near future.

However, true process diagnosis may only be done with better engineering analyses of the processes, which would enable the formulation of predictive mathematical models. There is still some time to go until we achieve that level of understanding of the oil mill processes.

8 Utility Requirement for Oil Mills

Utilities required for an oil mill are electricity, water and steam. Electricity is, of course, the main utility and contributes majorly to operation cost. Steam is the next expense item. Water is required not only to produce steam but also as cooling water and wash water. Compressed air is a utility which is produced in-house using electricity.

8.1 ELECTRICITY

Electricity requirement for various machines is discussed in Chapter 3. Additions of the requirement for the individual unit operation would give the total energy requirement for the oil mill. Of course, the energy for conveying elements and pumps would have to be added. Specific energy requirement, kWh/T of seed, depends not only on the oilseed to be crushed and the process to be followed (prepress or full press) but also on the plant capacity. In principle, the larger the capacity, the lower the specific energy requirement.

For example, the overall energy requirement of oil mill for prepress of rapeseed, including unloading, pre-cleaning and silo section, may range from 30 kWh/T for a 300 Tpd mill, to just about 20–22 for 1,000 Tpd.

Sunflower dehulling consumes significant amount of electrical energy. The press also consumes more energy than for rapeseed, due to the presence of hard hulls, left after partial dehulling. The overall consumption, therefore, is significantly more than that for rapeseed; and may range from 45 kWh/t for a 300 Tpd mill to 35 for 1,000 Tpd. While pressing whole sunflower seed, the requirement would be even higher, at nearly 50 and 40 kWh/T, respectively.

Copra is commonly pressed in two stages, in view of the very high oil content. Press throughputs are much lower than those for, say, rapeseed, hence specific energy requirement is much higher. For a 10 Tph (240 Tpd) copra crushing facility, overall consumption may be as high as 65 kWh/T. It may be noted that this process is a two-stage pressing to leave only 8% oil-in-cake, compared to the pre-pressing of rapeseed and sunflower seed mentioned above.

It may be noted that, due to the high consumption of electrical energy in oil mills, typically the electrical energy costs are double, or even triple, of the steam cost. This is in contrast with the cost structure in solvent extraction plants, where the contribution of steam, in overall utilities cost, is somewhat higher than the electrical energy.

Quality of electric power supply is of critical importance for efficiencies in, and upkeep of, oil mill. Fluctuations in the frequency of the AC power supply affect the speeds of various motors and directly affect machine performance. Voltage fluctuations may cause overheating of motors and affect their longevity.

DOI: 10.1201/9781003309475-9

8.2 STEAM

Steam requirement is easily computed from the pre-heat and cooking condition. Let us again take the example of rapeseed crushing 500 Tpd.

Basis: 1,000 kg rapeseed @ 20 C, 7% moisture
Process conditions:

Pre-heat to 50 C
Cook to 90 C
Moisture reduction to 5%.

So, initial moisture 70 kg; final moisture may be computed, by material balance, as 49 kg. Moisture reduction 21 kg. This figure may conveniently be expressed as equal to:

$$\Big(\big(\text{apparent percent reduction}(\text{APR}) + \text{final moisture percent} * \text{APR}/100\big)\%\Big)$$
$$* \text{ initial mass}$$

Let us now compute the heat load, required to raise the temperature as also for moisture evaporation. For now, we will not consider hot air drying and assume that all the energy is transferred from the steam heating of the mass.

$$\begin{aligned} Q &= \text{Initial mass} * \text{sp. heat} * \text{temp rise} + \text{initial mass} \\ &\quad * \big(\text{APR} + \text{final moisture percent} * \text{APR}/100\big)\% * \text{latent heat of evaporation} \\ &= 1{,}000 * 0.45 * (90 - 20) + 1{,}000 * (2 + 5 * 2 / 100)\% * 540 \\ &= 31{,}500 + 11{,}340 \text{ kCal} \\ &= 42{,}840 \text{ kCal} \end{aligned}$$

Considering saturated steam at 3 Barg, the latent heat is 510 kCal/kg. We will consider useful heat of 500, for the loss of some steam from steam traps.

Hence, steam requirement = Q/500 = 86 kg

We should add 5% for loss of heat through insulation, etc.

So, the required steam would be 90 kg/T!

When warm air is to be blown through the cooker for moisture reduction, the quantity of air and its temperature would be calculated using the air-water psychrometric chart. Heat requirement to heat the air would also be known from the calculation. Then, the load on heating from the cooker double-bottom, or jacket, would only be to raise the temperature.

In case of full press, cooking temperatures need to be higher, and so is the moisture reduction, since the target moisture is lower. Steam consumption then can be as high as 125 kg/T.

If oil were dried, the steam for heating the oil, from nearly 70 C to 95 C, as also the steam for vacuum ejector (if used), would have to be added to the above figure.

Quality of steam used for heating purposes is of critical importance. First, steam produced at boiler should be dry, with dryness factor above 98%. Moisture collects during the transport of steam from boiler house to the mill. This must be removed with a moisture separator installed at the inlet of the main steam header in the mill. Moisture separators are of different types, the simplest being the flap type, then the vertical partition type and finally the cyclonic separator. For high moisture content, cyclonic separator may be necessary, although it would involve higher pressure drop. The condensate collected in the separator should be removed with preferably a thermodynamic steam trap, which quickly flushes out the condensate, even at the expense of some steam.

A note on the boiler operation may not be out of place in this discussion, since it affects the steam quality. The steam boiler should be operated at pressure close to its rated pressure. Thus, if the boiler were rated at 10.5 Barg, and even if heating steam in the mill is to be at 3 Barg, the boiler should be operated as close to the rated pressure possible, say 8.5–9.5 kg, and not at a lower pressure. This is so because lower steam pressure in the steam drum of boiler means higher volume of steam, which increases the steam velocity out of the exit nozzle on the drum. Steam escaping at higher velocity invariably would carry moisture from the drum, thus the dryness fraction would reduce. Another factor to be noted is the 'total dissolved solids' (TDS) of the water in boiler. TDS higher than 2,500–3,000 ppm would cause foaming in the steam drum, again leading to carryover of water in steam. This case would be even more troublesome since the salts in the foam would then deposit in pipelines and on heating surfaces.

The major power consumption components in a steam boiler section are the FD and ID fans and the water feed pump. The overall power consumption in the section may contribute around 1 kWh, per ton of seed crushed, to the power requirement of oil mill.

8.3 WATER

Water is required for the following duties:

A. Cooling water, including:
- Press shaft cooling
- Press cage cooling
- Oil cooling
- Condensor cooling in oil drying section

B. Water for steam production

C. Floor wash water

Hardness of the cooling water in circulation should be maintained below 100 ppm, to prevent fouling of water channels. Make-up water for the Cooling Tower should be soft, with hardness below 5 ppm.

Cooling tower may be of forced ventilation, or natural ventilation, type. While the former comes fitted with an ID fan, the latter is fan-less. Forced ventilation towers are of compact design and are preferred over the fan-less type, for their performance, namely, the small 'approach' towards the wet-bulb temperature (difference between water-outlet temperature and wet-bulb temperature of the air). Among the fan-less towers, recently 'mist' towers have been launched, wherein water is sprayed up in the form of fine mist. Due to the superior contact with air, on the way up and again on the way down, these towers may be designed for low 'approach'.

Water for steam production should be soft and de-mineralised, either with 'DM' plant or with reverse osmosis (RO) membranes treatment, with TDS less than 5 ppm.

Floor wash water also may be soft, to help in effective cleaning of oily surfaces with soap.

Freshwater requirement for a rapeseed 300 Tpd oil mil may not exceed 40 kL per day. Thus, oil mills are not guzzlers of water, and the cost of water is always a very small portion of the overall utilities cost. However, for regions short on water, water usage should be carefully monitored, and efforts be made to minimise wastage.

8.4 COMPRESSED AIR

Compressed air is required for the operation of pneumatic control valves, pneumatic cylinders used in slide gates, air-vibrators, etc. as also for filter-blowing operation. Pressure required for these devices is usually in the range of 3–4 BarG. Generation at the compressor is typically at 7–8 BarG, to provide for pressure drops in pipelines and fittings. Compressors are either oil-lubricated (piston type) or oil-free (screw compressors). For piston compressors, oil filters need to be fitted in line. A dryer is provided downstream of the compressor to make the air bone-dry. For sensitive devices, air dryness should be very low (dew point –20 C). The compressor and the dryer may be mounted in an acoustic enclosure to minimise noise level to less than 70 dB.

Some of the air requirement may be erratic, e.g. for auto ON/OFF valves. To prevent large pressure fluctuations with erratic air consumption, a buffer should be provided. This vessel is called a Receiver, which stores compressed air, close to consumption area.

The main consumption of compressed air in an oil mill is for the operation of filter-blowing. The requirement may be as high as 0.5 m^3/kg of filter cake. Thus, for a filter of 60 m^2 (as required for an 800 Tpd rapeseed prepress operation), with cake thickness of say 15 mm, quantity of cake may be nearly 900 kg; hence, the compressed air requirement would be nearly 450 m^3. This requirement is required to be fulfilled within a short time, say 60 min. The compressor, therefore, would have to produce compressed air at the rate of 450 m^3/h. A screw compressor of this capacity would require drive motor of 45 kW. The overall consumption may be just one-third since filter blowing is required once per filtration cycle. Thus, the contribution of the air compressor to the power consumption of an oil mill may be around 0.5–1.0 kWh per T of seed crushed. This figure may be much higher for a copra oil mill (1.2–2.0), thanks to the much larger quantity of oil to be filtered.

The piping from compressor to receiver may be of galvanised iron. A GI/SS header from the receiver may be run along the length of the oil mill. Tappings to individual pneumatic components may be in nylon tubing.

9 Quality Issues

9.1 QUALITY OF PRESSCAKE

Oilseed must be well-cooked for efficient oil expression. The cooking process, while helping achieve beneficial changes in the cell and seed structure, also exposes the mass to elevated temperatures in presence of air. The air can act as a catalyst for protein-sugar reactions, collectively called the 'browning reactions'. Brown-coloured pigments are formed during the reactions. While some browning is accepted as indication of good cooking, excessive browning is an indication of overcooking, or charring, and is not desirable. The reactions are time-dependent. Thus long cooking durations would lead to more browning; hence, cooking time should also be controlled. This may be accomplished with an effective level control system.

Also while a 'cooked' flavour is readily acceptable, overcooked smell is not. Cooking temperatures, therefore, must be controlled closely.

Uneven surface of double-bottoms in cooker may also lead to charring of some portions of seed/grits. Plate straightness, and clearance of sweep arms, must be maintained tight to avoid such problems.

Small size particles are heated up quickly to the core and may get overheated over the time required to cook the larger particles. Hence, the quantity of fines must always be controlled during the operation of size reduction.

Charring may also take place within the screw press if high temperatures are generated. This may occur mainly due to sudden compression. As seen in the section on design of worms profile, the rate of change of volume determines the rate of pressure development. Hence too much compression (contraction of channel volume of a worm) over a short length may lead to pressure peaks and could lead to high temperatures.

Proper cooling of worm shaft is an effective means to minimise charring of cake and must be practised for all presses with long barrels, say, longer than 1.2 m.

If cake is to be stored, slow cooling with adequate air contact is necessary. Quick cooling may leave excess moisture in the core of cake and might lead to lumps formation, and in worse conditions fungus formation, during storage.

9.2 OIL QUALITY

Oil is used for human consumption. The world over there is a trend towards oils of native quality. While it is not possible to produce oil such as cold-pressed, with a large screw press, efforts must be made to minimise the change in oil quality.

The first parameter which should be focused on is the oil colour. While high cooking temperatures do not directly affect oil quality during cooking, thanks to the fact that oil is still within the cells, some of the pigments from overcooked cake do dissolve in oil and impart red colour to the oil (Vadke, 2015b). This is another reason

DOI: 10.1201/9781003309475-10

why cooking temperatures, as also press temperatures, must be controlled closely. Also, the presence of fines in grits is a sure recipe for producing charred particles which would impart red colour to the oil.

Cage cooling in presses is a good idea to reduce oil exit temperature. At present, only a few models offer this facility (Vavpot et al., 2014).

The oil expressed at the press contains foots, small particles of oilseed mass. These contain a concentration of red pigments, since smaller particles with their larger surface area are always prone to more 'browning'. It is, therefore, of utmost importance to separate the foots from oil as early as possible. As seen in the discussion on oil filtration, a screening conveyor, placed just below the presses, removes a major part of the foots from oil within a very short time, just a few minutes. Such a conveyor is thus to be preferred over the classical screening tank, with its long hold up, if only for the sake of oil quality.

Also, the oil hold up in the filter feed tank (FFT) should also be not excessive, since exposure to air at elevated temperatures would lead to some oxidation reactions of oil. As seen in the section on filtration, hold up of at least 60 min is necessary to allow oil recirculation at the beginning of the filtration cycle. But care must be taken to not have too big an FFT, with oil hold up of more than, say, 75 min. In fact, overall oil hold up in oil mill, prior to cooling, must be minimised. Therefore, it helps to size all oil tanks, FFT, Heel Tank and FOT, miserly.

10 Safety and Environment Aspects

10.1 SAFETY ISSUES

The importance of safety practices can never be overstated. Safety is not only to be followed during the erection of a process plant, but is a continuous process even during plant operation and maintenance. If a plant operation is trouble-free, it can sometimes lead to complacency and can seed habits which may cause risks in future. Many factories, therefore, have on-going safety training programs.

All process plants face some common safety issues. These may include mechanical and electrical risks. Common mechanical issues may be listed as follows:

- Maintenance of proper drive guards,
- Maintenance of lubrication of gearboxes and drive chains,
- Slippery plant floors, slippery staircases,
- Obstructions in pathways,
- Low overhead pipes,
- Low overhead structural members, installed during plant maintenance,
- Leakages in overhead pipelines of hot fluids (steam, vegetable oils, thermal oils)
- Gaskets in steam lines must be maintained well, to avoid steam bursts on gasket failure.
- Proper thermal insulation of all hot surfaces need to me maintained at all times, especially after plant maintenance during which some insulations may have to be removed for welding jobs.

Electrical issues commonly faced, such as:

- short-circuits, especially in motor control panels, due to failures of contacts over a period of time;
- failure, or by-passing, of drives interlocks which are in place to avoid damage to upstream motors, and or machines, when downstream machine stops;

Apart from common safety features, there are safety issues which are specific to oil mills. These include the following:

- Solids conveying elements are more prone to stoppages due to jamming. Repeated jamming may lead to mechanical failures. Installation of overload safety switch on a drag chain conveyor, or a screw conveyor provides good safety against damages.

DOI: 10.1201/9781003309475-11

- Metal-to-metal contact between the chain and casing, in a drag chain conveyor is a potential source of sparking. This can be avoided by using a plastic liner, such as UHMWPE (ultra-high molecular weight polyethylene). These liners are beneficial in two ways, due to low friction: they minimise the wear of metal parts, and, reduce power requirement.
- Failure of a link between the motor and the machine (or a conveyor), such as a flexible coupling or drive chain or belt, can cause jamming of upstream machinery because the running motor prevents the activation of drives interlock. A **zero-speed sensor**, with a switching action, which senses the rotation of the machine shaft, helps prevent such occurrences.
- An oil mill may face potential fire hazards. Overcooking may cause seed particles, especially fines, to catch fire. A temperature control system minimises such a risk. For a control system to work effectively, however, the sensing of temperature must be accurate. The sensor must be **calibrated** periodically. This applies to all sensors, and gauges, in the plant.
- Another potential fire hazard is at the screw press itself. Sometimes, if oil-in-cake rises, due to wear of worms and cage bars, there is a tendency to throttle the choke, as a first reaction. It must be noted that while the **choke** offers a convenient tool to arrest the OIC, it must be used judiciously. **Severe throttling** may lead to sudden temperature rise and may cause **fire** at cake discharge. This hazard is very real especially on 'full pressing' or 'second press' operations. A far better solution, to achieve lower OIC, is to reduce the speed of worm shaft.

10.2 HEALTH AND ENVIRONMENT

Luckily, there are no major environmental issues associated with the operation of an oil mill, which is inherently friendly to the environment compared to the solvent extraction plant (SEP). Oil refineries may have significant issues, even more than SEPs.

The only major hazard may come from the mis-operation of steam boiler. For boilers using coal as the fuel, coal grinding and boiler flue gases are both potential areas where fine dust may escape into the atmosphere. The coal grinding section must be properly ventilated, and all the dust trapped in an air bag filter. Air bag filters are also available for flue gas duties and are also not very expensive. Bag filters can control dust in flue gas below 100 ppm. For medium and large boilers, with steam capacity of more than 8 Tph, an electrostatic precipitator (ESP) system may be required in some countries. In the case of coals with high sulfur content, an alkaline absorption system may be required, to arrest SO_x gases, downstream of a bag filter.

The **ash** generated in coal-fired boilers maybe sprayed with cold water, and disposed off. Major consumers for the ash are brick-makers or cement kilns.

The seeds unloading area can be very **dusty**, depending on farm soil content. A proper aspiration system, with adequate aspiration hoods, is necessary to arrest the dust and keep the area workmen-friendly. This has been detailed in Chapter 3.

The area surrounding the screw press may have **oil mist** escaping from the press. The press chambers must be ventilated to suck out the mist. Before letting out the air outside the building, it may be passed through a baffle assembly and a demister pad to trap the mist.

The **floors** of the oil mill must be maintained **clean**, to prevent accidents from slippery surfaces. The wash water, used in floor cleaning, should be collected in an oil trap, before letting it out to an effluent treatment plant (ETP). This would be a compact ETP, designed basically for flocculation and settling.

Boiler water is bled-off regularly, to maintain the total dissolved solids (TDS) in water below 2,500 ppm. The blow-down water contains chemicals added to water for various purposes (anti-corrosion, anti-scaling). However, it does not contain suspended solids; so, it is not sent to the ETP. It may be collected, along with the treated water from the ETP and be used for spraying on coal and ash, as also for gardening. Extra water may be evaporated by means of fountain(s) in a pond.

Part 2

Solvent Extraction

11 Introduction

11.1 SCOPE

This part covers the principles and technology of extraction of vegetable oil by the method of dissolution in solvent; it also discusses design concepts of the equipment and the process line. The first chapter discusses the history of the solvent extraction industry briefly, throwing light on the development of technologies over the past century. It concludes with an overview of the present practice of solvent extraction. Chapter 12 starts with the theoretical aspects of solvent extraction, covering the cell structure and the resistance to oil extraction. The importance of various preparation operations to overcome the resistance is explained. It covers the technology of preparation of various oilseeds, as also an indication of the respective process parameters. Reasons for differences in preparation methods, and process parameters, for various seeds, are discussed. Also included are the preparation methods for rice bran, an oil-bearing material, which is an important raw material for VegOil extraction industry in India, China, Japan and south-east Asian countries.

Chapter 13 covers the design aspects and construction features of machinery used for oilseed preparation. Suitability for preparation of each oilseed, as also precautions to be taken during the operation of each, is discussed. Alternative preparation methods are discussed, with pros and cons w.r.t. the traditional methods of seed preparation. Readers may note that the preparation of oil-bearing materials, prior to solvent extraction, is critical, since the objective is to extract all the oil, leaving but a minute part in de-oiled meal. Any deficiency in preparation, therefore, can prove very costly. The concept of 'Milling Defect', a measure of shortcoming w.r.t. 'ideal' seed preparation, is discussed.

Chapter 14 discusses the theory and technology of solvent extraction. The main process sections in a solvent extraction plant (SEP) are discussed, with the principles of operation of each. The technology of solvent extraction of various oil-bearing materials, including the cakes obtained by prepressing, is discussed in detail. Difficulties in the processing of low-protein materials are explained. The discussion on process parameters for each material will be of help to plant operators and engineers to tune their processes, and help in trouble-shooting. Related technologies for soybean oil degumming, and miscella refining of cotton seed oil, process sections which are housed within the SEP, are also covered in this chapter.

Construction features, and design aspects, of each of the equipment in SEP are covered in Chapter 15. Discussion of optimum methods of operation of the equipment, and precautions to be observed, will help the plant operator in getting the best from each process equipment, and to help operate the plant with high efficiencies. Discussion of design aspects may help not only plant and equipment design engineers, but also provide sufficient insights to plant engineers to check the adequacy of, and to fine-tune, their process lines. Typical process control strategies and plant

DOI: 10.1201/9781003309475-13

automation philosophy are also discussed in this chapter. This chapter includes a section on 'Technologies of Future', including a different concept of extractor, and novel preparation methods, for low-protein, high-carb, materials which are difficult to process with existing technologies.

Chapter 16 gives an overview of the processing of de-oiled meal to produce various products. Edible products from de-oiled soybean are discussed briefly, and so is the possibility of in-house production of animal feed from cotton seed meal and by-products.

Meal bagging and storage are discussed in Chapter 17. Various methods of bag storage, and bags retrieval for dispatch, are outlined, apart from dust prevention systems.

The effects of processing on quality of products is discussed in Chapter 18. This should help plant operators tune their production processes and avoid pitfalls, so as to ensure product quality on a consistent basis.

The requirement of various utilities for the operation of seed preparation and SEP is discussed in Chapter 19. Steam requirement and steam pressure requirement are discussed. The possibility of re-design of plant, to enable operation with low-pressure steam, so as to allow the operation of a co-generation power plant, is discussed. Solvent extraction plants are ideal candidates for co-generation of power, which can help reduce the electricity bill to a great measure. The requirement of water for oil mill operation, while small, can be of critical importance in regions short on water. A brief discussion of saving of water is included. The quality of steam and water, usually a neglected area, may help plant operators get the best out of these utilities; this, however, has been covered in Part 1 and is not repeated in this part.

The last chapter of this part is dedicated to issues of plant and operator safety and environment. Safety of plant and operators is even more critical for SEPs, in view of the flammable and explosive nature of the solvent used. The importance of constant focus on these factors can never be over-emphasised.

Apart from the main text of the part, some important aspects of solvent extraction are covered in Appendices. Appendix 1 discusses the design and operation of batch plants, which are in use for special applications.

Appendix 2 gives a calculation method for the determination of the size of DT (Desolventiser Toaster) for a given material and plant capacity. It shows also the calculation for steam requirement, both indirect heating steam and 'open' steam, by means of heat and material balance equations.

Appendix 3 presents typical material balances for crushing of various oilseeds. This will surely be of help to plant engineers and managers.

Principles of plant start-up and plant shut-down, and the precautions required, are discussed in Appendix 4. A compilation of trouble-shooting methods should be of use to plant operators and engineers.

Appendix 5 gives the analytical procedure for the determination of 'Milling Defect'. This is an important parameter, often ignored, which can be of help in analyzing any fluctuations in residual oil in de-oiled meal. Appendix 6 gives a standard for 'food grade hexane' which is the common solvent used in solvent extraction plants all over the world.

The co-generation of power, using steam at high pressure, is outlined in Appendix 7. Readers may note that solvent extraction plants are one of the most suitable candidates for this technology, since the power and steam requirements are in 'ideal' proportions as required for co-generation.

Appendix 8 gives a typical layout of a solvent extraction plant. That should be useful to students of solvent extraction technology.

11.2 HISTORICAL PRACTICES

Historical practices used by humans for the recovery of vegetable oil were reviewed in Part 1 of this book. Earliest methods of seed maceration by stone, through boiling in water, stump presses, vertical jack-screw presses, to hydraulic presses, and finally screw presses, were covered in the discussion. The screw presses are the latest, and much superior, process for the mechanical expression of oil, as these allow continuous pressing operations, and hence allow the design of mills of large crushing capacities, with minimum manpower.

While the screw presses offer a versatile and easy method of oilseed crushing, they leave nearly 6–8% oil-in-cake; which may be 10–20% of the original oil in seed. Solvent extraction, developed since the 1920s, offers an alternative to enable extraction of up to 98% of the oil, leaving only 0.5–1.5% oil in the de-oiled cake (meal).

In this method, the seed is brought into contact with a solvent which dissolves the oil, the oil-rich solvent phase (miscella) is then separated from the oil-free seed mass (marc) by a simple filtration technique, and the oil is recovered by distilling off the low-boiling solvent. This method was first practiced, on a large scale, in Germany shortly after World War I. In the United States, solvent extraction was adopted on a major scale during the decade preceding World War II to extract oil from soybeans primarily. Solvent extraction is now used universally either as a 'finishing' unit operation after low-pressure expellers, or as the sole process for extraction of oil, from most oilseeds.

The earliest solvent extractors were 'batch' type and employed the 'immersion' technique, wherein mass of oilseed was dipped in a bath of solvent, soaked for sufficient time, with or without agitation, to allow the release of oil in the solvent (Norris, 1981). The batch process was then upgraded to continuous mode. One such design consisted of a belt fitted with perforated buckets full of oilseed moving continuously, the buckets travelling slowly thru the bath of solvent, before emerging out with de-oiled seed mass (Figure 11.1). Other designs of continuous immersion extractors included vertical multi-stage vessel, filled with solvent, where flakes were fed at the top, agitated by sweep arms attached to a central shaft, travelling to successive lower stages, and finally discharged from the bottom by a compression screw (Figure 11.2). Solvent was introduced at the bottom, just above the discharge screw, and miscella bled from the top, above the flakes feed level. These designs were in vogue for nearly three decades. The main difficulties encountered with these designs were, one the extractors were very tall, often two or three stories high, and two, the miscella required rigorous clarification for removal of fines (Karnofsky, 1949b).

To overcome these problems, a compact extractor, of 'percolation' type, wherein solvent was sprayed on slow-moving bed of flakes, it percolated through the bed, and collected with dissolved oil in hoppers below, was developed in US by Blaw-Knox Company (Karnofsky, 1949b). It was named the 'Rotocel' extractor. Similar

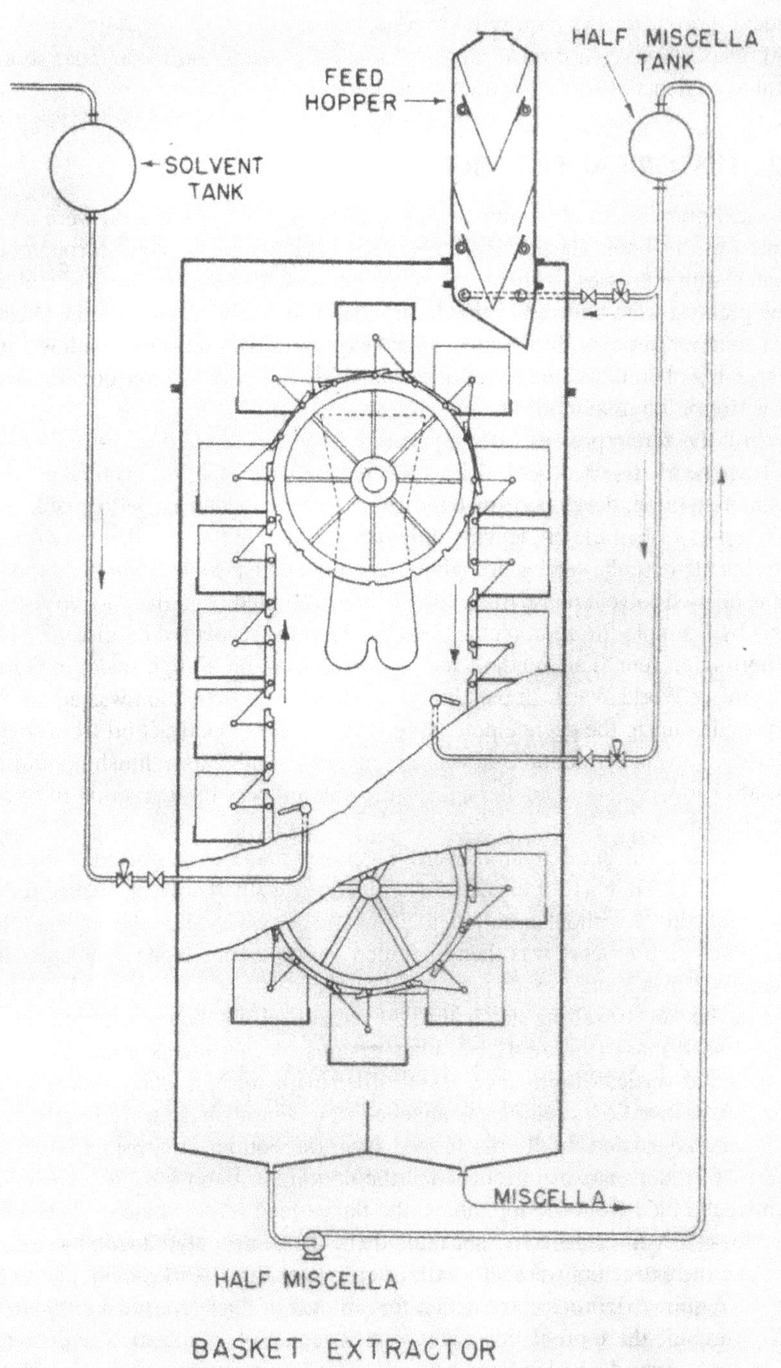

FIGURE 11.1 Basket Extractor.

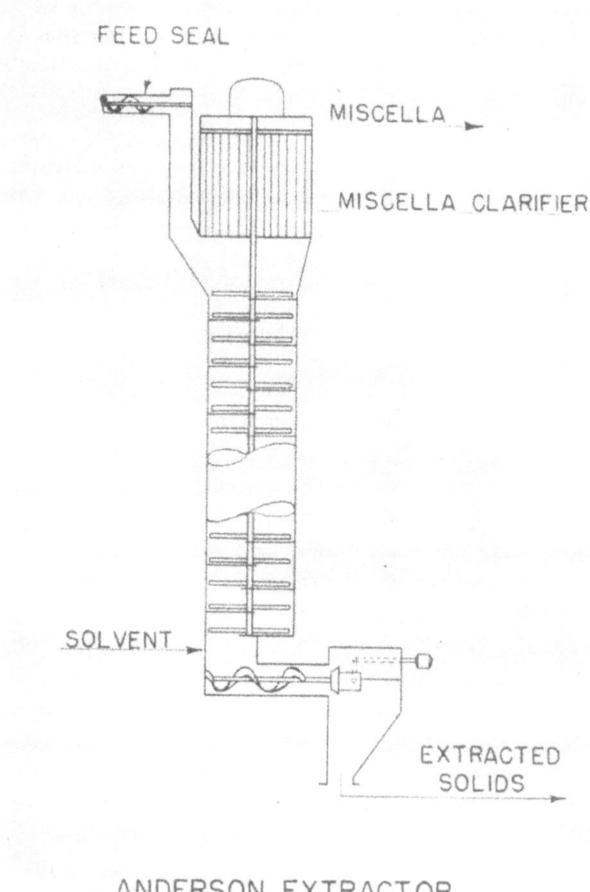

FIGURE 11.2 Anderson Extractor.

designs were being investigated in Europe around the same time, which led to a dispute on grant of patent in US (Shurtleff and Aoyagi, 2016). The case was finally settled in 1958, nine years after filing, to grant the patent to Blaw-Knox, who had to issue licenses to two companies in Europe as a compromise. Around the same time, a 'Linear' percolation extractor, with a slow-moving horizontal screen belt, was invented by DeSmet in Belgium (LeClef, 2020). A third type 'loop' extractor was developed by Crown Engg Works in US a bit later and is among the popular designs at the present time. In Germany, Lurgi developed a 'Frame Belt' extractor during the 1960s, which later got converted to 'Sliding Cell' extractor since the 1980s.

The percolation extractors proved to be much safer in operation, and produced clear miscella, avoiding the need for elaborate clarification. Continuous percolation-type extractors are now the norm all over the world; these require much less solvent hold-up and also cause less solvent losses, compared to the 'Immersion' extractors.

In all the designs of percolation extractors, bed of oilseed mass travels slowly through the extractor, while solvent is sprayed on top of bed at multiple locations. As the solvent flows down by gravity, it percolates thru the seed mass and dissolves the oil. The mixture of solvent and oil, called 'miscella' is collected in hoppers below. In all designs, a counter-current principle is employed, where the oilseed mass and solvent flow in opposite directions thru the multiple stages. The DeSmet design carries the mass on a slow-moving perforated screen belt, whereas in the Rotocel design, the screen is fixed, and mass is moved by vertical partitions, or 'buckets', attached to a central rotating shaft. In the Crown design too, the screen is fixed, and a drag chain conveyor moves the mass slowly. The Lurgi design is somewhat similar to Crown, but material is inside vertical cells (buckets), which move, on fixed screens, by means of chains on either sides.

Major changes have also taken place in the desolventisation process, to remove solvent from the de-oiled meal. In early continuous-mode plants, screw 'Dryers' were used for the purpose. A series of Dryers were mounted vertically, with material flowing from top dryer to lower ones. As plant capacities were increased it became difficult to manage the bigger and numerous dryers. The problem was solved in early 1940s by a vertical multi-stage vessel, somewhat similar to a vertical cooker (Kemper, 2005). Multiple drives were replaced with a single drive. Also, thanks to less openings for shaft ends compared to the dryers, the losses of solvent vapours to atmosphere were also controlled to a large degree. At that time, the toasting of soybean used to be carried out in a separate mixer vessel. A decade later, the process of toasting of soybean was accomplished within the equipment, by injecting live steam in the upper stages to increase moisture level to that required for toasting. Thus the multi-stage stacked equipment came to be known as the Desolventiser-Toaster, or simply the DT. The DTs were the mainstay of solvent plants during the 1950s through the 1970s. A major innovation in the design of the DT came from Dr. Heinz Schumacher, a German engineer during the early 1980s (Brueske, 1993). His design utilised the counter-current principle of stripping the solvent out by means of steam, and was able to minimise the residual solvent in meal, even while using much less steam for the purpose. The Schumacher DT, designed by several major machinery suppliers, is the DT of choice for most applications at the present time.

Distillation line was made more efficient over the decades, meaning less and less residual solvent was left in oil, while using less and less steam for the purpose. Distillation temperatures were also reduced over the decades, which resulted in better oil quality.

Solvent plant had to be, and still are, ventilated to keep the pressure, within the Extraction and Desolventising sections, a bit below atmospheric, so that solvent vapours did not leak out. As the ventilation air was vented, it carried solvent vapours along. The vapours were passed through a water-cooled condensor to trap a part of the vapour. Later, chilled water was circulated through the condensor, which improved the solvent recovery. During the late 1970s, vapour absorption system, using a mineral oil as absorbent, was designed. This system carries to the present time.

As a result of all the process improvements listed above, the solvent loss reduced significantly, from 30 to 50L per ton of material extracted in the 1960s to just about 1–2L at the present time. Steam consumptions reduced nearly by half to

200–280 kg/T in modern plants. The plants of yesteryears were not safe, and were prone to major accidents of fire and explosion; but modern plants are safe, thanks to the process improvements as also to safety features adopted over the decades.

Recent history of solvent extraction of oil from oilseeds will not be complete without a mention to the great engineer and researcher, Er. George Karnofsky, a chemical engineer, who worked for the Blaw-Knox company in Pittsburg, USA. He not only invented the Rotocel extractor but also put forth a theory of solvent extraction of oilseed flakes. He proposed a mechanism of extraction rate being controlled by 'undissolved oil' in the flakes, with diffusion being of consequence only in thick flakes (Karnofsky, 1949a). He followed it up in his later years, after retirement from active service, to present a mathematical method to calculate the residual oil (RO) as also miscella concentration at different stages of a continuous counter-current percolation extractor (Karnofsky, 1987). His final contribution was a comprehensive overview of engineering analysis of solvent extraction of solutes from solids, not limited to oil extraction, in an address to students in 2005 (Karnofsky, 2005), just three years before his final departure.

11.3 MODERN PRACTICE OF SOLVENT EXTRACTION

At present time, solvent extraction dominates the world of vegetable oil extraction. Almost all oil-bearing materials are subjected to solvent extraction. Soybean is the number one oilseed of the world, and soybean crushing plants are a very important part of the solvent extraction industry. The only major oil that is not extracted by solvent is palm oil, which happens to be the top produced vegetable oil in the world.

Oil-bearing materials are 'prepared' suitably, so as to yield maximum oil on solvent extraction. The preparation may or may not include part-oil recovery by mechanical expression, and is carried out in a separate building, at least 15 m away from the solvent extraction building, in most countries (Figure 11.3). Preparation machinery tend to be heavy-duty and require much electrical energy. The solvent extraction plant itself is designed to be an 'explosion-proof' environment.

The 'preparation' of materials may be viewed in three broad categories. The criterion is the oil content of the material. High oil content seeds, with more than 35% oil, such as rapeseed, sunflower seed, groundnut, palm kernels, etc. are pre-pressed to recover major part of the oil and produce cake with 16–20% oil, which is then solvent-extracted.

Low-oil content materials such as soybean and rice bran, with oil content less than 25%, are directly solvent-extracted without first recovery of oil by mechanical expression. Soybean is broken and softened and made into suitable form for solvent extraction; flakes or pellets/collets. Rice bran is steam-conditioned and formed into pellets. All the oil is extracted by solvent method.

Oilseeds in the mid-oil range, 25–35%, such as cottonseed (dehulled meat) and safflower, may either be directly solvent-extracted, or be subjected to mild pressing, to recover a small part of the oil, prior to solvent extraction. The later method offers advantage over the direct solvent extraction, in terms of higher total oil recovery and lower overall energy requirement. The part oil recovery may either be with Screw Presses or with 'drained' Expanders.

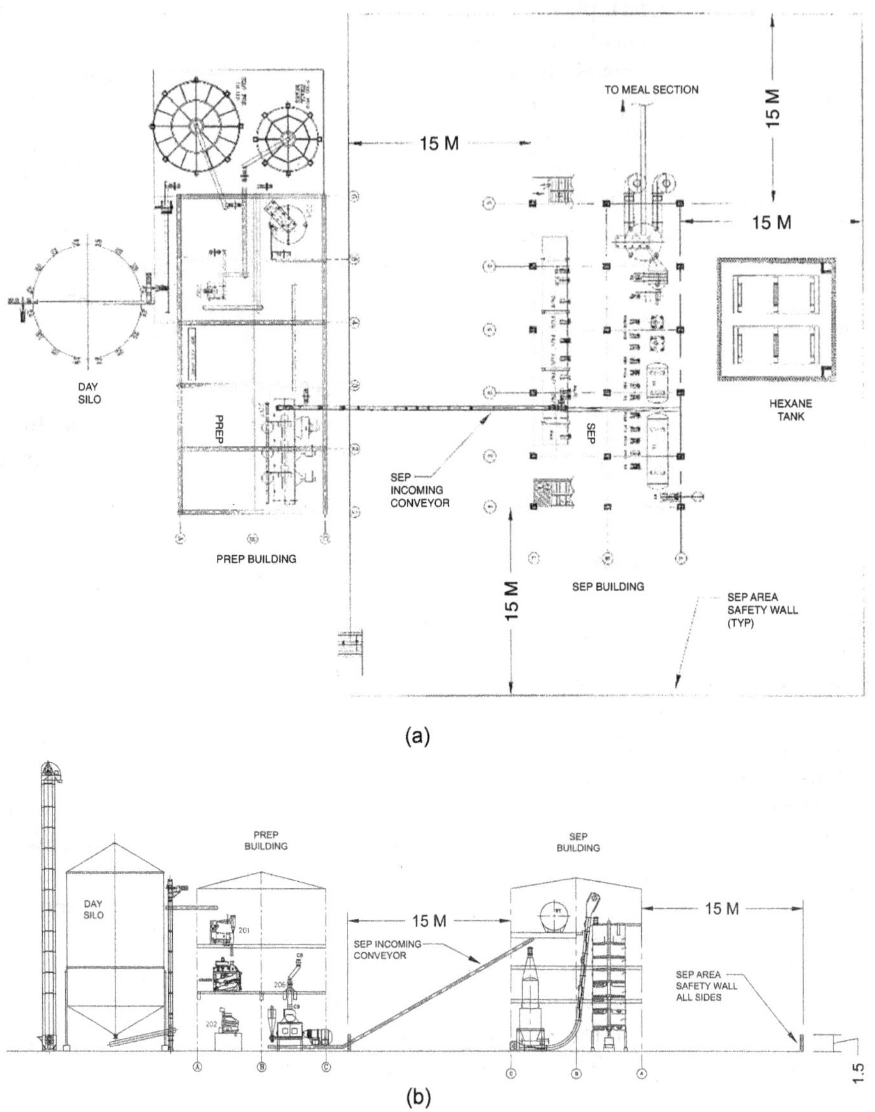

FIGURE 11.3 (a) Location of SEP and PREP Buildings (Plan). (b) Location of SEP and PREP Buildings (Elevation).

A schematic of main processes within a solvent extraction plant is shown in Figure 11.4. The plant has three main sections, namely, the Extractor, the Desolventiser and the distillation unit. There is a fourth, small section, for recovery of solvent vapours from vent air. These will be discussed in detail in later chapters.

Three main designs of Extractor are commonly used. One is the DeSmet type, with linear slow-moving screen, second the Rotocel type with fixed circular screen

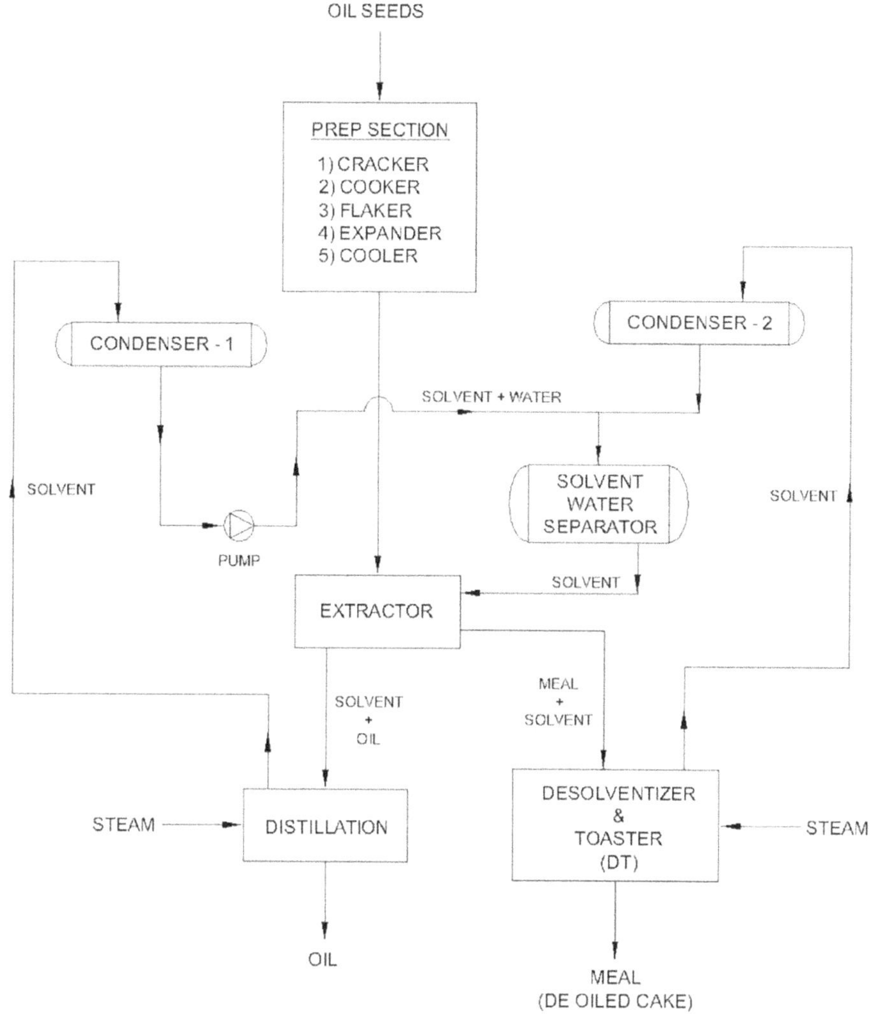

FIGURE 11.4 Schematic of Solvent Extraction Process.

with rotating vertical partitions, and the third the Crown design of a 'loop' extractor, with fixed screens in upper and lower casings, and material moved by a slow-moving heavy-duty drag chain conveyor.

The solvent of choice, the only one used for all vegetable oil except some minor ones, the world over, is food grade hexane, which consists of n-hexane as the main constituent with iso-hexane and some other associated compounds. Commercial solvents available from petroleum refineries are always a mixture of several compounds, mostly alkanes with 5–8 carbons, with boiling point range around 68–72C. A standard for 'food grade hexane' is given in Appendix 2.6. For some minor oils, such as neem (*Azadirachta Indica*) oil, ethanol may be the solvent of choice.

The desolventising operation, to remove the solvent from de-oiled meal, is carried out predominantly in Schumacher DT designs, modified by individual licensees and others. The drying and cooling operations are usually combined with the DT, in a DTDC.

Distillation lines are getting compact, and plate-type condensors have been introduced. Efforts are on to minimise the quantity of effluent water, carrying fines from beans, to approach the condition of 'zero effluent'.

Absorption systems are being refined to minimise solvent vapours escaping with vent air, in what may be called 'zero vent' system. These improvements positively impact not only the overall solvent loss from the process of solvent extraction but also the environment around the plant.

Plant capacities have increased over the decades in all parts of the world. During the 1960s, solvent plants designed for 60 Tpd, for soybean, were a common feature. At present time, in industrially well-developed countries, plant sizes in hinterland, near the farmlands, are sized between 600 and 2,000 Tpd, while those near the ports are much bigger, in the range of 2,000–15,000 Tpd. In developing countries, however, plant sizes range from 200 to 1,500 Tpd. Most of the port-based plants, in North and South America, work with soybean.

Batch solvent extraction plants, which were the mainstay up to the 1940s, are still in use for special applications, for capacities up to 100 Tpd, say for tree-borne seeds, such as Sal seed (*Shorea robusta*) and Neem seed (*Azadirachta Indica*), which are available in small quantities in a geographical area. However, the batch plants in use now are of percolation type and not the immersion type. The batch extractor is also utilised for desolventising of flakes. A distillation unit, batch or continuous, completes the plant. A typical batch plant is discussed in Appendix 2.1.

12 Preparation Prior to Solvent Extraction

12.1 PREPARATION – THEORETICAL ASPECTS

Oilseeds that are subjected to pre-pressing, to recover major portion of oil, have already been 'prepared' well for oil extraction and do not require much further preparation. However, those oil-bearing materials, containing less than 25% oil, which are directly solvent-extracted, require rigorous preparation. Oilseeds are broken, steam conditioned, and flaked, to enhance the oil recovery. Over the past two decades, an additional operation has been added to prepare the flakes further, namely, the expander treatment. It must be noted that the objective of solvent extraction is to extract almost the entire oil, so the seed preparation needs to be near-perfect.

12.1.1 Oil Cell Structure

As seen in Part 1, each unit operation plays an important role in facilitating the release of the oil globules that are embedded within individual cells of an oilseed (rapeseed – Figure 12.1 and soybean Figure 12.2), and thus helps in improving the oil yield on solvent extraction. It may be noted that the oil globules are dispersed in the interior matrix of an oil cell. The matrix is made up of various other bodies, such as cell nucleus, protein, etc. Thus, for the recovery of oil from individual oil cells, one objective is to bring all the oil globules towards the periphery.

When the oil is near the periphery, the next and most important objective is to get it past the cell wall. The cell wall is a rigid structure and offers the strongest resistance to oil extraction. As will be seen in the following discussion, most unit operations in the seed preparation are directed towards getting the cell wall softened and ruptured. The next objective is to flake the mass very thin so as to facilitate the contact of solvent with all cells. It further helps to make the oilseed mass, which consists of numerous cells, porous enough to facilitate the flow of solvent in and out of the mass. This objective is achieved in the Expander treatment, along with other benefits.

12.1.2 Seed Breaking and Cooking (Conditioning)

Cooking, or steam conditioning, is one of the most important unit operations in the preparation of oilseeds for solvent extraction. It not only helps overcome most of the resistances but also prepares the seed well for the final operation of flaking.

For the steam, and the heat, to get to the core of an oilseed, the seed coat must be broken, and a mid-sized seed such as soybean must be broken into smaller pieces. These objectives are met by an operation called 'Seed Cracking'. These pieces, or 'grits' are then subjected to cooking.

DOI: 10.1201/9781003309475-14

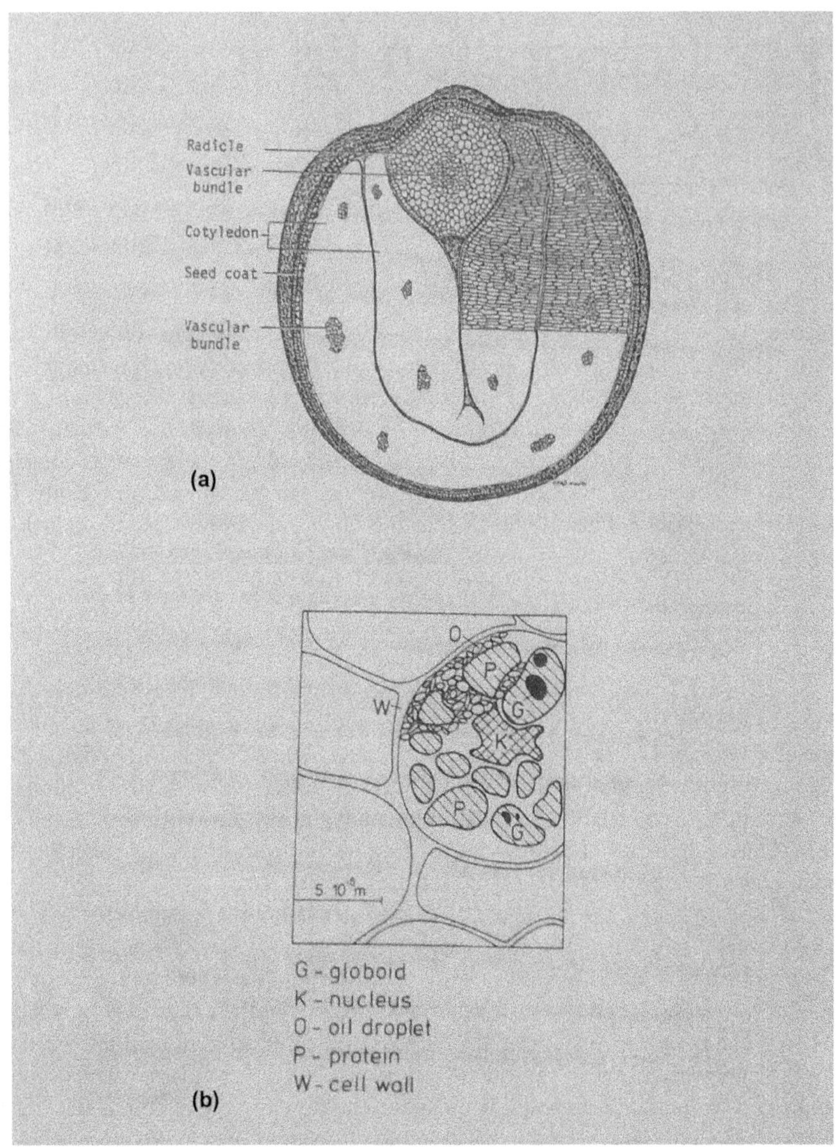

FIGURE 12.1 Rapeseed Oil Cell (Yiu et. al, 1983 and Eggers et. al, 1985).

A critical difference, between the preparation of seed for mechanical expression and that for solvent extraction, is that the presence of fines is to be strictly minimised for the latter. Fines hinder solvent percolation in the extractor. This requires that the sequence of unit operations as followed for mechanical expression be changed. In the mechanical method, cooking is the last operation prior to pressing. Flakes get broken during the cooking process. That is just not allowed for the solvent method.

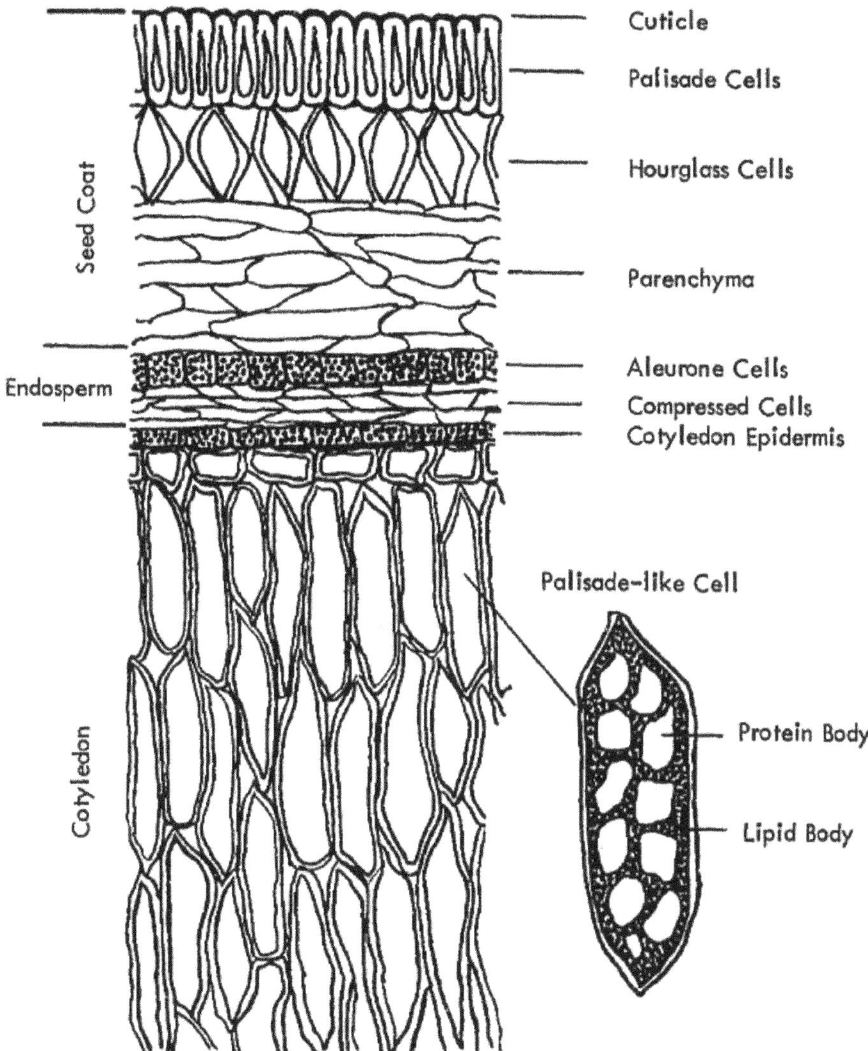

FIGURE 12.2 Soybean Oil Cell (Bair, 1979).

Here, flaking tends to be the final operation, and the flakes are gently conveyed into the extractor. So, the sequence of unit operations becomes breaking, cooking, and flaking. As an enhancing operation, flakes may be passed through an extruder, what is known in the industry as 'Expander'.

As discussed in Chapter 2 of Part 1 of this book, a cooking operation, post flaking, can achieve the following major objectives:

- Coagulate the protein within oil cells and release oil globules to the periphery of cell.

- Protein coagulation also helps to make the grits plastic enough to produce well-formed flakes, which would hold well under solvent sprays in Extractor.
- Reduce the viscosity of oil and improve its flow characteristics.
- Soften the cell wall.

However, when the cooking operation is conducted before flaking, it is done at lower temperature, and only the last effect is achieved to any major degree, along with some plasticity (Bair, 1979). There are no major changes in the cell structure, unlike in the case of cooking done post-flaking. Hence, in this case, the operation is better referred to as **Conditioning**.

For seed conditioning to be effective, adequate moisture level is necessary. For low-oil content seeds such as soybean, 10–11% moisture may be necessary, while for medium-oil content seeds such as cottonseed, 8–9% moisture may be adequate.

12.1.3 Flaking

The cooked grits are given an instantaneous, strong impact between two smooth rolls to crush the cell structure within the grits. Should the cell walls have been softened well enough during cooking, the walls may be ruptured to overcome the most important resistance to oil extraction. Thin flakes also expose most cells to direct contact with solvent. A good soybean flaking operation, post steam conditioning, was shown to have ruptured most cell walls (Bair, 1979).

Even the strong impact, within a fraction of a second, does not cause the grits to shatter, thanks to the plasticity achieved by cooking. The flakes are then to be handled gently, to retain the shape and not to produce fines, during transport to the extractor.

While the thickness of flakes is an important factor, the **Degree of deformation** (DoD) undergone by grits during flaking is of critical importance. As discussed in Part 1, it is the DoD (difference between grits thickness and flakes thickness) that determines the damage to structure of seed and cells. It may be noted that the size of an individual oil cell in most oilseeds is in the range of 20 μm. Thus, even within a flake 0.3 mm thick, there are many layers of cells. Thus, the DoD becomes one of the most crucial factors in preparation of the seed.

Karnofsky (1949b) calls the DoD, the 'distortion'. He has presented comparative data on solvent extraction of soybean flakes, all of the same thickness and prepared from the same batch of soybeans, but of different sizes, separated by sieving. Bigger flakes, which had undergone larger distortion, extracted much faster, and had significantly lower ROC after 60 minutes than smaller flakes.

12.1.4 Expander Enhancement

Extrusion enhancement of soybean was introduced during the early 1980s (Bredeson, 1983). The treatment offered three major enhancements to soybean flakes. One, it aided further agglomeration of protein, thereby releasing more oil globules, and producing a denser mass. Second, as the mass was forced out through a die-plate, the pressure was released, and the moisture within the mass evaporated instantaneously, causing numerous micro-paths through the collets. The mass remained relatively dense, even after the expansion on exit.

The abundance of micro-paths would help with the percolation of solvent through the mass and contact with oil cells. Release of more oil globules would reduce the residual oil content (ROC) in meal after solvent extraction. Third, The denser mass allowed to pack more material in an extractor. Apart from the recovery of more oil, these factors helped increase the processing capacity of the extractor. In fact, the capacity increase could be as high as 25–30%.

As the collets drained the solvent faster, at the end of extraction cycle, the carry-over solvent to the DT was less. This helped increase the capacity of the DT as well.

Since the micro-paths increased the percolation rate, it was possible to use richer miscella (oil-solvent mixture – with a higher oil concentration) for the initial wash of fresh collets fed to the extractor. This meant that the outgoing miscella had a higher concentration of oil. In other words, the quantity of solvent going with oil to distillation was less. Thus, existing lines of distillation were able to handle more oil.

As the capacities of all the three sections of solvent extraction plant (SEP) increased, the overall plant capacity increased. All this, just by adding one operation to the preparation plant! Compared to this, additional investment in extra equipment in the SEP to achieve higher capacity would be huge. Thus, expanders were, and are, seen mainly as capacity boosters for solvent extraction plants.

Should there be any shortcomings in the SEP, such as higher ROC (residual oil content), or high solvent losses, an expander may help overcome these shortcomings too. Expanded collets are not only easily extractable, compared to flakes, but are also easier to desolventise, thanks to the well-cooked protein mass that would release the solvent quicker. Steam consumption may also be reduced, mainly due to less solvent in DT, and to distillation. The extra electrical energy required for the expander may be offset by the increased plant capacity with unchanged energy input to other machines.

Because of the advantages offered by expanders, more and more soybean crushing plants are now fitted with these machines.

Injection of steam into the flakes inside the expander helps further the coagulation reactions. The steam remains in liquid form within the pressurized barrel and flashes off on exit from the die-plate. Thus, steam injection boosts the expansion of collets and the creation of micro-paths.

The introduction of expander also allows the pre-cooking of flakes, since the integrity of flakes is of no concern. Cooking at a temperature of 80–85C for even just 5 minutes, upstream of expander, helps further the desirable actions within the oil cells, and aids the extrusion operation.

Collets are discharged from the expander at around 90–95C, and typically have 11–12% moisture, after a part of the moisture has flashed off. Moisture must be reduced to below 10.5% to achieve good solvent extraction. Also, collets need to be cooled down to 60C. The cooling should be achieved slowly to firm up the pellets, which are quite 'plastic' at expander discharge.

It may be noted that, while the machines (expander) used are very similar to those used for extrusion enhancement prior to mechanical expression (Part 1, Chapter 2), the treatment is much lighter; the speed of rotation of shaft is much higher, and the residence time is very short, less than 15 seconds! Power consumption is also much less compared to the extruders used prior to the press. Thus an 'Expander' is designed for much higher throughput compared to an 'Extruder' of the same size.

Rice bran, as obtained from rice milling, is in powdery form, and does not lend itself to 'percolation' type extraction. When treated in an expander, it can form porous pellets, which are easily extractable. This treatment forms an attractive alternative to the traditional pelleting operation.

Expanders are also now used for oilseeds of mid-oil-content range, such as cottonseed meat and safflower, but fitted with a 'drained cage', to recover a part of the oil. These accomplish all the advantages of a normal expander and also reduce the oil content to less than 25%, to produce strong collets.

12.1.5 The Degree of Preparation – Milling Defect

The degree of preparation of oilseed, such as soybean, from cracking through to flaking, and may be further through expanding, as it reflects in performance in solvent extraction, may be determined by a 'Milling Defect' (MD) test. If the preparation were 'perfect', the MD would be zero. And such perfect flakes, or cake, could be extracted fully, to leave no residual oil, in an 'ideal' extractor. A laboratory extractor, where a cake sample is subjected to a continuous solvent wash for a long time, may be considered an ideal extractor. So, if a cake sample were extracted for 3 hours in a lab extractor, any oil still remaining in the cake would be the 'milling defect'.

In practice, an MD of 0.20% is achievable, with controlled operations of cracking, conditioning and flaking. This means that even an 'ideal' extractor would not be able to extract oil in flakes down to less than 0.2%. In practice, the extractor may be considered to have performed well if the ROC is down to 0.5% when MD was 0.2%. The difference between MD and the ROC may be referred to as the 'Extraction Defect'. Thus, an extraction defect (ED) of 0.3% is quite acceptable for commercial extractors. Further reduction in ED would require much longer extraction times, which would make the operation impractical. The laboratory method for determination of MD is given in Appendix 2.5.

The same measure of MD is applicable to the pre-pressed cake of various oilseeds. In this case, an additional operation, pressing, has been performed on the seed, to prepare it for solvent extraction. Quite often, machinery suppliers for oil mill and solvent extraction plant may be different. In such cases, the solvent extraction plant supplier may guarantee ROC based on MD in the cake. For example, the guarantee would quite often be MD + 0.3%.

12.2 PREPARATION OF PRESSCAKES

Presscakes from prepress, or sometimes from full press too, are subjected to solvent extraction for recovery of balance oil. As discussed in Part 1 of this book, these cakes should be prepared well for solvent extraction. The preparation steps include breaking and cooling. In some oil mills, cakes are first cooled and then broken. However, if cakes are broken first, they may be cooled uniformly.

One issue, in breaking of warm cakes, is they may be somewhat plastic and may clog the rolls of a breaker. This may especially be true of cakes with high protein content, 30% or more, such as those of rapeseed, groundnut, cotton seed, dehulled sunflower seed, as also castor (Table 2.1, Chapter 2). To overcome the issue of plasticity, the conveyor carrying the cake to the breaker may be ventilated; air contact over a minute, as with a slow-moving conveyor, can make these cakes sufficiently firm for effective breaking.

Such firm cakes with high protein content, may be broken down to 10–20mm size. This small size would expose a lot of surface area for the action of solvent. It may be noted that the most solvent percolation occurs through the broken sides of cake, rather than the flat surfaces where mass has been well-compacted. With the plasticity of the cake, there is little risk of creating much fines. This major size reduction may be accomplished with 4-rolls, i.e. 2-pass, machine,

Cakes with intermediate protein level, between 20% and 25%, such as those of whole sunflower seed, sesame, and corn germ, may be safely broken to 20–30mm size, but not smaller. This may be achieved with a single-pass, 2-roll machine

Cake of palm kernel, with low protein content, around 14–15%, is not firm, and may be gently broken only to 30–50mm, to avoid the creation of too much fines. This is also true for cake of neem seed.

Shea nut cake, containing just about 10% protein, is too fragile and usually crumbles during conveying between the press, cooler and extractor. As it is, fines are created during the conveying. Hence, it should not be passed through a breaker. As discussed in Chapters 2 and 4 (Part 1), on the practice of shea nut preparation and pressing, the recently modified preparation process may help form a somewhat firmer cake. Even then, breaking of the cake would not be recommended.

12.3 SOYBEAN PREPARATION

Soybean crushing industry is a major part of the solvent extraction industry worldwide. In fact, soybean is the major oilseed of the world. Nearly 300 million ton of soybean is crushed for oil, compared to less than 200m ton of all the other oilseeds put together. Since most other oilseeds first pass through the mechanical expression phase, the cakes subjected to solvent extraction may be just about 2/3rd of the latter figure; that means the quantity of soybean crushed in solvent extraction plants would be more than double the quantity of cakes from all other oilseeds.

Soybean is mid-size seed, 7–10mm long and 5–7mm dia. Although high in protein (34–38%), it is harder than groundnut and rapeseed, due mainly to low oil content (18–22%), much lower than the two. Beans moisture is typically 11–13% in western countries (U.S., Brazil, Argentina, Paraguay) and 8–10% in warmer eastern countries (India, southern China, west Africa). The beans in eastern countries tend to be smaller in size, and higher in protein, @ 36–38% against 34–35% in western countries.

Soybeans are prepared for extraction either whole or after dehulling. Dehulling helps increase the protein content in de-oiled meal beyond 46%, and reduces fibre content which helps its usage in poultry feed.

The preparation steps for whole soybeans include cracking, conditioning and flaking, and may be 'expanding' (Figure 12.3). For dehulled soybeans, a hulls separation step is introduced.

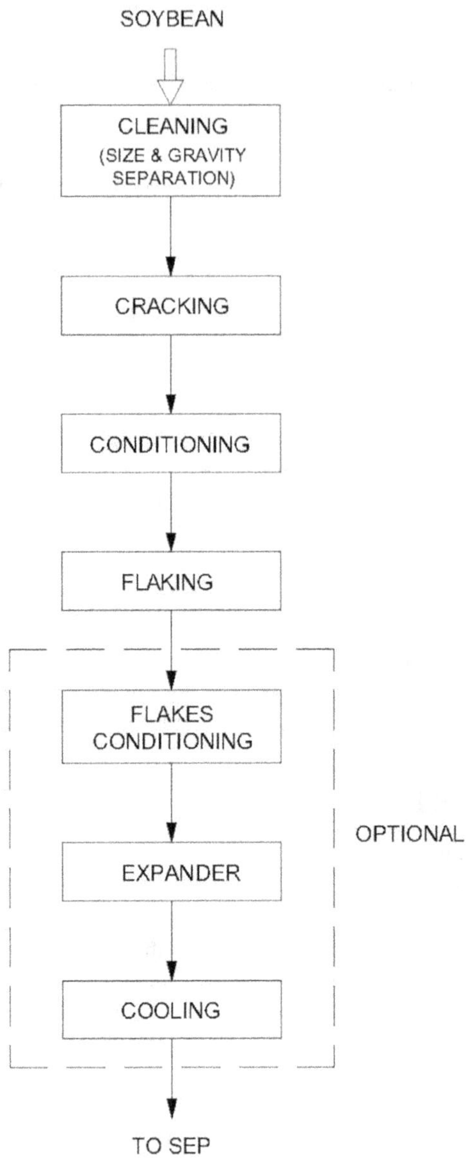

FIGURE 12.3 Schematic of Soybean Preparation (w/o Dehulling).

12.3.1 Beans Cleaning

As beans are received in factory, it is unloaded, pre-cleaned to remove chaff, large impurities and fine dust; and stored in silos. The operations of seed unloading, pre-cleaning, and silo storage, have been discussed in Part 1 of this book.

If the beans are moist, with moisture more than 11.5%, they need to be dried, to allow cracking downstream. Drying may be accomplished in vertical silo-type dryers, followed by tempering for a couple of days. Alternatively, if the beans are to be dehulled, drying may be accomplished in a conditioner-dryer. These operations are discussed in Chapter 13.

As the beans are taken for preparation, the first objective is to clean the beans off all impurities. These include farm impurities, such as chaff, bag threads, and soil particles. Metallic impurities may also get in during multiple handling. The beans are first subjected to cleaning by size separation, with vibrating sieves. Metallic (iron) impurities are removed by magnetic separation. Stones and mud balls of size similar to beans are removed by gravity separation. The beans may then be weighed online as they are conveyed continuously to the next operations.

12.3.2 Beans Cracking

Beans have to be broken down into smaller pieces, and the hull to be opened, for effective steam treatment. The breaking operation is accomplished with Beans Cracker, a machine with one or two pairs of corrugated rolls. The objective is to make small pieces ('grits') of 3–4 mm, without creating too much fines. Cracker rolls are configured in holding-cutting mode for the purpose – this is discussed in Chapter 13. Larger grits would require long cooking time, during which the surface would get overcooked. Very small particles do not lend themselves to sufficient disruption of cell structure during flaking. As discussed in the earlier section, the degree of deformation (DoD) achieved during flaking depends not only on the flake thickness but also on the grit size. The larger the DoD, the greater the work done on cell walls.

The larger western soybean, typically US grade 2, is cracked into 7–8 pieces, whereas the smaller Asian and African beans may be cracked into only 4–5 pieces. Typically, 4-roll machines are used for soybean cracking. While on the lower pair rolls, the pitch is nearly 3.6 mm (7 corrugations per inch) for all beans, the pitch on the upper rolls may vary depending on the beans size. For western beans, a wide pitch is used, nearly 8.5 mm (3 corrugations per inch); whereas for smaller beans, it may be between 5 and 6 mm (4–5 corrugations per inch).

Effectiveness of the cracking operation should be ascertained by sieve analysis. The first step may be to separate the uncracked beans and splits manually. These should not be more than 3% and 5%, respectively. The grits may then be sieved carefully on a 2 mm sieve. The fines, passing through 2 mm, should not exceed 5%.

If beans are to be processed for high-protein meal, it is critical that all beans are cracked effectively. This requires that the beans be of similar size. Should the beans lot have wide size distribution, it may be subjected to a 'grading' operation, wherein 'small' beans may be separated by a sieving operation.

Beans containing high moisture, more than 11.5%, are difficult to crack. Such beans must be dried prior to cracking. The drying may either be accomplished in large silos, with warm air, or in vertical conditioners. These processes are discussed in the following section on 'dehulling'.

12.3.3 Conditioning

This operation is similar to cooking, used for seed preparation for mechanical expression, but milder. The objective here is not to achieve full cooking, to rupture cell walls by the explosion of water vapours within the cells, but only to prepare the grits adequately for an effective flaking action. The reason for the change of sequence of cooking and flaking operations, compared to that for mechanical expression, has been explained in earlier Section 12.1 – theoretical aspects of seed preparation. The objectives of this operation also have been listed in the section.

In most plants, soybean is heated to 72–75C, and conditioned for 18–20min. However, longer time (25–30min) at lower temperature (70–72C) helps in better uniform conditioning. In fact, there is now a tendency to opt for even slower conditioning for 40min at 65C. Optimum moisture range for flaking is 10–11%. Should the grits be dry, 'open' steam may be injected to raise the moisture to 10–10.5%. For very dry beans, with moisture content of 6–7%, as encountered in Asia and Africa during the off-season, hot water may be sprayed in the first compartment of conditioner=cooker, up to 2%, and steam be injected in the second compartment, to raise the moisture to the required range. Water addition of more than 2% is not desirable, as some moisture may remain as surface moisture, and may cause problems in flaking operation. Also, care must be taken to spray the water evenly on the entire seed bed.

The conditioning time may be reduced by 1/4th if flakes are to be treated further in expander.

12.3.4 Flaking

This may usually be the last operation of soybean preparation, and may affect the performance of the solvent extraction plant to a large extent. The steam-conditioned grits are flaked hard to 0.28–0.30mm thickness. The objective is to totally disrupt the beans' structure and rupture as many cell walls as possible. Thinner flakes would mean more damage to cell walls, but these would get compacted too much within the extractor; the excessive compaction would hinder solvent percolation and would be counter-productive to extraction. Thinner flakes would also require more energy input to flakers. In fact, for deep bed extractors, having beds deeper than 2.5m, flake thickness may be restricted to minimum of 0.35mm, to avoid too much compaction. Typical energy consumption, for flake thickness 0.28–0.3mm, may be as high as 8–9 kWh/T.

As discussed before, a good preparation of soybean should achieve milling defect (MD) of 0.2%. As the flakes are conveyed to the extraction plant, a gentle air aspiration should be applied to remove surface moisture, and to cool the flakes to the extraction temperature of 58–60C.

Should the flakes be further treated in expander, thicker flakes upto 0.4mm, with MD of 0.3%, may be allowed. These flakes should then be further conditioned for at least 5min at 80–85C, prior to passage through expander.

12.3.5 Expanding

Because of the advantages offered by Expanders, more and more soybean crushing plants are now fitted with these machines. The advantages are discussed in Section 12.1.

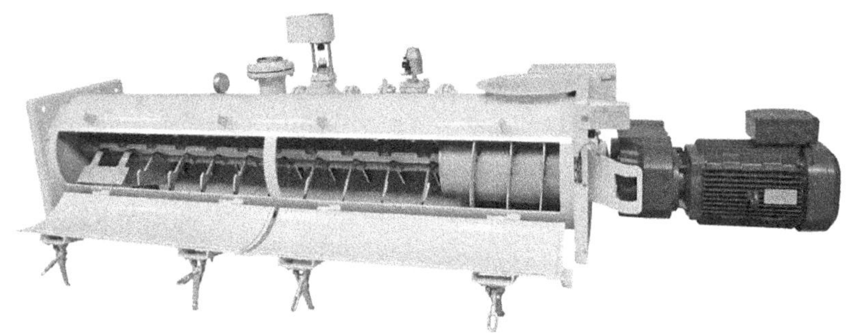

FIGURE 12.4 Paddle Mixer Conditioner. Courtesy: Van Aarsen International.

Flakes are heated to 80–85 C and conditioned for 5 min, usually in a Paddle Mixer-conditioner, with steam jacket (Figure 12.4). Alternatively, a Rotary Conditioner may be used. Since heating requirement is low, from 70 C to 85 C, and so is the holdup time, these conditioners are relatively compact. Vertical multi-tray cookers are not preferred, since those require higher electric energy.

Expander wormshaft rotates at relatively high speeds (200–250 rpm). The high shearing action takes the heat quickly to cell components and facilitates protein coagulation. Steam is injected into the expander to aid the coagulation process, and to aid expansion, as also to reduce the load on the drive motor. Exit temperature is maintained at 95 C minimum, by regulating the steam input and the choke pressure, to ensure good flash off at exit, which in turn ensures good expansion of collets.

Expander may be fitted either with a die-plate or with an annular 'floating' choke. The latter allows uniform pressure on the exiting mass, as it balances any pressure variations, due to variations in feed, with the hydraulic pressure applied on the choke. In case of die-plates, taper dyes may be used, with outlet dia of 16–18 mm; after expansion, the collets dia may become 18–20 mm.

Energy input into an expander may be considered an extra. Typically, for soybean expansion, energy requirement is 4–4.5 kWh/T. However, this is partly offset by saving in energy input to flakers. In absence of expander, soybean grits need to be flaked to 0.28–0.3 mm, using nearly 8–9 kWh/T energy. With expander, however, thicker flakes, 0.4 mm, are allowed, requiring only about 6–7 kWh/T. Thus, energy saving up to 2 kWh/T is achieved on flakers, which offsets nearly half the energy input to expander. Add to this, the savings in steam in solvent plant, and expander may be seen as an energy-neutral device.

Should the MD of flakes be low, within 0.2–0.25%, expander is not able to reduce it further. However, should it be higher, say 0.3–0.4%, an expander may help reduce it by a 0.05–0.1 percentage point.

12.3.6 Cooling of Collets

Collets are discharged from an expander at around 95–100 C and have moisture in the range of 12–13%. Nearly a percent moisture flashes off on discharge, reducing the temperature by about 10C. These must be cooled further to around 60C, and moisture be reduced to less than 10%, before being conveyed to the SEP.

At expander discharge, collets are 'soft' and must be handled gently, to not produce fines. A slow-moving 'dwell' conveyor, with wide cross-section, to hold the collets for a minute, helps firm those up prior to being fed to cooler. This process prevents the disintegration of collets as they fall onto the cooler screen.

The most common design of coolers used for collet cooling is the horizontal, slow moving screen belt cooler (Figure 5.3, Chapter 5). It is of modular design, with multiple modules of dryer and cooler assembled together. A dryer module has an FD blower (forced draft), an air heater, a suction (ID – induced draft) fan, and a cyclone, all mounted on the module. Air may be blown at 130C, to effectively remove moisture. A cooler module may just have a suction fan and a cyclone. An FD fan may sometimes be added, to balance the pressure below the bed. An oscillating feeder helps in the even distribution of collets on the screen.

The soybean collet cooler is typically designed for 4–6 min hold up. Some designs use only cold air, but it requires nearly double the air quantity, so the electrical energy requirement for fans goes up considerably.

The cooled firm collets are conveyed gently to SEP, by means of en-masse conveyor, with chain moving at a speed of 22–25 m/min max.

12.3.7 Dehulling

Soybean crushing gives just about 18–19% oil, and nearly 80–82% meal. Thus, soybean meal is the more important product, and its quality is of paramount importance. When the whole soybean is crushed, the meal contains nearly 6% fibre. This level of fibre is not attractive for its usage in poultry feed. Therefore, much of the soybean is crushed after dehulling to remove much of the hulls, to have a product with 3.5% fibre max.

Western soybean varieties typically produce meal, when whole beans are crushed, which have around 43–44% protein. The hull content of beans is nearly 8%. If 7–7.5% hulls were removed, the protein content goes above 47%. This is the minimum protein level required for a product called 'Hi-Pro' soybean meal. Most of the Asian and African varieties have higher protein content, and the meal from whole beans itself may contain up to 45–46% protein. These varieties also tend to have smaller beans with more hulls, nearly 9–10%. Should 8–9% hulls be removed, the protein content may increase up to 50% or even more.

12.3.7.1 Cold Dehulling

Classical dehulling method involves the separation of hulls, which have come loose on cracking of beans. The separation may be accomplished either just by air aspiration, or by a combination of size separation and air aspiration (Figure 12.5). When only air aspiration is used, much of small grits also are carried with hulls; the former have to be recovered by beating and size separation.

To ensure that all beans are effectively cracked, beans should be of similar size. If the beans lot has wide size distribution, it may be subjected to grading by size. Small beans that are separated, may be cracked separately; or may be stored, to be

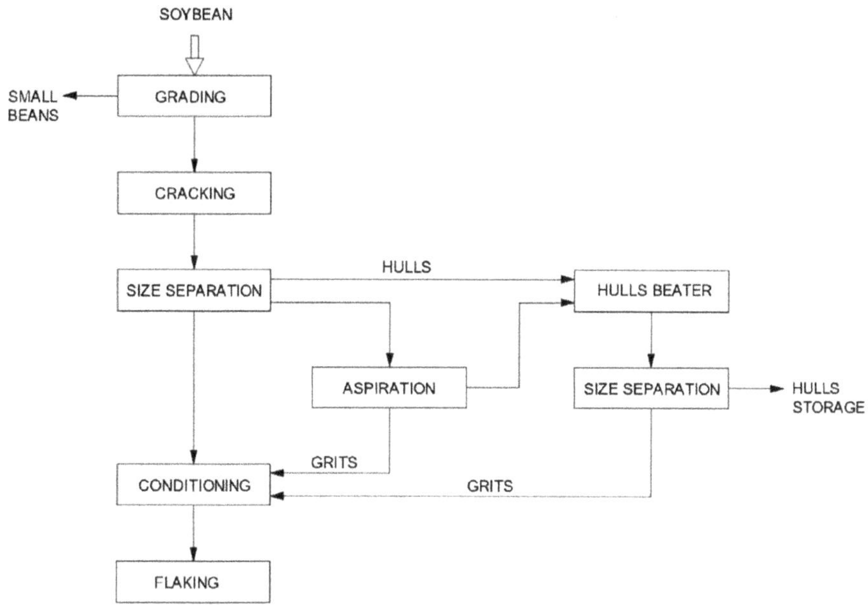

FIGURE 12.5 Schematic of Cold Dehulling of Soybean (Single Step).

mixed later with lots that are not dehulled. Grading is achieved by size separation with vibrating sieves.

For western beans, where due to the cold growing season, the beans moisture is relatively high, the hulls are tight on the cotyledon, and a two-step cracking process gives better separation (Figure 12.6). The first cracker is a two-roll machine, which splits the beans. The cracked splits are subjected to gentle multiple impact, as during passage through an 'impact aspirator (wide chute with internal pipes arrays), to loosen the hulls further. The splits are then cracked in a four-roll cracker. The grits, produced in the second cracker, are also subjected to multiple impact, and hulls are removed by aspiration. Hulls from both aspirators are screened separately to recover fine particles of the cotyledon.

After the dehulling process, hulls remaining in the meat should not be more than 1.5%. Also, hulls purification, for recovery of fines, should ensure that the oil content of hulls fraction should not be more than 1% above botanical oil content (0.8–1.0%) of pure hulls. Should soybeans be processed for edible products, hulls must be reduced to below 1%.

Should the beans moisture be high, more than 11.5%, cracking is not effective; the beans come out just pinched and flattened, instead of in multiple pieces. Therefore, high-moisture beans must be dried and tempered and then cracked. In fact, for proper separation of hulls after cracking, beans moisture should be reduced to below 10%.

The heat energy requirement for the drying operation is high, and the cooled beans will again be heated for conditioning prior to flaking. Also, big silos are required for tempering the beans over one or two days. These issues may be overcome with an

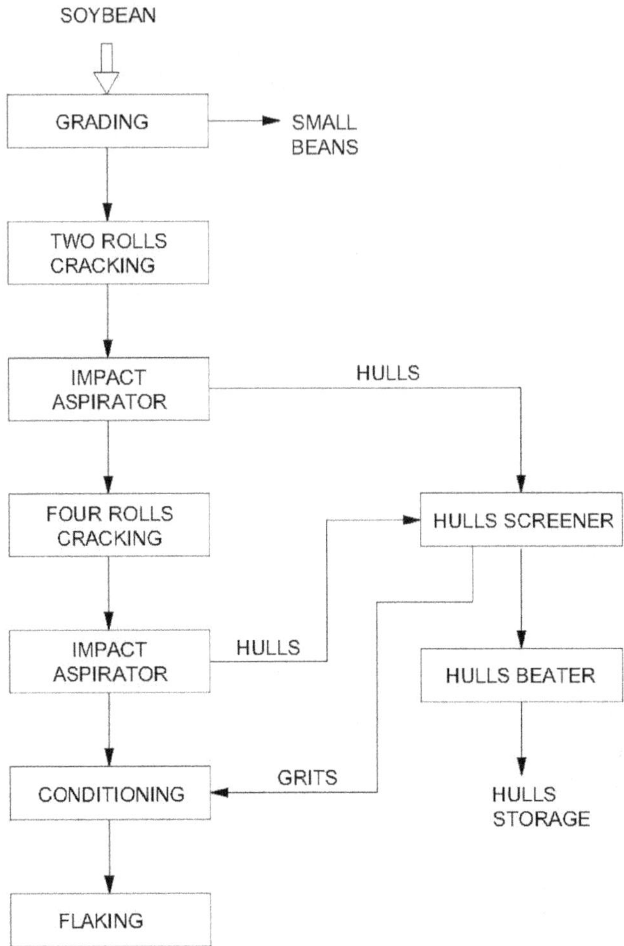

FIGURE 12.6 Schematic of Cold Dehulling of Soybean (Double Step).

alternative process of steam conditioning-cum-drying, followed by cracking in hot condition. This process is known as 'warm dehulling' or 'hot dehulling', and was introduced by Buhler group around the early 1990s.

12.3.7.2 Warm/Hot Dehulling

In Warm Dehulling system, which works with soybean moisture up to 13%, a vertical conditioner-dryer is the heart of the process (Figure 12.7). As soybean flows down by gravity, it comes in contact with hot pipe surfaces, and get heated to 60–70C. Rows of pipes, either circular or elliptical cross-section, placed at triangular pitch, help ensure effective contact. Some of the bean moisture comes to the surface and accumulates inside the hull. As the beans 'sweat', the kernel softens. The pressure of the water vapour loosens the hull from the kernel. Thereafter, hot air at 120–130C is blown through the beans to quickly dry out the hull and the accumulated vapour. In

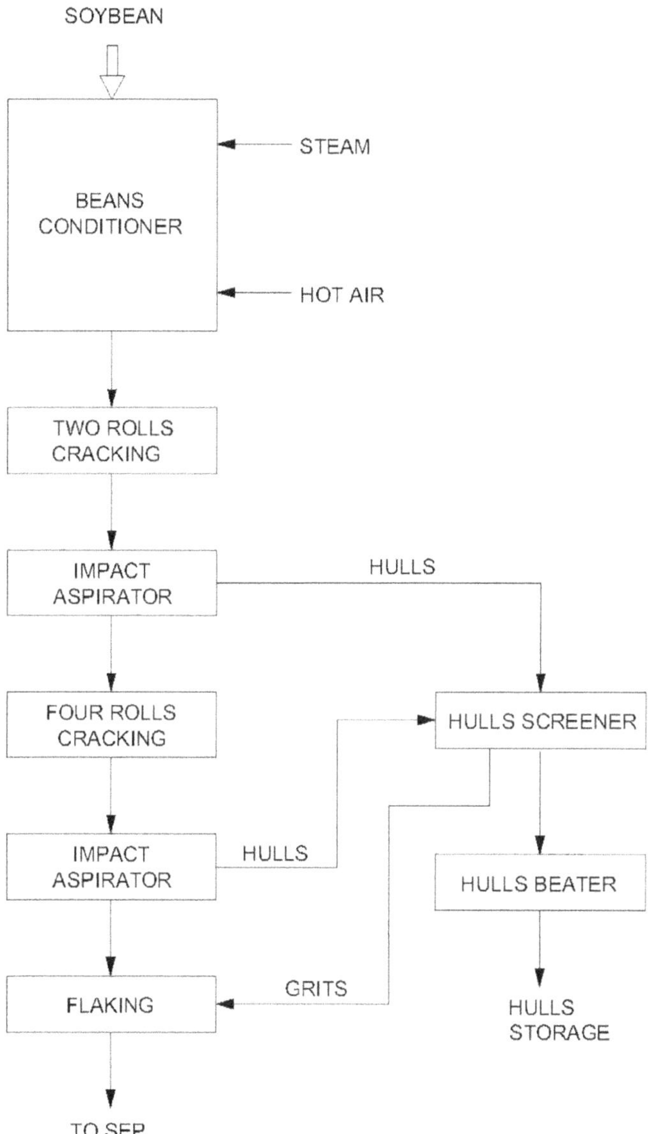

FIGURE 12.7 Schematic of Warm Dehulling of Soybean.

some designs, the heat-dry cycle is repeated two or three times. Thus, three objectives are met:

- Conditioning of the beans
- Moisture reduction, and
- Loosening the hulls

The conditioner is usually built with a square cross-section. Such a cross-section facilitates the fitment of rectangular tube sheets for the steam-heated tubes. It has a 'tempering' section at the bottom. The beans discharge is by means of a row of long rotary dischargers, with a common drive, or a Rake discharger; the discharge rate is controlled with the speed variation of the Discharger. Temperature of the beans at discharge is close to 70C (Ziemann, 2020). These are then cracked, and hulls removal is accomplished by a similar process as with dry dehulling.

The total residence time may be more than 40–50 min in single-cycle conditioners, but may be 60–70 min in multi-cycle design.

In another process, known as Hot Dehulling, in addition to the cycle(s) of 'sweating' and drying, a separate dryer is provided downstream for final moisture reduction (Heinrich, 2020). This may be a fluidised bed dryer, with hot air at 140–150C. The separate dryer may allow the intake of beans with moisture up to 14%.

Steam consumption for warm dehulling process is around 90–95 kg/T, compared to the total steam requirement for drying and beans conditioning of nearly 140–150 kg/T for dry dehulling. Also, big tempering silos are avoided in the hot dehulling process. For these reasons, the hot dehulling process is now the norm, particularly in new crushing plants, for dehulling of soybeans containing more than 11% moisture.

12.4 RICE BRAN PREPARATION

Rice bran is another oil-bearing material which is directly extracted by solvent. India is the leading country in rice bran (RB) processing, as also in the production of refined RB oil, followed by China, Thailand and Vietnam.

Rice bran may contain between 15% and 25% oil, depending on the method of rice milling. It is very low in protein, less than 10%.

12.4.1 Raw Rice Bran

Rice bran is mainly of two varieties, raw rice bran and parboiled rice bran, obtained from milling of raw rice and parboiled rice, respectively. Raw RB must be processed within 24 hours of rice milling. As the milling disrupts the oil cells, lipase enzyme comes in contact with oil globules, and the attack on tri-glyceride chains starts. The free fatty acid content increases every hour and may go up to 15% within 24 hours, and another 3–4% with each passing day. Should the FFA rise beyond 20%, the oil may not be refined economically for edible purpose, and would be sold as industrial grade oil. Oil content in raw RB ranges from 15% to 20%, depending on the milling process.

Because of the requirement of processing within 24 hours of milling, (raw) bran may only be collected from nearby rice mills, say from a radius of 100 km. For this reason, it is difficult to find rice bran processing plants bigger than 500 Tpd capacity.

12.4.2 Parboiled Rice Bran

Bran from parboiled rice contains more oil, 22–25%, and is more stable, as the enzyme has mostly been inactivated during the parboiling process. It is possible to get parboiled rice bran with 3–4% FFA, increasing @ just 1% every day. Thus, it may be processed within a week of milling and still have FFA below 10%. Colour of oil produced from parboiled RB is darker, with more 'red'. The availability of parboiled RB is limited, as compared to the raw RB.

As obtained from rice mills, RB is in powdery form. If fed to (percolation) extractor in that form, solvent percolation is severely restricted. That results in a very low capacity of extractor, as also higher residual oil. For this reason, the bran is to be moulded in a form which would allow good solvent percolation, as also good contact of solvent with oil cells.

12.4.3 RB Preparation Process Steps

Traditionally, bran is steam-conditioned and pelleted (Figure 12.8). The pellets are cooled and sent for solvent extraction. This preparation allows almost complete oil extraction, leaving just about 0.5% residual oil in meal.

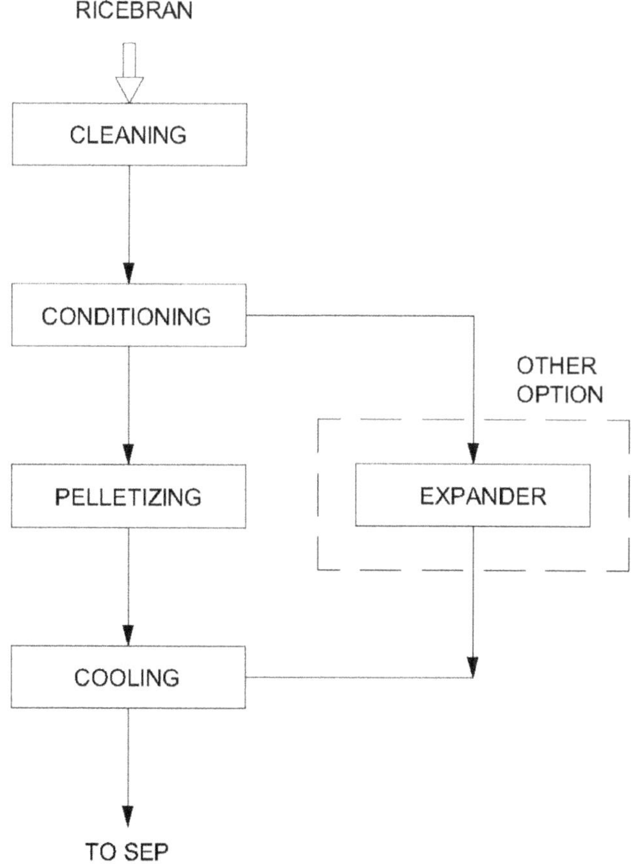

FIGURE 12.8 Schematic of Rice Bran Preparation.

However, the pellets do not hold well under solvent spray for long. If the extraction time is, say, 90 minutes, the pellets start disintegrating after half an hour, and there are too much fines on top of the extractor bed by the middle of the extraction zone. This layer of fines affects solvent percolation and causes increase in extraction time requirement. Also, the pellets at the feed point are very compact (hard) and do not allow the use of rich miscella for the first wash. Thus, miscella concentration out of the extractor has to be maintained low. This means higher distillation load and higher steam requirement.

These problems may be partially overcome by the use of an expander in place of a pellet mill. The collets, out of the expander, are porous enough to allow the use of rich miscella, and are strong enough to withstand solvent spray.

12.4.4 Cleaning

A single vibrating screen, with opening of 4 mm (6 mesh), is sufficient to remove all oversize impurities from rice bran. Undersize, such as soil particles, cannot be separated, since the bran itself is in powdery form. High throughput capacity is achieved because of the big difference in size between RB and impurities. A rotary drum sieve may also do the job. Aspiration is not used, since it would take away the bran which is very light.

Size separation should be followed by magnetic separation, to ensure there is no damage to the pelleting machinery downstream.

12.4.5 Conditioning

Traditionally, bran is conditioned in small paddle-screw conditioners, with the residence time of just 2–3 minutes. The paddle screw is placed above a pelleting mill. The screw is jacketed for L.P. steam; also, live steam may be injected into the bran, to improve the binding in pellets. Bran moisture, as received, ranges from as high as 15% at the start of the season (October-November), through nearly 12% (Dec-Feb), to a low of 9–10% during the off-season. Nearly 3% live steam is injected for the purpose of binding, except at the beginning of the season, when just about 1% live steam may suffice for the purpose. The outlet temperature is maintained at around 70–75 C. The seasonality of bran moisture can be quite different in Vietnam, where multiple paddy crops are a norm.

Pellet mills capacities are low. So, a plant designed for 400 Tpd, would have a line of 3–5 pellet mills, each with its conditioner. An overflow bin, at the end of line, is recommended, to ensure that even the last mill is fully loaded. The bran from overflow bin may be conveyed back to the feed bin, or to a common conditioner as described below.

Vertical multi-tray type conditioners are used in some plants. These are usually designed as a single conditioner, common for all pellet mills. A distribution conveyor may feed all the pellet mills, and also maintain a small overflow of material to an overflow bin. Alternatively, a short discharge screw conveyor may be fitted at conditioner outlet, and its speed may be varied, so as to maintain a constant height of bran in a small hopper placed on top of the last mill.

A vertical conditioner may be designed for longer conditioning time, of nearly 10 min, to achieve better conditioning, which results in better binding of pellets at lower moisture content. A lower moisture content of pellets results in less energy input in pellet drying-cooling operation. Conditioning time may be much shorter than for soybean (20–30 min), because of the small particle size of rice bran.

12.4.6 Pelleting

Conditioned bran, at elevated moisture content, is fed to a pelleting mill. As the rotating rolls force the bran through a die plate, the bran gets compacted into hard pellets. Pellet size has reduced over the decades. From what used to be dia 8 mm nearly two decades back, the standard size now is 5 mm. Taper dies 6 mm/5 mm are used for the purpose, for better compaction. The smaller size pellets not only are stronger but also are easily extractable.

Processing of parboiled bran is always a tricky matter. It is difficult to form strong pellets with parboiled bran. This is so because the water absorption capacity of the starch and protein in parboiled bran is reduced, due to the parboiling process. The common practice is to mix some amount of raw bran, say 10–15%, upstream of conditioner. Thinner pellets, dia 4 mm, are produced to help achieve better strength.

In locations where raw bran may not be available, some quantity of ground rice husk is mixed, to give additional strength to the pellets. However, these pellets do not hold well under solvent spray. It would be better to work the bran more vigorously instead, like passing through high-taper die of 6 / 4 mm.

The largest pellet mills available are with Die-Plate dia 900 mm, driven by 75 hp motor. The mill can process up to 125 Tpd of raw RB and 150 Tpd of parboiled RB.

12.4.7 Expander Treatment

A few plants have utilised expanders in place of pellet mills. Pellets appeared softer compared to those from pelleting mills. However, these held well under the spray of solvent in the extractor, thanks to superior protein-starch coagulation due to continuous shearing action. Also, due to the high porosity, achieved with expansion at exit, it was possible to wash the fresh collets with rich miscella; that meant the miscella concentration out of extractor was higher, and solvent load to distillation was lower. Thus, steam consumption in SEP was lower.

Expander may also be built for higher capacities; so, a single expander may cater to the capacity of the plant, in place of multiple pellet mills. In fact, a single vertical conditioner with a single expander would simplify the entire operation; and would also enable good process control. However, higher electrical energy input to the expander, almost twice in fact, compared to pelleting mills, has held back its wide application in the industry.

Since rice bran has very low protein content, the material has to be worked more, compared to soybean, to achieve protein starch coagulation. So, the dies in the die-plate must be much smaller dia than those for soybean. This means the electrical energy consumption is nearly twice that for soybean.

12.4.8 Pellet Cooling

Pellets must be cooled before being fed to the extractor. Cooling helps make the pellets firmer, and helps improve integrity under spray of solvent to some extent. Also, moisture reduction to 10% is required for effective extraction.

Normally, to achieve 2% moisture reduction, hot air drying would be utilised. This is the norm for cooling of collets of soybean, for example. However, in case of rice barn pellets, the use of hot air, temperature higher than 60 C, is prohibited. It has been observed that pellets subjected to drying by hot air, though they hold well under solvent spray, are difficult to extract (Modi, 2003). This is so perhaps because, with hot air, there may be a layer of coagulated protein-starch on the pellet surface, which hinders solvent penetration into the pellet core.

Since all the moisture removal is to be accomplished by passing cold air, the air quantity is substantial. It also takes longer than usual. Hence dwell time in the cooler are up to 8–10 minutes, compared to 5–6 minutes for most other materials.

Collets from expander, on the other hand, are very porous and may be cooled in a shorter time.

Pellets must be handled gently, to avoid breakage, during conveying to solvent plant. En-masse conveyor, with a chain speed of 20 m/min max, may be used for the purpose. The speed may be a bit higher, up to 25 m/min, for expanded collets.

12.5 COTTON SEED PREPARATION (ALSO SAFFLOWER SEED AND CORN GERM)

As seen in Part 1 of this book, cotton seed may either be pressed to produce a cake of 16–18% oil content or may be treated with expander, prior to solvent extraction. The preparation steps for prepressing, and the prepress operation itself, have been discussed in Part 1. Here we will discuss the expander process.

Cotton seed meats, obtained from delinting and dehulling process, should be conditioned, flaked, cooked and passed through the expander (Figure 12.9). The individual operations are discussed below.

12.5.1 Pre-heat and Flaking

Cotton seed meat is conditioned at around 70 C, for 20 min. The operation is much more rigorous than that for the prepressing process. The temperature is higher, and the time is longer. This is so because the preparation has to be stronger, and the flaking should achieve adequate disruption of cell wall structure, to compensate for the loss of strong treatment in a press, the expander being a light-duty machine relatively.

The conditioned meat is flaked to 0.3–0.35 mm, a bit thinner than for prepress.

12.5.2 Cooking

The flakes are usually cooked at 80–85 C for 4–5 min, as for soybean. The optimum moisture level at conditioner outlet is 7–8%. Should the flakes moisture be higher, ventilation should be provided for moisture removal.

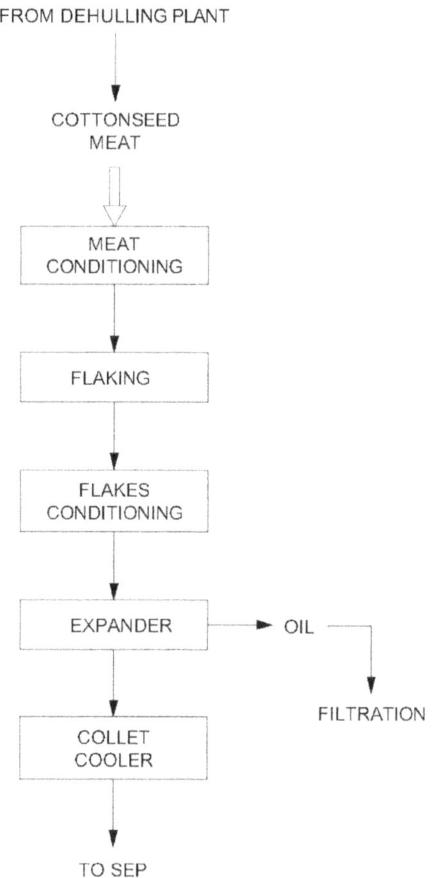

FIGURE 12.9 Schematic of Cottonseed Meat Preparation.

However, it helps to cook for a longer duration of 15–20 min, somewhat similar to the conditions prior to prepress. Longer cooking duration helps achieve low ROC in the final de-oiled meal after solvent extraction. As will be seen in the discussion below, a lower ROC also means lower 'free gossypol' in meal, which is desirable.

12.5.3 Expander

Expander processing started in the US during the 1970s. The conditioned flakes were passed through the expander, shaft rotating at lower speeds than for soybean. Red-coloured collets emerged through the die-plate at discharge, coated with released oil. The oil got absorbed as the collets fell into the chute to the cooler below.

The expander process saved substantial amount of electrical energy that was spent on pre-press. The cost of an expander was just a fraction of the press.

The collets were readily extractable, although the ROC in meal was higher @ 0.8–0.9% compared to 0.5–0.6% in meal after prepress plus extraction. Steam

TABLE 12.1
Relative Utility Consumptions[a] in Alternative Processes for Preparation of Cotton Seed Meat

Utility and ROC	Pre-Press	Expander	Expander With Drain Cage
Electricity in Prep, kWh/T	35	20	24
Electricity in SEP	14	16	15
Total electricity	**49**	**36**	**39**
Steam in Prep, kg/T	100	90	100
Steam in SEP	190	270	230
Total steam	**290**	**360**	**330**
ROC in Meal	**0.5–0.6%**	**0.8–0.9%**	**0.6–0.7%**

Basis: Plant to process meat (30% oil) derived from 500 Tpd white cotton seed.
[a] Electricity and steam figures on per ton meat basis.

consumption was somewhat higher, due to the extra solvent, going with the extra oil, to distillation. However, the overall energy balance favored the expander process.

The practice of fitting **drain cage** to the expander was started in India nearly a decade back (Vadke, 2017). The released oil was let out through the drain cage so that the collets had just about 22–24% oil. The collets were much stronger than those with full oil. The steam requirement in distillation reduced nearly to the level of the prepressed cake process. The ROC in meal reduced to 0.6–0.7%. The electrical energy requirement was only marginally higher than that for expander without drain cage. This practice is now catching up in many countries.

The oil is screened and filtered, as discussed in Chapter 6. The filtered oil is sent to solvent extraction plant, for mixing with miscella (oil-solvent mixture), prior to miscella refining.

Relative utilities requirement is given below for the three processes (Table 12.1).

It may be seen that the process of expander, without drain cage, would require significantly more steam, and lose more oil in meal, compared to the Pre-Press process. Thus it would offset the benefit of saving electricity. However, expanders were still preferred for much lower investment, and ease of operation.

On the other hand, there is a clear benefit in adopting the process of expander with drain cage. The electricity saving is maintained, while steam and ROC figures are also closer to those of Pre-Press option. Combined with lower investment, and ease of operation, expander with drain cage is the clear winning option.

12.5.4 Cooling of Collets

The collets should be cooled to 58–60 C, slowly over 5–6 minutes, to make these firm. Ambient air cooling is sufficient, since no significant moisture removal is necessary.

The collets may be conveyed gently in en-masse conveyor, with maximum chain speed of 25 m/min.

12.5.5 Safflower Seed Preparation

Safflower seed contains nearly 32–36% oil, has a soft seed coat, with soft kernel inside. It may be broken in a cracker mill, with a single passage (2-roll), then prepared in a similar fashion as the cottonseed meat, and processed through expander with drain cage. All the process parameters described for cotton seed meat are applicable for safflower seed. Power and steam requirements are also very similar to those for cottonseed meat. Collets, with 22–24% oil, may be cooled and sent for solvent extraction.

12.5.6 Corn Germ

Corn germ may either be prepressed prior to extraction; or be subjected to direct solvent extraction. The first option has been discussed in Part 1. Here we discuss the preparation for direct extraction. It should be noted that corn germ contains more than 40% oil, and is preferably prepressed prior to solvent extraction. Direct extraction route may only be utilised when small quantities are to be crushed in a solvent extraction facility that has no presses.

The germ may be broken in a cracker mill, with a single passage (2-roll). The grits are conditioned to 70C for nearly 20–25 min, and then flaked to 0.3 mm. Thus, the preparation is very similar to that for soybean. However, the flakes do not hold as well, because of lower protein content, giving poor performance in the SEP. The extraction performance may be greatly enhanced by second cooking prior to passage through a drained expander, like that for cottonseed.

12.6 SUNFLOWER SEED AND PALM KERNEL

Sunflower seed and palm kernel (PK) may be directly solvent-extracted, without the step of prepressing. The flakes may be handled well within the extractor, and final ROC of 1.0% (sunseed) or 1.2% (PK) may be achieved. However, the overall energy balance is totally in favour of the prepress plus solvent process. This is mainly due to the much higher steam requirement in SEP to handle four to five times the oil, and the associated solvent. The electrical energy saving in the preparation, without the press, does not balance the much higher steam cost, and the higher oil loss in meal.

Although the utility balance is not in favour of the direct solvent extraction process, it is sometimes adopted, mainly to save on the investment. In some countries, very low utilities cost (availability of cheap gas to produce steam) may also make the process of direct extraction attractive. Also, should a plant be operating on soybean for most of the months, but may have sunseed available on and off, this process may be of interest, since they would use essentially the same preparation line, without any additional investment. However, plants may operate solely on direct extraction of PK, because PressCake of PK can be powdery, which is difficult to desolventise.

The preparation of sunflower seed, and palm kernel, for direct extraction, follows similar path as that for the prepressing operation, except there is no cooking of flakes. In the absence of cooking and pre-pressing, the other preparation steps must be vigorous than those prior to pre-pressing. Sunseed should be cracked in 4-roll crackers, conditioned at 70–75 C for 20–25 min, and flaked to 0.3–0.35 mm. Flaker rolls should be operated at lower speeds compared to soybeans, at around 200 rpm. Resultant capacity may only be 2/3rd that for soybeans.

Should expander be available in the plant, a passage of flakes through the expander, after steam-conditioning, would help enhance the preparation. Even a better option would be to fit a drain cage on the expander, to recover part of the oil; the collets would then be much stronger, and extraction performance would approach that of cakes. The collets would have to be cooled to 58–60 C slowly.

The grits of **palm kernel**, as obtained by passage through hammer mill to create large pieces, and another passage through a seed cracker, must be conditioned prior to flaking. The conditioning step is usually avoided in many plants, due to the perceived fear of oil release in pre-heater. But that is a mistake and can lead to high ROC in meal, as also higher losses of solvent with meal. Conditioning of ground kernel at 60–65 C for 30 min helps soften up the seed structure, helping structural distortion on flaking. Also, due to slight coagulation of protein, the flakes (0.35 mm) hold somewhat better and do not disintegrate easily in Extractor.

12.7 MINOR SEEDS: SAL SEED AND MANGO KERNEL

12.7.1 Sal Seed

Sal (*Shorea Robusta*) seed is a minor oilseed, collected from Sal forests in India. Contains around 15–20% fat, which, in refined form, is used as a replacement for cocoa butter.

This is a large seed, oval shaped, nearly 24–28 mm long, and 15–18 mm wide (Figure 12.10). Hard shell is 1/4th by weight of seed, and kernel the 3/4th. De-shelled

FIGURE 12.10 Sal Seed Kernel.

kernels are received at crushing plants. Kernels contain nearly 20–25% oil, and are low in protein, less than 10%, high in starch. Kernels are hard and must be prepared well before extraction.

The kernels received at factory are usually in 3–4 pieces. These are broken either with Seed Cracker or with hammer mill, steam-conditioned, and flaked. The flakes are sent for solvent extraction. Steam conditioning parameters tend to be 75–80C for 15–20min. Flakes are usually around 0.5mm thick. It would be better if the conditioning time be increased to 35–40min, so as to achieve conditioning to the core, which would allow producing thinner flakes of 0.3–0.35mm. Also, size reduction with Cracker (4-rolls) mill should be preferred to minimise fines. That would improve extraction, and reduce the ROC to below 1.0%.

Some plants have now started using expander to prepare the flakes further. This is a good technique to overcome deficiencies in conditioning and, to some extent, in flaking. In this case, grinding of seed using hammer mill may be acceptable, since fines production would be of no matter; all the fines would get coagulated in the expander.

12.7.2 Mango Kernel

Mango kernel is received in factory without the shell and pre-dried to less than 10% moisture (Figure 12.11). Oil content is in the range of 12–15%.

The kernel is cleaned of impurities by size separation and magnetic separation. Clean kernel is broken with hammer mill to 8–10mm size. The kernels being a bit plastic, either the hammers should have twin edges, or single-edge knives be used.

In most factories, the grits are sent for direct extraction. However, conditioning and flaking could help reduce the extraction time significantly. Also, a second passage through another hammer mill, to create smaller pieces, 3–5mm, would help the subsequent preparation.

FIGURE 12.11 Mango Kernel.

13 Preparation Equipment *Construction Features and Design Aspects*

Most machinery used in the preparation of oilseeds has been covered in Part 1 of this book. Storage silos, day silo, pre-cleaners, cleaners, destoners, crackers, cookers, flakers, as also delinters, dehullers and grinders have been covered in Chapter 3. Cake processing machinery, namely, cake breaker and cake cooler, have been discussed in Chapter 5.

Machines that are utilised for the preparation of oil-bearing materials for solvent extraction, but are not required in preparation for mechanical expression, are only a few. These machines, namely, grader, pellet mill and expander, especially with drain cage, as well as flakes conditioner, are discussed in this chapter.

13.1 GRADER

This machine is used to remove 'small' beans from a lot of soybeans, to ensure that all the 'normal' sized beans are effectively cracked. The machine is a simpler form of a seed cleaner. It contains deck(s) of vibrating sieves, like in a cleaner. No scalping screens, for the removal of large particles, are included. Also, quite often, aspiration channel, for the removal of light particles, is also not necessary, as all 'lights' have already been removed in seed cleaner and destoner.

Since the size separation is to be achieved between beans which are not very different in size, the amplitude of vibration is maintained low. This means a machine with the same sieve surface area has a much lower throughput capacity compared to a cleaner. Typically, the grader machine would have half the capacity of a cleaner of equal sieve area.

13.2 PELLET MILL

Pellet mills are mainly of two types, vertical and horizontal. The former is commonly used to make pellets of meals for animal feed, while the latter is commonly used on rice bran. The heart of a pellet mill is heavy rollers moving on a thick die-plate, which rotates at 60–75 rpm, forcing material through the dyes (Figure 13.1). The casing wall rotates at a lower speed in opposite direction, thereby churning the bran or the meal mixture. Pellets emerge from the other side of the plate. In vertical mills, the die-plate is mounted vertically, and in horizontal mills, horizontally.

DOI: 10.1201/9781003309475-15

(a)

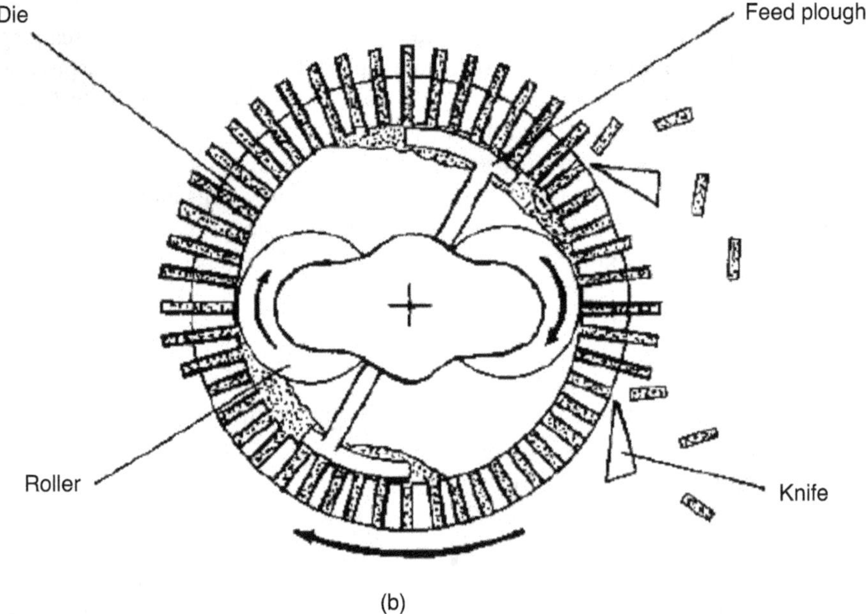

(b)

FIGURE 13.1 (a) Vertical Pellet Mill. Courtesy – Spectoms Engineering Pvt. Ltd. (b) Vertical Pellet Mill – Roller Mechanism. Courtesy – Spectoms Engineering Pvt. Ltd.

Traditionally, in rice bran processing plants in Asia, horizontal mills have been used. Two decades back, the common machine had a die-plate of 600 mm dia and was driven by a 50 hp motor. The die outlet used to be 7 or 8 mm dia. Over the past decade, the most common size has been 750 dia plate, with drive motor of 60 hp, capable of processing nearly 90–100 Tpd rice bran.

The substantial increase in capacity was achieved even with smaller pellets, dia 6 mm, thanks to thinner die-plates. Recently, bigger machines have become available, with die-plate of dia 900 mm. These are driven by a 75 hp motor and can process 125–150 Tpd bran. With bigger machines, specific energy requirement has been reducing over the years, and is down to just about 7–8 kWh/T from nearly 10–11 units earlier.

The die-plate is made either of mild steel or cast steel; it is flame-hardened and tempered. Life of a die-plate is, typically, 50–60 days. Special wear-resistant materials have not been tried by the rice bran industry. Plate thickness, for the mild steel plates, is typically 40 mm, which suffices to produce strong pellets out of raw rice bran. More compaction is required for parboiled rice, hence die-plates are thicker, around 55 mm; these thicker plates are usually made of cast steel.

Many of the horizontal mills are operated in 'open' condition, without a shroud. Water vapours emerge continuously from the mill, carrying some fines. Thus, the surrounding is always warm and dirty. To improve the working, proper hood should be provided, which may be gently vented to remove the vapours.

13.3 EXPANDER

Expander (Figure 13.2) is similar in construction to the extruder machine used in the preparation of oilseeds, especially for the full press process. However, it is used for much lighter duty, with just about 1/4th the drive power. Expanders are operated with high speeds of shaft rotation, typically in the range of 200–250 rpm. The residence time is of the order of just 15 seconds max.

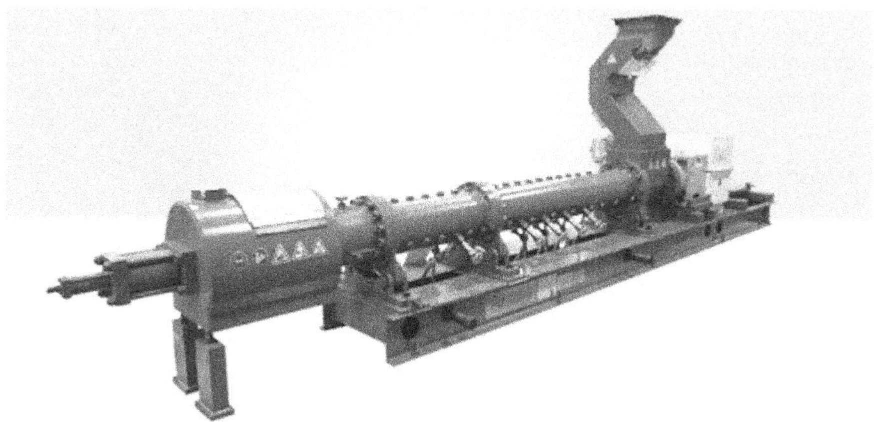

FIGURE 13.2 Expander. Courtesy – Anderson International.

The main objectives of the expander operation are as follows:

- Coagulate protein further, so that collets retain shape under solvent spray;
- Further release the oil globules within oil cells, to collect these near the cell wall; and,
- Cause sudden evaporation of moisture at exit, to create numerous micro-paths within collets, so as to improve solvent contact.

Expander process is not meant to rupture cell walls much. Thus, generally it is not able to reduce the milling defect. However, should the preparation have been inadequate, and if MD was high, say more than 0.3%, the expander would then help overcome the deficiency, to reduce the MD by around 0.1 percentage point. This is mainly achieved by the release of oil globules within individual oil cells.

The worm shaft is assembled with many worm pieces. The feed worm is designed for long pitch, to accept the feed with low bulk density, due to the presence of air between flakes. The main pressure worms are of much lower pitch, as these are designed for the volume of mass without air.

This principle is explained in detail in Chapter 4. Care should be taken to have a transition worm, between feed worm and pressure worms, of reducing pitch if possible, so as to minimise load on drive motor due to sudden compression. Cutting 'bolts' are inserted from the barrel, to cut and mix the mass inside; this action prevents slippage of the moist mass in the worm channels and increases throughput capacity.

Expanders are available in various sizes, with barrel dia 200–400 mm, barrel length 1.8–2.5 m, driven by motors rated for 100–300 kW, and may process up to 1,600 Tpd soybean. Energy consumption is much lower than for extruders used for prepress, from 8 kWh/T for 200 mm machines to just about 4 kWh/T for the large 400 mm expander. A comparative chart of various size machines is given below (Table 13.1).

Expander worms are of cast steel, case-hardened to a depth of nearly 2 mm. Good case-hardening materials may be EN31B, EN31C, MnCr5, etc. Last 4–5 worms, which are subjected to maximum pressure, and hence maximum wear, may be hard-faced with wear-resistant material such as chromium steel (Cr > 15%) to a depth of 3–4 mm. The barrel is lined with replaceable sleeves, which should be replaced after wear of 2–2.5 mm. Wear on cutting bolts also should be monitored; bolts should be replaced on wear of 5 mm a side.

TABLE 13.1
Expander Capacities and Energy for Soybean Processing

Barrel dia	200 mm	250 mm	300 mm	350 mm	400 mm
Capacity on Soybean, Tpd	250 T	600 T	900 T	1,200 T	1,600 T
Drive motor kW	90	132	200	250	315
Energy consumption, kWh/T	7.5	4.6	4.3	4.2	4.0

Since the shaft is supported only at feed end, and free on the other, the machine should not be run empty for any length of time. If the machine has to stop in interlock, the feeder should stop first, and the machine may be stopped after, say 15 seconds, when it is almost empty. This is to avoid leaving the hot moist mass inside the barrel, to become a compact hard cake within minutes.

Die-plate at exit has circular dies. Die size for soybean may be 16–18 mm; with a die of 18 mm dia, soybean collets may expand up to 20–21 mm. For rice bran, smaller dies should be used, as more pressure must be put on material which has much less protein than soybean, dies of 12 mm do produce good quality collets.

For soybean, which has high protein content, coagulation of protein does not require too much shearing and mixing. Thus light pressure at exit is allowable. For this reason, an annular choke maybe used in place of die-plate. It further helps to have the choke floating, supported by hydraulic pressure, to maintain the pressure on exiting material constant.

Steam injection is an important aspect of expander operation. Steam injection not only reduces the motor load, but also helps in protein coagulation and in the expansion of collet at exit. Injection nozzles are provided at various locations along the length of expander barrel. These nozzles must be aligned properly to inject steam in the direction of material flow, which is determined by the angle of worm channel. Should care not be taken to align the nozzles properly, these might clog with oilseed mass.

13.4 EXPANDER WITH DRAIN CAGE

When expanders are to be used for mid-oil content seeds such as cotton seed (meat) and safflower, it is much beneficial to fit a drain cage to allow draining of oil released within the barrel. A part of the barrel, towards discharge end, is replaced with a drain cage, with flanges bolted to the barrel segments. The cage is built much lighter compared to a cage on a screw press (Figure 13.3). Length of the cage may typically be 600–900 mm. A 360-degree hood should be fitted to avoid oil mist to escape to surroundings. The hood may be conical at the bottom, to collect the oil which may flow by gravity to a screening conveyor below.

A 'standard' expander, without drain cage, has pressure worms with uniform pitch and constant boss dia. For a drain cage, however, since oil is to be released, it helps to have worm profile with reducing pitch and/or reducing channel depth, so as to slowly increase the pressure. A pressure 'Hump', as explained in Chapter 4, at the end of the pressure section, would help further to squeeze out the oil. Two precautions are necessary to make optimum use of the short drain cage. One, the pressure worm segment should start just before the drain cage, so that oil may be released from the beginning of the cage itself. Second, the 'Hump' may be fitted at least 100 mm before the end of drain cage, to avoid carryover of the oil into the undrained segment.

Power requirement for drained expanders is slightly higher than that for standard undrained expanders. Table 13.2 gives the energy requirement for various size expanders for cottonseed processing duty, to produce collets containing 22–24% oil.

It may be seen that the energy consumption is much higher than that for soybean processing with standard expander. This is due to two reasons; one, cotton seed meats

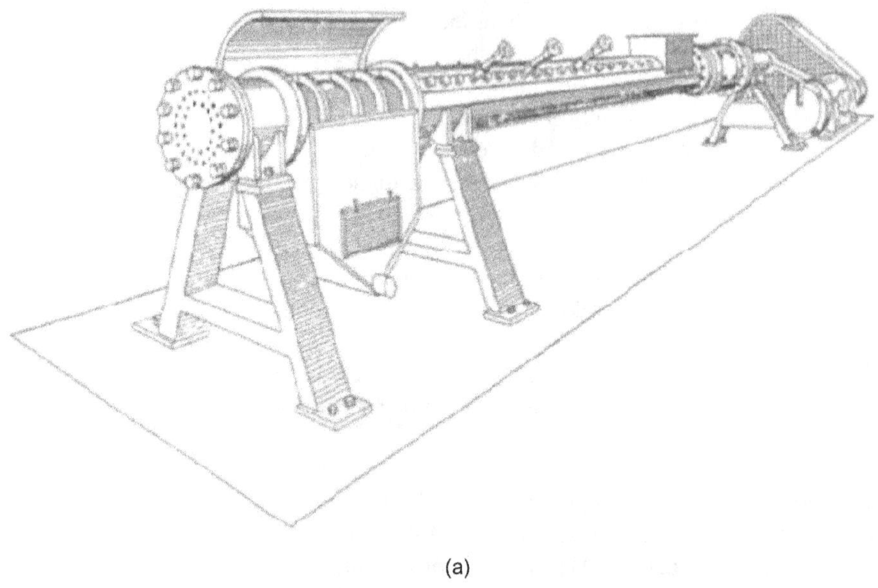

(a)

(b)

FIGURE 13.3 Expander with Drain Cage. (a) Machine Outline; (b) In Operation (Cotton seed).

Source: (a) Courtesy – Anderson International; (b) Courtesy: Thakurji Solvex, Jalna, India

contain part of hulls and lint, both with high fiber content, which put more frictional force on worm surfaces; second, higher pressures are required for squeezing out part of the oil. It should be noted, however, that the energy consumption is just about half of what is required for prepress operation (Table 4.2, Chapter 4).

TABLE 13.2
Expander (Drained) Capacities and Energy for Cottonseed Meat Processing

Barrel dia	200 mm	250 mm	300 mm
Cotton seed Meats capacity, Tpd	150 T	280 T	460 T
Drive motor kW	110	150	225
Energy consumption, kWh/T	15.0	11.0	10.0

The worms and the cage bars, for the drain cage, should be built to the same specifications as those in screw presses, as discussed in Part 1. Design of the worm profile has also been discussed in Part 1. For a cage 900 mm long, just three numbers knife bars may be used, with as many knives, for cutting and mixing of the mass in the cage. More 'knives' are not necessary, since there is not much compaction of the mass, unlike in a Pre-press. Dies of dia 12–14 mm can produce good collets.

The oil should be filtered prior to mixing with miscella in solvent plant, for miscella refining; this process is discussed in Chapter 14. The oil is first pre-filtered on Screening Conveyor placed below the expander. Vertical PLF is preferably used for filtration, so that manual work is minimised. Both these equipment are discussed in Chapter 6.

14 Solvent Extraction *Technology and Plant Operation*

14.1 SOLVENT EXTRACTION PROCESS

14.1.1 Process Overview

A solvent extraction plant has three main sections. First is, naturally, the extractor, where flakes/cakes/collets are contacted with solvent multiple times to extract the oil. Two product streams emerge continuously from the extractor. One is the mixture of oil with solvent, called 'miscella'; the other is the de-oiled meal laden with solvent, called 'marc' or 'spent flakes'. Miscella is sent to the distillation section, where solvent is evaporated by heating and flashing under vacuum in multiple vessels operating with increasing vacuum. Oil, free of solvent, is pumped out to storage tank; this product is called the 'crude oil'. The 'marc' is desolventised in a multi-stage vessel where each stage has heated bottom. The desolventised 'meal' is dried and/or cooled with hot/cold air and sent for bagging. The solvent vapours from both the sections, distillation and desolventiser, are condensed in water-cooled condensors, liquid solvent separated from condensed water, and recirculated back to extractor – see Figure 11.4.

Apart from the main three sections, there is a fourth small unit, for the recovery of solvent vapours from vented air; this is known as the Vent Absorption unit or the Recuperation unit.

Apart from the main sections, there are support equipment, such as vapour scrubber, condensors, solvent-water separator and water heater; these are shown in a typical flow sheet of a solvent extraction plant (Figure 14.1).

14.1.2 Extraction

This is the heart of a solvent extraction plant, not only because the oil is extracted here, but also because the performance of the extractor directly affects the performance of the other two sections, namely, desolventisation and distillation. The extractor must be designed and operated well, to achieve good performance of the plant.

14.1.2.1 Nomenclature

Some nomenclature need to be fixed for the following discussion of extractor. The oil-bearing material fed to extractor maybe either in the form of flakes or in the form of collets or cakes. For simplicity, we will refer to it as flakes. The de-oiled material may be referred to as 'spent flakes'. The concentration of oil in miscella is called

DOI: 10.1201/9781003309475-16

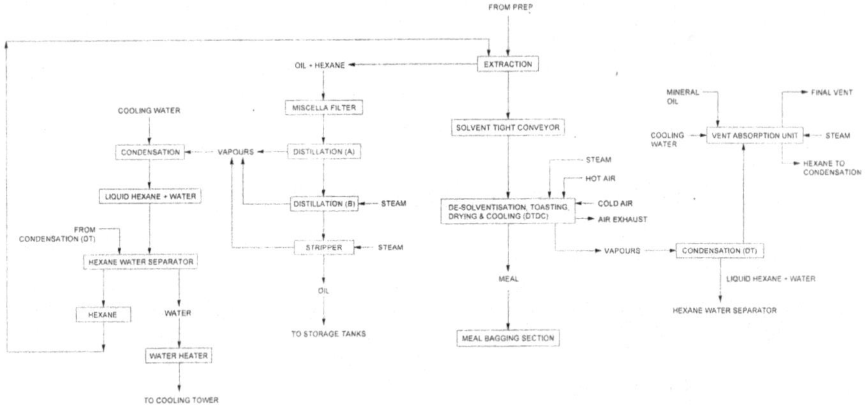

FIGURE 14.1 Flow Sheet for Solvent Extraction Plant.

the 'miscella concentration'. The oil content in spent flakes, on solvent-free basis, is called the Residual Oil Content (ROC). The figures of ROC discussed for various materials hereafter are based on sample taken at extractor discharge. These figures may be different from the oil content analysed on sample at bagging point, due to what is known as the 'DT Jump'; this term is explained in the next section, on soybean processing.

14.1.2.2 Extractor Types

Three main types of extractors are in practice the world over, linear moving screen belt (promoted by DeSmet), circular motion baskets (promoted by Blaw-Knox, Rosedowns, Krupp-Extechnik and French – now DeSmet), and the loop extractor (promoted by Crown); we will refer to these as 'Linear', 'Rotary', and 'Loop', respectively. All three are percolation extractors with counter-current multi-stage design. In the first design, the screen moves and the material bed moves with it (Figure 14.2). In the other two, the screen is fixed, and the material is moved either by vertical partition (Figure 14.3) or by an en-masse conveyor (Figure 14.4), respectively. Fourth type of Extractor, 'Sliding Cell' type, is in use mostly in some European countries, especially Germany (Figure 14.5).

14.1.2.3 Flow of Flakes and Solvent

Flakes are fed at one end of the extractor, and fresh solvent is sprayed on top of bed near the discharge end. The solvent percolates through the bed, giving final wash to the outgoing material, and extracting final oil as possible. It collects in hopper below the bed. A pump recirculates the solvent on top of bed, some distance before the fresh hexane spray. As more solvent is spread continuously, the solvent from hopper overflows to the adjacent hopper. This way the flow of solvent occurs in the direction opposite to the movement of the bed. Each hopper has a pump for the recirculation of miscella on to the top of bed. Each succeeding spray takes more and more oil from the bed. Thus, the miscella in succeeding hoppers gets richer and richer. The last rich

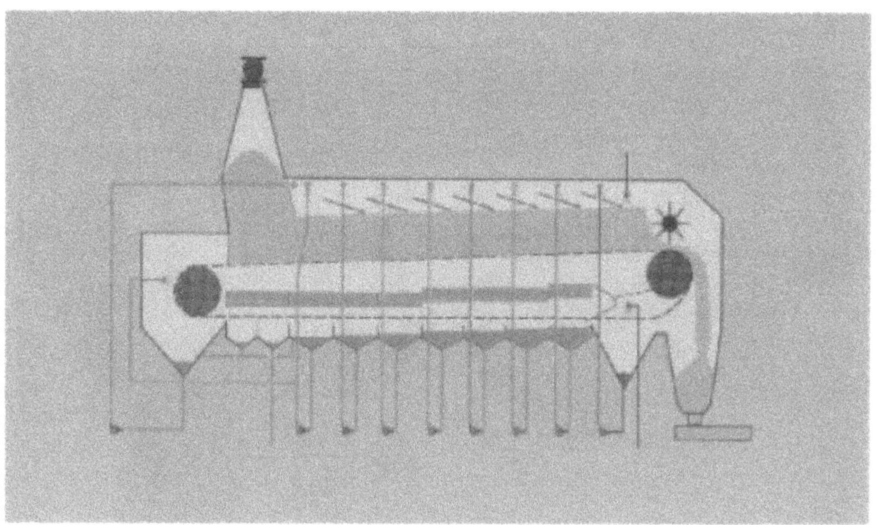

FIGURE 14.2 Linear (DeSmet) type Extractor.

FIGURE 14.3 Rotary Extractor.

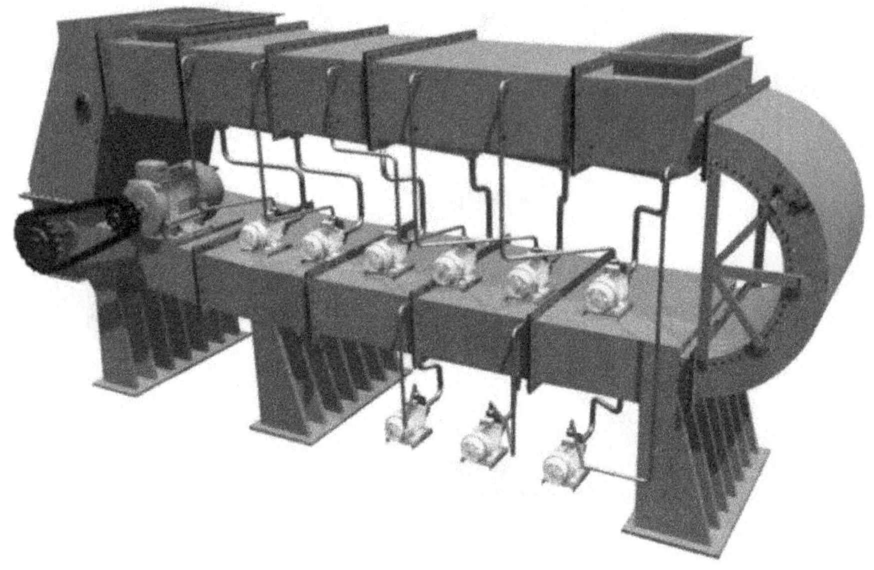

FIGURE 14.4 Loop (Crown) Extractor.

FIGURE14.5 Sliding Cell (Lurgi) Extractor.

miscella, in the hopper towards feed end of extractor, is sprayed on the fresh flakes. Since fresh flakes have maximum oil, much of the oil, nearly half in most cases, is extracted with this wash itself.

14.1.2.4 Number of Extraction Stages

Number of sprays depend on the extraction characteristics of the oil-bearing material, as also on plant throughput capacity. The minimum number of sprays to obtain the benefit of counter-current principle and to obtain rich miscella is four. It should be understood that richer miscella at extractor outlet means less solvent evaporation load on distillation. Typically, extractors are designed with five to nine sprays, apart from the last spray of fresh solvent (Kemper, 2005).

14.1.2.5 Residual Oil in Spent Flakes

The main control point for the efficacy of extraction is that the oil content in last spray, just upstream of the fresh solvent spray, should not be more than 0.2%. If it is higher in concentration, it indicates that more oil has passed to the spent flakes, and that the extraction is incomplete. There may be cases where the final spray concentration is within the required 0.2%, yet the spent flakes have higher residual oil content (ROC) than expected. This would indicate that the preparation of flakes was incomplete. In such a situation, the milling defect test must be performed in laboratory, to determine the MD. If it shows higher value than desired, steps need to be taken to improve the preparation of material.

14.1.2.6 Outlet Miscella Concentration

The miscella concentration in miscella going out of the extractor depends on structure of fresh flakes. If the structure is such that it allows high rate of miscella percolation, then the miscella concentration may be maintained high. This is achieved by feeding less fresh solvent to the extractor, which in turn is achieved by lower flow rate of miscella to distillation. On the other hand, for materials which do not allow high percolation rates, e.g. fragile cake of shea nut, or pellets of rice bran, more solvent must be fed to the extractor, to recover balance oil in the last sprays; this causes low miscella concentration out of the extractor, and more solvent load on distillation.

14.1.2.7 Solvent Carryover in Spent Flakes

Solvent retention in spent flakes at extractor discharge is a function of several factors and determines the load on the downstream operation of desolventising. An adequate drainage time should be provided at the end of the extraction zone, so that maximum solvent drains out of the flakes into hopper below. This drainage time differs for different seeds (oil-bearing materials). It also depends on the kind and quality of seed preparation. For example, soybean flakes would take longer to drain out, and also drain to higher residual solvent, as compared to soybean collets. Cakes with fines would also require longer drainage and retain more solvent. In principle, materials with higher solvent percolation rates drain faster and retain less solvent, and vice versa.

It may also be noted that **higher moisture** in flakes also causes **higher solvent retention**. Thus, higher moisture content in flakes at extractor inlet would have direct impact on the performance of desolventisation unit.

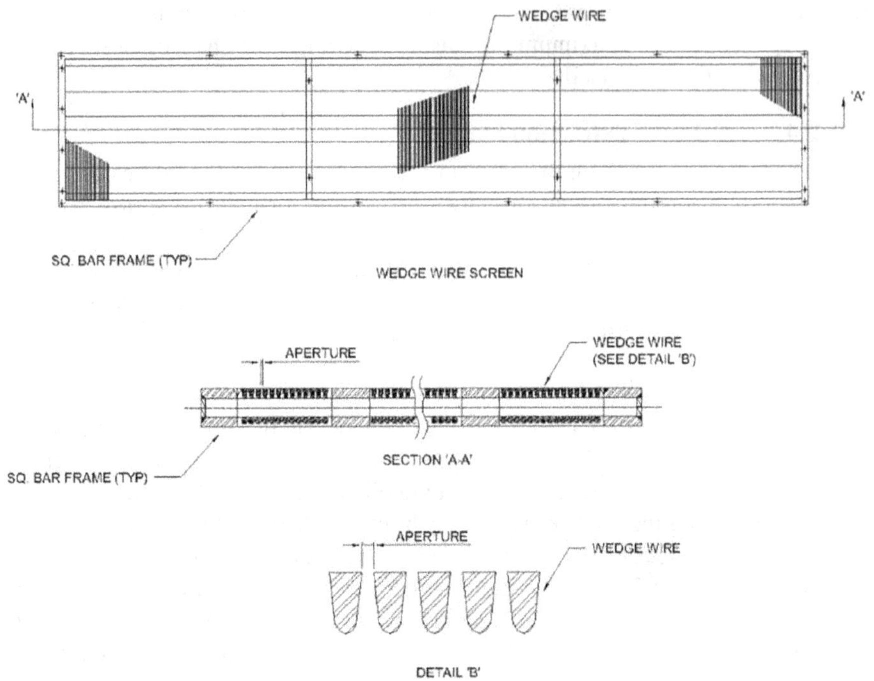

FIGURE 14.6 Extractor Screen.

14.1.2.8 Extractor Screen

The screen, whether stationary or moving, is made from wedge-wire elements. This has self-cleaning properties, thanks to the tapering wires, with increasing clearance towards the bottom side (Figure 14.6). In recent times, these screens have replaced all other types, wire-mesh or Hollander mesh, which were common until the turn of the century, especially in Linear extractors. In smaller extractors of Linear type, with capacities up to 500 Tpd, however, wire-mesh screens are still in use, especially in Asian and African countries.

Wire mesh screens tend to clog quickly, especially with high-protein materials such as soybean, rapeseed and sunflower seed (dehulled). A strong rinsing device is necessary on the return path of screen after the spent flakes have been discharged. This, of course, is relevant only to the Linear extractor design. Even wedge-wire screens must be rinsed, although these tend to clog less. For the extractors of the other two types, Rotary and Loop, this requirement is not so stringent, since the screens are wiped clean by movement of flakes. However, recently a similar screen cleaning system, with solvent sprayed upwards from below the screen, has been patented for the Loop extractor too.

14.1.2.9 Intermediate Drain Zones

In the Rotary design, there is a drain zone between two successive wash zones. This is achieved by spacing the sprays such that there is no solvent spray over alternate buckets. During the drain zones, most of the solvent drains out. Thus, when the net spray starts, there is not much mixing with the solvent from previous spray in the bed. Smaller drain zones exist in the Linear design, where bed scraper arms create small mounds of flakes on top of bed between adjacent extraction zones, to prevent the miscella from adjacent sprays to flow into each other. This is further explained in Chapter 15, on construction features of Extractors.

In the Linear and the Loop type, care is taken, by adjustment of sprays, to not have a layer of miscella on top of the bed, to prevent intermixing of adjacent sprays. The intermixing of miscella sprays, within the bed, should be minimised, since the two sprays are of different concentration, and mixing would affect their extraction potential. It is the extraction potential which primarily determines the rate of oil extraction – see section below.

14.1.2.10 Combination of Percolation and Immersion Principles

For optimum operation of an extractor, it is necessary that the bed is filled with solvent in the extraction zone(s). Coats and Karnofsky (1950) showed that soaking of flakes helps hasten the dissolution of oil. This condition helps achieve the benefits of an 'immersion' extraction mechanism in a percolation extractor, and is the only way to ensure contact of solvent with all parts of the bed. This is especially true with materials which allow high percolation rates. With such materials, even if circulation pumping rates are high, the miscella might pass though the bed quickly, by-passing parts of the bed.

One way to ensure a filled bed is to flood the bed with solvent, with a layer of solvent on top of the bed. This may be the practice in Rotary extractor. However, this method is not allowed in the other two types, Linear and Loop; since there is no physical partition between two spraying zones, a flooded bed will always cause mixing of miscella from two adjacent zones. Intermixing of miscella from adjacent zones would disturb the miscella concentration profile and affect the overall oil extraction adversely. The other method, and the only way for the Linear and Loop extractors, is to observe a slight spread of miscella on the bed, as it strikes from the spray above. This is the correct way the flow of circulation pump should be adjusted. Too much flow would cause flooding, and too little flow would cause by-passing parts of the bed. Correct adjustment of flow allows us to obtain benefits of both, the 'immersion' and 'percolation' techniques.

14.1.2.11 Extractor Safety

Extractors should be ventilated to maintain slight suction, to the tune of 1–3 mmWc, so as to prevent leakage of solvent vapours from large flanges and manholes. Higher suction would suck too much air from surrounding into the extractor and would overload the vent absorption system. Air lock is maintained at the feed end by means of a rotary airlock, to prevent the escape of solvent vapours and also to minimise air intake. A pressure safety valve may also be provided to prevent pressure build up in the extractor.

14.1.2.12 Extractor Operation

The feed rate of flakes from the preparation plant is usually subject to some fluctuations. In order that the Extractor operates continuously, and also no material is held back, the speed of movement of flakes' bed in extractor is auto-regulated. This is discussed in detail in Chapter 15. Discharge of spent flakes is also regulated, so that the load on the downstream equipment, the DT, is maintained constant.

14.1.2.13 Extraction Potential

Extraction potential at any point in the extractor may be seen as the difference between the diffusivity of oil from 'flakes' (or collets/cakes) and the 'solubility deficit' of the contacting miscella; the 'solubility deficit' being the difference between solubility in fresh solvent and solubility in miscella. This may be written in the form below:

$$\text{Extraction Potential} = \text{Diffusivity of oil from flakes} - \text{Solubility Deficit} \quad (14.1)$$

where, Solubility Deficit = Solubility in fresh solvent − solubility in miscella.

Since the solubility in miscella up to 20% concentration is not significantly different from the solubility in fresh solvent (Coats and Karnofsky, 1950), the solubility deficit (SD) is very small in the later part of extractor and is significant only for rich miscella near the feed end. In a counter-current operation, since the rich miscella contacts the fresh flakes, whose diffusivity is high, thanks to the high oil content, the extraction potential (EP) is still high (Figure 14.7). As the extraction progresses, and oil content in collets reduces, its diffusivity reduces as well, but thankfully it is contacted by leaner miscella whose SD is very small; thus, the EP remains large. Near the discharge end, where the diffusivity is quite low, due to very low oil content, flakes are contacted by fresh solvent, whose SD is zero; thus, the EP remains positive. Hence, a counter-current extractor is able to achieve the extraction in shortest duration. Also, it is able to achieve very low ROC (where diffusivity is very low), thanks to its final contact with fresh solvent (solubility deficit zero), keeping the extraction potential positive. Contrast this with a co-current process, where the potential reduces drastically towards the discharge end, becoming zero at a much higher ROC.

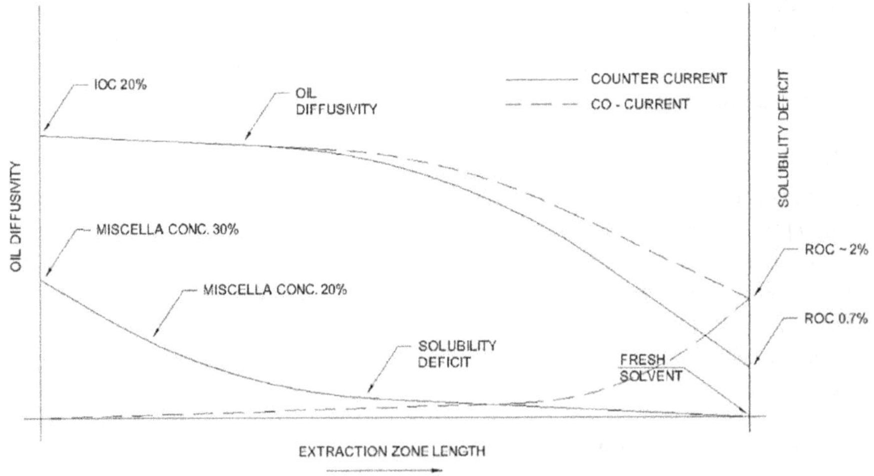

FIGURE 14.7 Extraction Potential, effect of Counter Current Process.

14.1.2.14 Effect of Flake Thickness and Temperature

As discussed before, flake thickness is an important parameter that determines the milling defect, and hence also the final ROC after solvent extraction. Thinner flakes have been shown to extract much faster, and to lower final ROC, than thicker flakes (Karnofsky, 1949b). In fact, the flake thickness has been shown to impact the extraction time, required to reach a particular ROC, inversely to the power 2.2 (Becker, 1978):

$$Y_a / Y_b = [X_b / X_a]^{2.2} \tag{14.2}$$

where, Y_a is the time required to reach final ROC for flakes of thickness X_a, and Y_b is the time required to reach the same ROC for flakes of thickness X_b.

Whenever the importance of flake thickness is stressed, it must be remembered that equally important is the DoD or distortion, which is the difference between the thickness of grits and flakes. Thus, the above relation is valid for flakes only of equal DoD.

It should be noted that flakes of different oilseeds, with the same thickness, would extract differently. For example, it was shown that soybean flakes extracted much faster, and to a much lower final ROC, than cottonseed flakes, which fared better than flakes of flaxseed, all of nearly the same thickness and prepared in nearly the same fashion (Karnofsky, 1949b).

It may also be noted, that while thinner flakes extract faster in a laboratory extractor, flakes thinner than 0.25 mm tend to pack in an extractor bed, affecting the solvent percolation, and disturbing the extraction rate adversely. Hence, the optimum thickness of flakes is 0.28–0.3 mm for most oilseeds.

Temperature of extraction also affects the extraction rate significantly. Becker (1978) claimed that the extraction rate varied inversely to the square of extraction temperature, where temperature was measured in Fahrenheit. If temperature were expressed in Celsius instead, the power may be approximately 1.5, for the range of interest, 50–65 C:

$$Y_a / Y_b \sim [T_b / T_a]^{1.5} \tag{14.3}$$

where, Y_a is the time required to reach final ROC at extraction temperature T_a, and Y_b is the time required to reach the same ROC at temperature T_b; T_a and T_b both expressed in Celsius.

14.1.2.15 Mathematical Analysis of Extraction Process

A mathematical model of percolation extractor is available, thanks to George Karnofsky (Karnofsky, 1987), the inventor of the Rotocel extractor. The model can predict extraction time required to attain a specified final ROC, intermediate ROs as also the miscella concentrations at various time intervals during continuous extraction. Essential input for the model is the laboratory extraction data of RO and the miscella concentration over time, for flakes of the thickness used in commercial extractor. The model is based on a mechanism of the extraction rate being dependent on the fraction of 'undissolved oil', as also diffusion of oil from flakes, which becomes significant only towards the end of the extraction period, when residual oil is low. However, the model predicts shorter extraction times than actually required in percolation extractors. Nevertheless, it is a good start that may enable the development of a more appropriate mathematical model in the future.

14.1.2.16 Batch Extraction

While continuous extractors dominate the vegetable oil extraction industry, batch extractors are still used for special applications, including extraction of minor seeds which may be available only in small quantities, as also for special products such as soybean protein concentrate which have a limited market. A batch extraction system is discussed in Appendix 2.1.

14.1.3 Desolventising of Spent Flakes

Spent flakes from percolation extractors contain between 25% and 33% solvent on wet basis, i.e. 33–50% on solvent-free basis. All the solvent has to be recovered entirely, and residual solvent in meal has to be brought to just a 'trace' level. Requirement for the residual solvent used to be 500 ppm until a decade ago. But, now customers insist on much lower residual, to 100–200 ppm. Desolventised flakes of soybean may either be used for animal feed or for human food. All other meals are only used for animal feed up to the present time.

While the extractor is the heart of the plant, the desolventiser is the most critical equipment operationally; this is so because this equipment consumes maximum steam, maximum electric energy, and also may be responsible for maximum solvent loss. For these reasons, desolventiser is equipped with the most sophisticated control systems, and grabs attention of the plant operator most of the time.

Primary method of desolventising is a combination of indirect heating by steam, and direct steam injection, to evaporate the solvent. Until the 1950s, indirect heating was the main method, aided by live steam injection only for the removal of the last part of solvent. As the importance of more live steam injection, especially for toasting of soybean was realised, the use of live steam increased.

Soybean flakes must be 'toasted' before they may be used for animal feed. This is necessary, so that enzymes which would bind proteins and make those unavailable for birds' digestive systems, must be denatured and inactivated. In early years, the 'toasting' operation used to be performed in a separate equipment, downstream of desolventising. However, by the early 1950s, both the operations were successfully accomplished simultaneously in the desolventiser, by judicious use of live steam to adjust the moisture, which would help accomplish simultaneous toasting. Thence, the equipment came to be known as the Desolventiser-Toaster, or simply, the DT. This name has stuck even for other oilseeds which do not require toasting per se.

In early decades of the industry, jacketed horizontal paddle conveyors, of large diameters, were employed for desolventisation. Heat was supplied by steam given to the jackets. These were commonly known as the Dryers. Later, a Steaming Pan was added, downstream of the dryers, for live steam injection, for the final solvent removal.

The paddle conveyors were replaced, around the 1940s, with vertical multi-stage desolventisers, where heating was accomplished by steam given to double-plates at the bottom of each stage. Material was swept in each stage by sweeping arms attached to central shaft, and flowed from the top stage to bottom stage through gates in the stage bottom plates. Desolventised flakes were discharged from the Desolventiser through a vapour-tight gate fitted into the bottom plate of the last stage. By the early 1950s, a process was patented by French Oil (Shurtleff and Aoyagi, 2016) to combine the operations of desolventising and toasting within the same equipment, by means of injecting sufficient quantity of live steam into the meal. Thus, the equipment came to be known as the Desolventise-Toaster, or the DT. This design continues to be in operation till date.

In the 1980s, the desolventisation operation was revolutionised by the Schumacher DT, with its special design of steam and vapour passage through holes in double-bottoms of the DT. The holes, distributed over the entire area of double-bottom, improved the contact between live steam and flakes so effectively, that the time required for desolventisation reduced significantly. Also, it became possible to achieve very low residual solvent in meal. Licensee companies of the Schumacher design improved upon the principle in later years to optimise steam usage for specific applications.

Unlike distillation, desolventising is accomplished principally at atmospheric pressure condition. This is so, mainly because the attraction between solvent and meal is less compared to that between solvent and oil; it is also very difficult to keep the present design of the DT leak-proof under vacuum. Recently, the last stage of the DT is being subjected to a slight vacuum, for safety against a risk of higher solvent escape during any process disturbance. This operation also helps evaporate some moisture, and reduce the load on the drying operation downstream.

The desolventising time, required for different oilseeds, is determined primarily by experimentation. In theory, the time may be calculated if the 'diffusivity' coefficients, at various concentrations of solvent in flakes, are known. Principles of mass transfer may then be applied to calculate the time required to reach a particular final concentration. Such a model was presented recently (Cardarelli et al., 2002) and is discussed briefly in Chapter 15.

The DT gases have very high heat content, which is beneficially utilised to evaporate solvent from miscella, in the first evaporator of the distillation section. The gases contain some fine particles, which must be scrubbed off, before the gas is let in to the evaporator, to prevent fines deposition on the tubes. Usually, scrubbing medium is hot water. The water is circulated in closed circuit, with slight bleed-off, maybe once a shift, followed by addition of make-up water, to maintain the fines concentration. Another system design uses a cyclone, at the outlet of DT, to remove part of the solids; the balance fines are scrubbed off with a spray of solvent, which is then sent to the DT. This system eliminates effluent water stream, but requires higher energy to re-evaporate the solvent in DT; and is not very common.

The desolventised meal may contain high moisture, at times above 17%, as in case of soybean. Warm air is blown through the meal to remove moisture. The dry meal is then cooled by blowing cold air. For meals derived from presscakes, moisture content is not very high, and just a cooling operation suffices. When the dryer and cooler stage is added below the DT, the equipment is referred to as the DTDC. This is a standard feature in modern soybean plants.

Elaborate control systems are employed on the DTDC. Level of flakes is maintained in each stage, so as to regulate the desolventising, as also drying-cooling, time. The constant levels also help ensure optimum contact with stripping steam, to effectively remove the solvent. Temperature of meal is controlled, as also the temperature of DT gases. These control systems are discussed in detail in Chapter 15.

14.1.4 Miscella Distillation

Distillation to separate the solvent from oil is performed fully under vacuum. Also, unlike the DT, this process is carried out in multiple vessels, each performing a distinct task. It cannot be performed under atmospheric pressure condition because of high affinity of oil to the solvent. This affinity is much more than that between spent flakes and solvent. The high affinity means that the separation, at atmospheric pressure, would have to be carried out at very high temperatures, at which the quality of oil would suffer.

The major part of distillation is accomplished by indirect heating. For the removal of final portion of solvent, indirect heat alone is not sufficient. This is so because at very low concentration of solvent, the binding force from the surrounding oil is very high, so volatility of solvent reduces drastically. This volatility is measured by vapour pressure. For example, the vapour pressure of solvent n-hexane at 0.1% concentration (oil being 99.9%) at 90C may only be 6 Torr (approx. 9 mBar), compared to nearly 1,400 Torr of free n-hexane, at the same temperature. Thus, to boil off the solvent at the low concentration of 0.1% by indirect heat alone, the system pressure would have to be reduced to 6 TorrA. If the final residual solvent desired was only 200 ppm (0.02%), then the vapour pressure would be even lower, just 1.2 Torr. These are very high vacuum levels and could be created only at enormous energy cost.

A much cheaper method is to inject a carrier gas, non-reacting with either oil or solvent, to create additional pressure, so that the combined pressure of solvent and the gas would be equal to the system pressure of say 50 TorrA (equivalent to 710 Torr vacuum at sea level). It may be noted that energy requirement to create and maintain vacuum @ 50 TorrA is just a fraction of that required to maintain 6 TorrA. Luckily, such a non-reacting carrier gas is readily, and cheaply, available in factories, namely steam. Hence, steam is injected to boil off the final portion of solvent from oil. This operation is known as 'steam stripping', and the equipment in which it is carried out is called the 'Stripper'. Striping operation itself may be carried out in multiple stages, to optimise on steam requirement, and thus reduce the subsequent condensation requirement. Thus, quite often a distillation line consists of multiple equipment for indirect heating, followed by stripping column(s).

The first part, evaporation by indirect heating, is usually carried out in two stages (Figure 14.8). The first stage uses heat from the hot gases from DT, comprising mainly of solvent vapours, with a portion of water vapour. Major evaporation is achieved in the first stage itself, ranging from 60% to as much as 90%, for different oilseeds, depending on solvent content in spent flakes and the temperature of DT gases. Since steam is not used in the first stage, the equipment is called the steam economiser, or

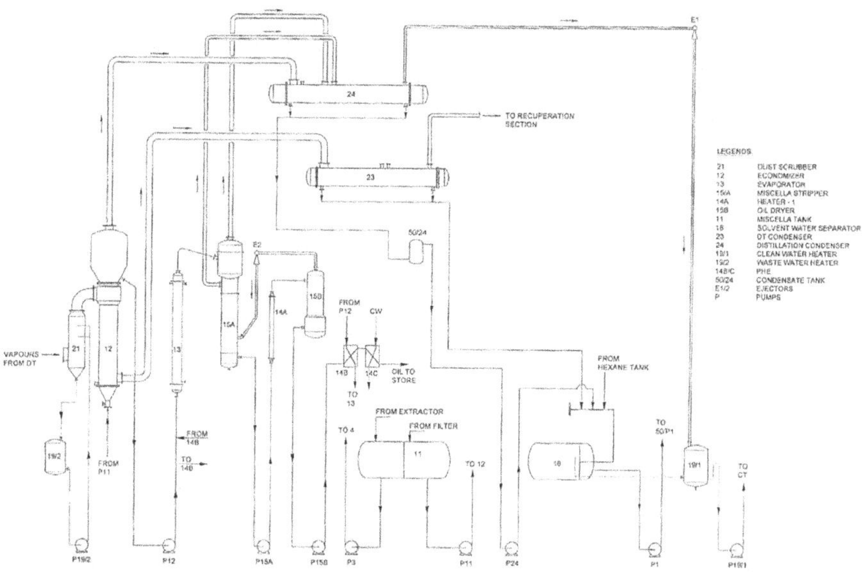

FIGURE 14.8 Flow Sheet for Distillation Section.

simply the 'Economiser'. Since major part of solvent evaporation is achieved without using any steam, the overall steam requirement for distillation section is relatively small. It is for this reason then that, even though the quantity of solvent going with spent flakes to the DT is much less than that going with oil to distillation, the steam required in DT is more than that required in distillation.

Bulk of the remaining evaporation is achieved in the second stage Evaporator, where the heating medium is steam, leaving just about 2% of the total solvent to be separated by the technique of steam stripping. The specific parameters, of temperature and vacuum, are discussed in Chapter 15.

Vapour pressures of solvents at various temperatures are well-documented. Data of vapour pressures at different solvent concentrations in miscella, at different temperatures, may be calculated by empirical methods available; but, these are only approximate, and it is better to generate this data by experimentation in laboratory. Once this data is available, all the distillation stages may individually be calculated, to know the heating steam required, the stripping steam, as also the load on vacuum system. These methods are discussed in Chapter 15.

A preparatory step, at the beginning of distillation, is miscella filtration. It is important to remove insolubles, coming from seed or soil, from the miscella, to prevent sedimentation in heater tubes or Stripper surfaces. Such sediments, of proteinaceous matter, get charred over time, and impart dark colour to oil. The most common equipment, used for the purpose, is a hydro-cyclone. Over the past two decades, the use of wedge-wire filters has increased. These are able to screen out particles above 30 microns. Construction and design features of both these equipment is discussed in Chapter 15.

The solvent vapours emerging from evaporators and stripper(s) are condensed by indirect cooling with cooling water. The condensors are usually shell-n-tube type. Recently, plate heat exchangers have been introduced for the purpose. The PHEs are compact, but maintenance of the gaskets is somewhat tricky; improper closing of the PHEs may result in losses of solvent vapours to the surrounding. For this reason, the usage of PHEs is still in infancy.

Vacuum is maintained in condensors, and in distillation equipment, by means of steam ejectors. As long as the steam going to ejectors is dry (>98% dryness), and the steam pressure is maintained, the ejectors are capable of maintaining vacuum constant, provided there is no large fluctuation in miscella flow or concentration. The vacuum systems are so designed that the ejector steam is beneficially used in the process, either as live steam in upstream equipment or as heating steam for water. The ejectors are maintenance-free and hence are preferred over vacuum pumps.

The condensed solvent from the distillation section contains some water, coming from live steam injected into the Stripper. Condensed solvent from the DT has significant amount of water, part of the live steam that was injected into the flakes for stripping of solvent. The water must be separated from solvent before the solvent is circulated back to extractor. Luckily, the common solvent, hexane, is immiscible with water and can be separated by a simple process of decantation, i.e. gravity separation. The decantation equipment is discussed in Chapter 15.

Another equipment for additional heat recovery from DT vapours may also be used. The vapours at exit of Economisers are at 62C and may still contain some transferrable heat. This heat may be utilised for heating the solvent, which is at around 38C, going to the extractor. Such a heat recovery may save up to 10–15kg steam per ton of soybean. That is a significant saving and is very attractive in view of the low cost of the additional equipment required.

14.1.5 Solvent Recovery from Vent

As discussed earlier, apart from the three main sections of a solvent extraction plant, there is a fourth, albeit small, section. The job here is to recover solvent from vent air. The air comes mainly from two sources. One is the leakage from various flanges on the plant equipment. The leakage into the equipment happens because the equipment must be ventilated to maintain a slight suction, or negative pressure, to prevent leakage of solvent vapours out of the equipment. As the equipment, mainly the extractor and the DT, are ventilated, and the air is let out of the vent, the air carries solvent vapours along with it. The second source of air intake is the air contained in gaps between flakes fed to extractor. This air quantity may be reduced significantly by having a proper rotary airlock at extractor feed point. This principle is explained in Chapter 15.

Fraction of solvent vapours in the air depends on air temperature. The air temperature depends on the condensation temperature of the condensor used for DT gases. Spray of oil on the air, aided by close contact in a packed bed, allows absorption of most of the solvent vapours. The absorption section is designed to minimise the solvent going out to the level of 50 g/m^3 or less. In recent times, absorption systems are being designed for just about 10–20 g/m^3 solvent in vent air. The absorber

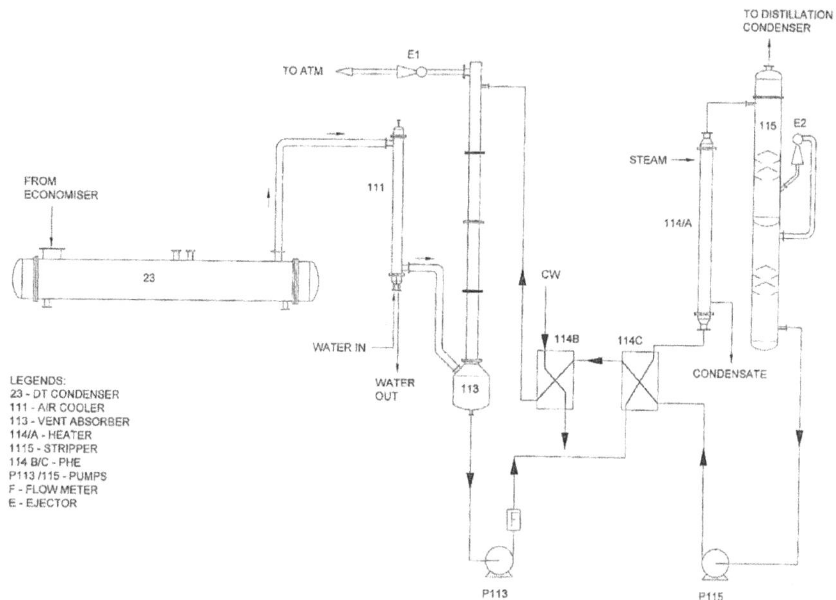

FIGURE 14.9 Flow Sheet for Vent Absorption System.

is typically a tall column, incorporating multiple stages for air-oil contact, aided by packing medium.

The solvent-rich oil is then heated and steam-stripped to recover solvent vapours which are sent for condensation. The lean oil is cooled and circulated back to the absorption column (Figure 14.9). An oil-oil heat exchanger is used in between the oil cooler and oil heater, to gain heat for cold oil, and partially cool the hot oil; it helps save steam on one hand, and helps make the oil cooler compact.

Mineral oil is used for the purpose of absorption of vapours. Vegetable oil is not preferred, as those contain fractions with lower boiling point, which are subject to evaporation; vegetable oils are also prone to polymerisation over multiple cycles of cooling and heating.

Absorber design models are available in published literature. It only requires the data on vapour pressure of solvent in air and in oil. While the former is available in literature, the latter should be determined in the laboratory.

14.1.6 Explosion-Proof Environment

The solvent used in vegetable oil extraction, hexane, is highly volatile and carries an explosion risk. It is classified as flammable class IB, moderately high-risk explosive solvent. Therefore, all electrical motors and fittings in SEP must be explosion-proof. Construction of equipment also must adhere to spark-free requirements. The electric panel is installed in a safe room. This is discussed further in Chapter 15.

In most countries, space of 15 m all around the plant is treated as 'safe zone', bounded by wall 1.5 m high, to accommodate all hexane vapours in case of accidental escape from the plant. The preparation plant, and meal bagging section, are located outside the safe zone. Moreover, access doors or windows, at ground level, to nearby buildings, have to be located such that the access path is at least 30 m from the plant.

Auxiliaries required for plant operation, cooling tower and solvent storage are allowed within the safe zone, with required precautions of explosion-proof electric motors and fittings. Hexane is usually stored underground, although in cold countries, it may be stored over-ground under a shade.

All the tools used in the plant must be spark-proof. The key is to make these out of soft metals, such as brass, so that any impact with carbon steels or stainless steels does not produce sparks. Torches used in plant need to be explosion-proof. Mobile phones may either be deposited at the entry gate to the safe area or need to be switched off.

An automatic sprinkler arrangement should be provided on top of all equipment, for water spray in case of fire. Heat melts the glass bulbs of sprinkler nozzles, and water spray starts. The sprinkler system should be connected to high-pressure hydrant system of the factory.

In case of power failures, all the hot vapours need to be condensed, to avoid pressure build-up. An overhead water tank of sufficient capacity is kept full at all times, for the purpose. A one-way valve, fitted below the tank and kept closed by pump pressure, opens when water pumping stops, and enables water circulation to all the condensors.

High-pressure water hydrants are located all around the plant, for safety against accidental fire. Access roads should be provided on at least two sides around the safety wall.

14.1.7 Alternative Solvents

While hexane is the solvent of choice in vegetable oil industry around the world, efforts are on to find a less hazardous solvent. Apart from the risk of explosion, hexane is also categorised as a 'hazardous air pollutant' (HAP). Various solvents have been tried, in laboratory studies, for the extraction of oil. Apart from hexane, iso-hexane, heptane and ethanol have shown some promise. Heptane would be attractive because of lower volatility, and lower explosion risk, but requires much more heat for evaporation, hence is not attractive commercially.

Iso-hexane also offers somewhat lower explosion risk and is not categorised as an HAP in USA. It is currently used in some plants, especially in old plants which are now in the middle of human habitation in the US.

Ethanol has very high latent heat and cannot replace hexane as the solvent of choice. But ethanol has the added advantage that it can dissolve sugars, hence is commercially used for the production of protein concentrates of soybean, and now, from some other legumes.

14.2 SOYBEAN PROCESSING

As mentioned before, prepared soybean has been fed to the extractor traditionally in the form of flakes. As the importance of thin flakes was realised, thin flakes 0.2 mm thick were fed to immersion extractors. With the advent of percolation extractors, as the bed became deeper, it became apparent that thin flakes compacted too much. So over the last six to seven decades, flakes thickness of 0.3 mm has been the norm.

Since the 1980s, with the advent of expanders, first for the preparation of cotton seed and then for rice bran, its utility for the preparation of soybean flakes was realised. Over the past three decades, expanders have become very popular. At the present time, maybe half of the soybean crushing plants are equipped with expanders. In Asia, and Africa, expanders are even more popular, and it is difficult to find a soybean plant without expander.

14.2.1 Extraction

Soybean, when prepared well, is a relatively simple material to extract oil from. It particularly lends well to percolation extraction. Flakes 0.3 mm thick do not compact too much, even with deep beds. The bulk density increases from about 0.32–0.35 kg/L in air to say 0.55 in extractor bed, leaving enough voids for flow of solvent. Expander collets, cooled properly, resist compaction even better. Both flakes and collets retain their form under repeated solvent sprays and do not crumble into fines; thus, the percolation rates are maintained even towards the discharge end. With shallow beds, as encountered in the 'loop' extractors, the compaction is even less. This shows in the difference in extraction time for flakes, to achieve final ROC of MD + 0.3%; if it is 50–55 minutes in deep beds, it may only be 40–45 minutes in shallow bed extractor.

Percolation rates may be around $20\,m^3/m^2/h$ for flakes, of moisture content 10.5–11%, in medium bed Linear extractor. This improves to nearly $30\,m^3/m^2/h$ for flakes containing lower moisture, less than 9.5%; this is the norm in most Asian and African countries. Collets have even higher rates and allow miscella pumping rates up to $35–40\,m^3/m^2/h$. These percolation rates may be maintained up to a bed height of 2.5 m, beyond which the rate decreases, on account of packing of flakes, especially at the bottom. It may be noted that higher pumping rates mean faster extraction. The percolation rates are slightly higher in shallow-bed extractors, thanks to less packing of material in bottom layers.

After completion of extraction, the flakes have to be drained of the solvent, to minimise the solvent carryover to the DT. The drainage time required is a function of not only the voids but also the bed depth. Thus while deep bed of soybean flakes may have to be drained for almost 20 minutes, shallow bed may be drained adequately within just 10 minutes. Also, collets drain faster than flakes. Table 14.1 shows comparative extraction performance of flakes and collets.

Soybean has high protein content. Hence flakes tend to stick to the screen in the Linear extractor. A good screen rinsing arrangement is necessary on the return path, just after the spent flakes have been discharged. Normally the screen is rinsed with fresh solvent, which is then collected in hopper below and pumped on to the bed. The rinsing liquid has to strike the screen at high velocity and cover the entire width

TABLE 14.1
Extraction Performance of Soy Flakes and Collets (Linear Extractor)

Parameter	Soy Flakes	Soy Collets
Solvent percolation rate, $m^3/m^2/h$	20–25	35–40
Extraction time	50–60 min	40–50 min
ROC (MD = 0.25%)	0.5–0.65%	0.5–0.55%
Miscella concentration	24–26%	28–30%
Solvent in Marc, wet basis	28–30%	25–26%

of the screen. This requirement is not so rigorous for wedge-wire screens, as it is for wire meshes. For the Rotary and Loop extractors, of course, there is practically no such requirement, thanks to the sweeping movement of flakes on the screen.

Soybean contains significant amounts of phosphatides associated with oil. Phosphatides are extracted slowly compared to oil, and in fact, slow down the rate of oil extraction in later stages (Karnofsky, 1949a). In a study on phosphatide (PS) content of lipid fractions extracted on successive extraction cycles from a batch of soybean flakes, it was shown that while initial cycles which extracted maximum oil had very low PS content, it increased quickly in successive trials (Bull and Hopper, 1941). Out of total five extraction cycles, after three cycles the ROC reduced to 0.68%, and PS in extracted oil was just 1.75%. In the fourth cycle, which yielded only 1.1% of the total oil, and reduced ROC to 0.56%, PS increased steeply to 14%, taking the cumulative PS content to 1.9%. And in the last cycle, which yielded further 1.1% of the total oil, and reduced ROC to 0.43%, PS content was almost 19%, taking the cumulative PS content to 2.1%. The final lipid fractions were very dark in colour. Also, as seen earlier, the extraction time required to go below 0.5% ROC would increase greatly, requiring large extractors and higher energy input. For these reasons, the usual practice is to limit oil extraction to final ROC of 0.5–0.55%.

The ROC is quite often analysed on samples collected after desolventising and drying-cooling. The results can be quite different from those on samples collected at extractor discharge. The reason is as follows. In the sample at extractor discharge, there is still some oil trapped deep inside oil cells, which may not be extracted in the standard analytical test. In the DT, due to vigorous heating and steaming operation, that oil also becomes available for laboratory extraction. This difference in analytical ROC values is known as the **'DT Jump'**. For soybean, the DT Jump may be as high as 0.35%. Thus, a sample analysed to have ROC of 0.5% at extractor discharge, may show ROC of 0.8–0.85% at DT discharge. Since it is much easier to collect sample, and analyse, after DT, many plant operators are happy if their meal samples, at bagging, shows ROC of less than 0.9% or even 1.0%.

The extent of DT Jump appears partly dependent on protein content of the meal. The lower the protein content, the lower the DT Jump. Thus while it is 0.3% or higher for soybean, it is around 0.1% for rapeseed, sunflower seed and cotton seed, and practically nil for rice bran, which is very low in protein. Rice Bran is also a product of fine milling of rice, so most cells have already been damaged. Another reason for the high DT Jump in soybean may be the presence of high phospholipid content in oil. As part of the phospholipids remain un-extracted in the meal, they might release oil into solvent very slowly. But as the cells are further damaged during desolventising, thanks to hexane vapours bursting out of the cells at the desolventisation temperature; the oil would become available easily for analysis.

14.2.2 Desolventising Toasting

Just as soybean oil is one of the easiest materials to extract, the spent flakes are also relatively easy to desolventise. The high protein content in soybean allows good coagulation of the protein starch mass, during cooking and again during desolventisation. As the proteins are denatured, their affinity to solvent has also reduced. Thus, if good contact of steam with the flakes is established, all the solvent may readily be evaporated, leaving very low residual solvent in meal.

Solvent carryover with spent flakes of soybean is typically between 28% and 30% on wet basis, equivalent to 39–43% w/w of solvent-free flakes. The variation in the solvent carryover depends on the adequacy of soybean preparation, especially the extent of protein coagulation achieved during cooking, as also on bed depth and drainage time, apart from flakes moisture. When collets are fed to extractor, the drainage is somewhat superior, leaving only 24–26% solvent in spent collets. Thus, an expander helps reduce the load on the DT. It may be noted that a reduction of 5% solvent (solvent-free basis) in flakes would mean a reduction of nearly 10 kg steam, per ton of soybean processed. This is demonstrated by calculations in Chapter 15.

A good desolventising operation should reduce the residual solvent in soybean meal to less than 300 ppm. This 'standard' used to be 500 ppm two decades back. If the 'flash' stage under vacuum is operated well, the residual may even be reduced to 200 ppm. In any case, some more solvent is carried out with air in downstream drying and cooling operations, reducing the residual to less than 100 ppm.

Toasting is a special requirement for soybean meal when it is used for animal feed. Toasting, i.e. heating in presence of high moisture, destroys Trypsin Inhibitor (TI), which otherwise binds the Trypsin enzyme required for protein digestion. Desired result can be accomplished by toasting at 100C for 15–20 min at moisture level of 17.5%. At lower moisture levels, longer duration becomes necessary. Residual TI may be determined in laboratory, albeit with a very lengthy procedure. Usually, another component, Urease, is measured to reflect the level of damage to TI. The Urease Activity (UA) may be determined quickly, and readily in any small laboratory, in terms of increase of pH after reaction with urea for a certain amount of time. Desired range of UA of 0.05–0.15 pH rise indicates good toasting (corresponding level in raw soybean is around 2.3–2.4). The lower the pH rise, the better toasting.

As mentioned above, toasting is a function of three parameters, temperature, moisture and time. At a given temperature, longer duration is necessary at lower moisture. So, if less live steam be used for desolventising, longer desolventising time, or higher temperature, would be necessary to achieve sufficient toasting. Both the factors, either a longer heating duration, or higher temperature, would damage other nutrients in the soybean meal, and lower the meal quality. Typically, protein and starch components would undergo 'browning' reactions, reducing the protein availability. Hence, it is important that sufficient live steam be used in the desolventising operation, to achieve optimum toasting within shortest time. Methods to calculate the steam injection requirement and moisture content in meal will be discussed in Chapter 15.

Since the Schumacher DT, with holes in double bottom for steam injection, was introduced in the early 1980s, it has become the equipment of choice for soybean DT. Licensee companies to the design introduced their own improvements to the design. One issue with the design was that the high quantity of live steam required to drive out the entire solvent led to increase in moisture content in meal to 19% and higher. This level was not really required for good toasting and put unnecessary high load on drying system downstream. An important improvement was the introduction of the upper pre-desolventising stage(s), to reduce solvent content, so that the remaining solvent could be evaporated with just enough steam to raise moisture in meal to 17.5–18%. This improved the overall energy balance of the DT and drying system. This point will be clear from calculations presented in Chapter 15.

An indication of 'sufficient' steam injection is the temperature of gases exiting the DT from top. For soybean, DT gas temperature between 70 and 72C is a good process indicator, for effective desolventising. With previous designs, prior to the Schumacher DT, DT gas temperature had to be at least 75C for effective desolventising, meaning more steam had to be injected. This was so, because of the poor contact of steam with meal. The extra 3C would mean additional steam usage of nearly 10kg per ton of soybean.

Later, drying and cooling stages were added below the DT, to a system popularly known as the DTDC (Figure 14.10). This is now the most common, and important, piece of equipment in soybean extraction plants all over the world. As mentioned before, the DTDC is the most energy-consuming equipment, both electricity and steam, and is also the equipment where most solvent losses may occur. For these reasons, the DTDC is the centre of attention by plant operator and is subjected to elaborate control systems. These will be discussed in Chapter 15.

14.2.3 Meal Drying and Cooling

Since soybean is toasted at a moisture level of 17–18% in the DT, it has to be dried before storage. For storage, moisture level of 12–12.5% max.is desirable. Drying is accomplished by blowing of hot air through the meal. Until the 1970s, Drying used to be effected in a separate Dryer away from the DT. Generally, these dryers used to be placed in a separate building, where meal was bagged. With the advent of the DTDC, the dryer and cooler operations are combined, with the DT, in a single equipment. Drying alone is not sufficient for safe storage. If the warm meal is bagged, it may form lumps, and turn dark, during storage. Cooling operation, therefore, is also required along with drying.

When the Dryer-Cooler (DC) were built as a separate unit, these were mostly of two types; one, the slow-moving screen as used in collet cooling; or, a vertical unit with two or three stages. The construction of vertical unit was similar to a vertical DT; equipped with a sweep arms in each stage connected to a slow-rotating central shaft. The hot moist meal would enter the top stage and move by gravity to lower

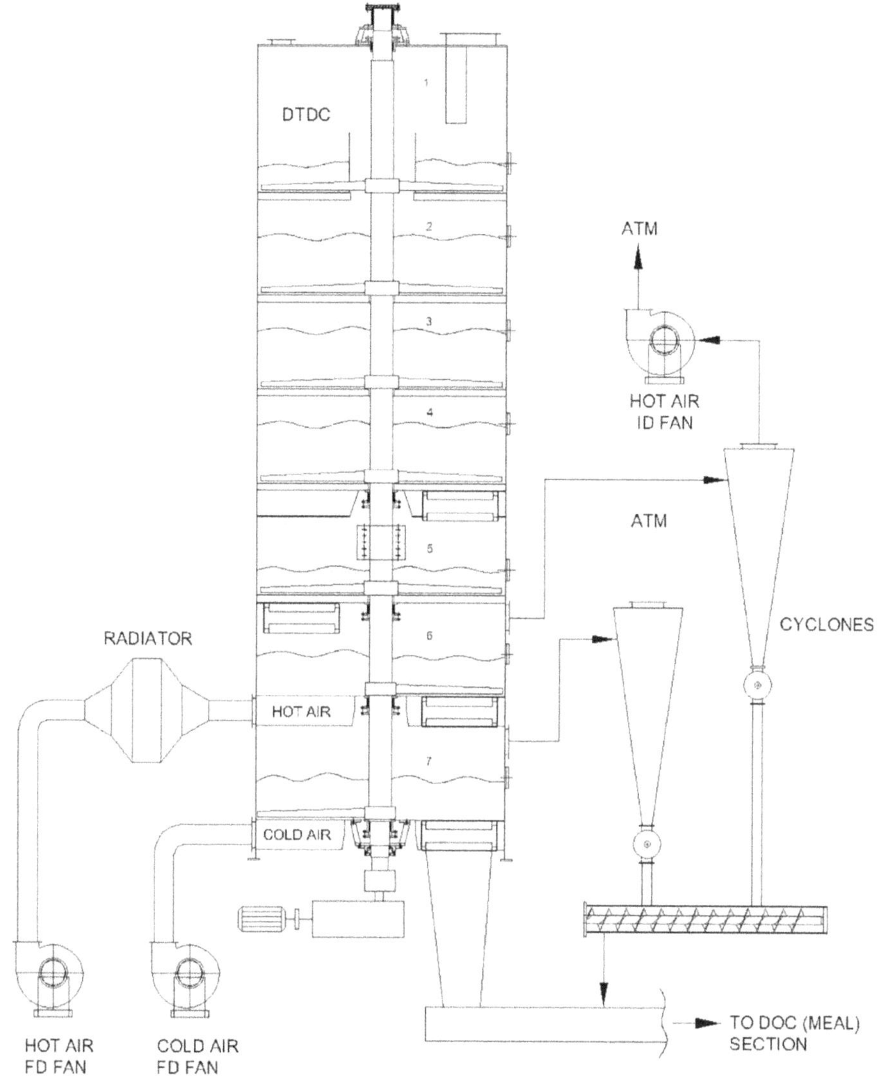

FIGURE 14.10 GA of DTDC with Fans, Radiator and Cyclones.

stages. Each stage would have stirred meal bed of 500–600 mm height, through which air, hot or cold, would be blown from the bottom. It is these stages that are combined with the DT in a DTDC.

The DTDC offers several advantages over the separate DT and DC. One, the conveyor that carried the hot, moist meal faced operational issues. One was the corrosion issue, which necessitated construction in SS. Second was the elongation of the drag chain due to the temperature; the elongation caused slackness, resulting in slipping the chain off the wheel, and hence jamming of conveyor. Third, as some moisture evaporated due to the temperature, it condensed on the cover, together with fines. These wet fines deteriorated over time and had to be cleaned manually. If, on the other hand, the conveyor was ventilated, to prevent condensation and deposition, it meant some fines went out with the vapour. Fourth issue was higher electricity requirement for the two separate equipment. Another important issue with the separate DC was it required presence of an additional operator in the meal bagging area. Thus, a DTDC not only saves energy, and saves an operator, but provides for a clean operation, with a cooled dry meal being discharged. It does mean, however, that the motors for the air fans of the DC have to be of flameproof construction, which is a small price to pay for the major benefits.

The DC operation too is auto-controlled, for best consistent results. These systems are described in Chapter 15.

The air requirement, as also the air temperature, is determined with the help of air-water psychrometric charts. These are discussed in Chapter 15. Typically, about 25–30 kg steam is required for the drying operation per ton of soybean meal. The energy requirement for air blowing is typically within the range of 1.5–2.2 kWh/T.

As air is blown through the meal, warm air for drying and ambient air for cooling, it carries some fines with it. The fines must be separated from the exhaust air. Typically, air cyclones are used for the purpose. Much of the very fine particles, smaller than 20 microns, escape into the atmosphere. The fines escaping to atmosphere not only constitute material loss, but more importantly, cause air pollution.

Solution to arrest the fines would be to use air bag filters in place of cyclones; however, their use is limited as the wet fines stick to the air bags and tend to clog those. The better option is to go for 'high-efficiency' cyclones, which can remove particles up to 10 microns; these entail somewhat higher pressure drop and hence require higher energy input to blower. For the cooler section, however, the use of air bag filter is now increasing. The fines collected at the bottom of cyclones, or bag filters, are conveyed to the meal conveyor.

14.2.4 Meal Sizing and Bagging

The meal, dried to less than 12%, and cooled to nearly 40C, is conveyed to the bagging section. In many countries, meal is first subjected to lump separation and breaking operations. These lumps are either from the collets or formed out of flakes in the DT. In the DT, with the heat treatment in presence of moisture, and also due to the steam condensation on the surface of flakes, some lumps are formed. During a normal operation, lumps may be up to 15–20 mm size. Should the cooking operation, during soybean preparation, not be adequate, bigger lumps might form. Often, the lumps are required to be broken to smaller size, say 6 mm, before the meal is bagged (Figure 14.11).

Lump separation may be achieved, by size separation, either in vibrating sieves or in rotary drum sieves. The separated lumps are then broken with a hammer mill and mixed with the meal.

The meal is mostly bagged in polythene bags to 50 kg. Auto-bagging machines are used for the purpose; these operations are discussed in Chapter 15.

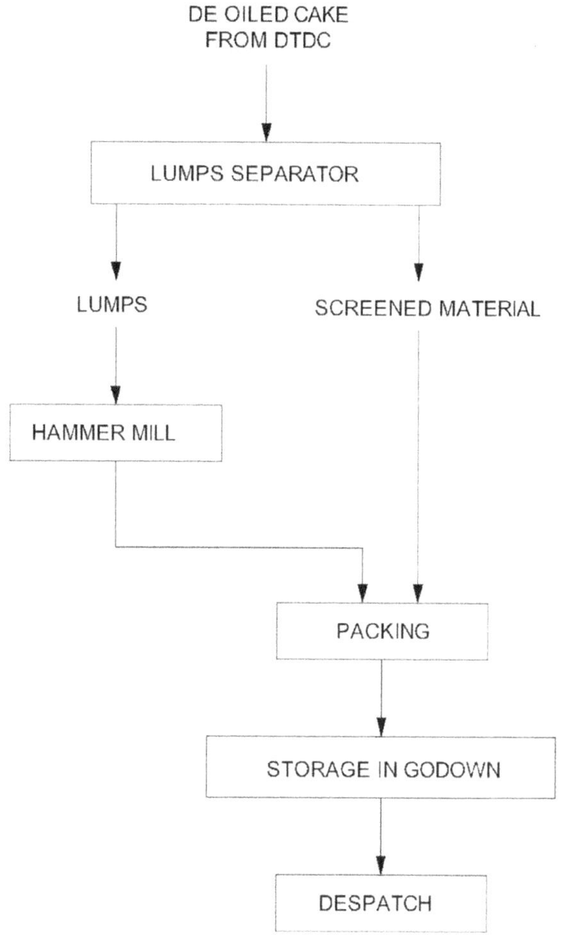

FIGURE 14.11 Flow Sheet for Meal Sizing and Bagging.

14.2.5 Distillation

As discussed before, miscella filtration is an important preparation step for distillation. The step prevents choking of heater tubes and also helps minimise dis-colouration of oil. This operation is critical for soybean, since the insolubles always get into the gums and thence into the soy lecithin, affecting the lecithin quality. After filtration, the clean miscella is sent for distillation.

Miscella obtained from soybean oil extraction contains typically 24–26% oil when flakes are extracted. On the other hand, with collets from expander, concentration may be increased to 28–30% without compromising on the final ROC in spent collets. For miscella of 25% conc., if 18.5% oil were extracted from soybeans, 555 kg of solvent would be handled in the distillation section per ton of soybean. On the other hand, if miscella concentration were 30%, only 432 kg solvent would need to be distilled off. However, this significant reduction in solvent quantity does not reflect in proportional reduction in steam consumption, as seen below.

As mentioned before, the first equipment in the distillation line is the Economiser, which uses the heat in DT gases to distill off most of the solvent. It is possible to reach 70–75% concentration at Economiser outlet, starting from 25% at inlet, depending on the temperature of DT gas, leaving only 60–80 kg solvent, with 185 kg oil, in miscella per ton of soybean processed. This balance quantity of solvent is required to be distilled using steam. If the inlet concentration were 30% instead, the Economiser maybe designed to enrich miscella to 80–82% concentration. It may be noted that concentrating the miscella above this level is not economical, since the increasing miscella concentration is accompanied by rising temperature; higher vacuum would then be required to maintain temperature difference from the condensing gas; high vacuum would mean higher costs. Concentration of 82% means 40 kg of solvent with 185 kg oil. Thus, effectively only 20–40 kg less solvent would have to be distilled, compared to that from 75% miscella, using steam; this is much less than the 125 kg/T difference in solvent quantities of inlet miscella in the two cases; the steam saving then would be around 4–8 kg/T only.

The first evaporations, in the Economiser and the next Evaporator, are typically carried out at a vacuum of 400–425 Torr (absolute pressure of 325–350 mm Hg), as it simplifies the subsequent condensation of solvent vapours using cooling water at 30–32 C. If evaporation were conducted at higher vacuum, i.e. lower absolute pressure, the condensation operation would require cooling water at lower temperatures, which is not always economical. If the miscella is heated up to 90 C and flashed at 350 TorrA, outlet concentration is 94%. Thus, only 12 kg solvent (per ton of soybean) remains with oil, to be treated in the next stage, namely the Miscella Stripper, where live steam is used as carrier vapour to boil off the solvent.

Stripper may be designed to leave 100–700 ppm solvent (0.01–0.07%), depending on the relevant legislative requirement in various countries. Stripping is usually accomplished with multi-stage equipment, the last stage being under high vacuum, usually around 50 TorrA. Live steam must be injected to aid the boiling of solvent at very low concentrations. The trick is to inject adequate steam to boost the combined vapour pressure, of solvent and of steam, to the system pressure. For example, the vapour pressure of hexane at 400 ppm concentration in oil, at 90 C, is just about 2.5 Torr. This means adequate steam should be injected to create the balance 47.5 Torr pressure, so that combined vapour pressure is 50 Torr, equal to the system pressure. The quantity of steam, required for the purpose, may be calculated using the laws of physical chemistry. Should the requirement of residual solvent be lower than 400 ppm, steam requirement would increase correspondingly.

In recent times, there is a move towards carrying out complete stripping operation in a single column maintained at moderate vacuum, say 300–350 TorrA similar as for evaporation. For this purpose, the Stripper is designed with a larger number of contact stages, so as to accomplish the complete removal of solvent with minimum steam. Here, a final Oil Dryer is additionally required, wherein oil is flashed under higher vacuum, say 50 TorrA, for removal of any excess moisture in oil.

Because of the high phosphatides content in soybean oil, there is a risk of foaming in the flasher of Evaporator and in Stripper. This is especially so at the beginning of soybean season (November-December), as more phosphatides are readily extracted with oil. The risk multiplies in presence of fines in oil. Should there be fines deposited in evaporator tubes, or on stripper surfaces, occasionally a part of the sediment is flushed into the oil, and it can create foaming which may persist for hours, until the fines are discharged with the crude oil. The foaming may cause carryover of miscella in to condensors and may contaminate the solvent. When the contaminated solvent is circulated back to extractor, it may be unable to extract the last oil; the ROC would increase as a result. The main precaution, to avoid foaming, is to minimise the fines in miscella by good filtration, before it is received in the miscella tank which feeds the distillation section. As a safety, nozzles may be provided on top of flasher and strippers, to let in a blast of steam to break the foam. The quantity of 'foam-break' steam must be regulated, as too much steam would affect the vacuum.

With distillation temperature below 100C, light colour oil may be produced. It should be noted that the yellow colour is native to soybean oil, whereas red colour is produced during processing. Thus, a good processing line should produce oil with low red colour. Typically, most plants produce oils with red colour ranging between 1.2 and 2.0 units (in 1/4 inch cell, on Lovibond scale). However, recent results in some plants show that, if the distillation temperature is maintained below 100C at all stages of distillation, and if sedimentation of gums in evaporator tubes, and on Stripper surfaces, is avoided by appropriate designs, it is possible to produce crude oil of red units below 1.0 (Tiwari, 2018). With these conditions, while processing dehulled soybeans, even lighter oil may be produced, with 0.8–0.9 red units.

Readers may note this colour improvement can be of significance during oil refining, as lower red would mean lower refining costs. Also, the colour of neutralised oil, i.e. after the operations of degumming, neutralising and water wash, is generally around 0.6 Red. Thus, the low red colour in crude oil could allow simplification of the entire refining process.

The oil must be cooled to 40C prior to storage, to avoid any possibility of oxidation. This, again, may be accomplished in two stages, one by heat exchange with the miscella from Economiser which is at 55–60C, followed by cooling with water from cooling tower.

14.2.6 Solvent Recovery from Vent

The air intake into the plant, with soybean flakes, maybe estimated at 2 m^3 per ton of soybean. However, if a good rotary airlock be used at extractor inlet, the air intake may be reduced substantially, close to zero in fact. For safety, however, it may be considered at 1 m^3 per ton, and 10–20 m^3/h air leakage added to the figure, depending on the size and tightness of the plant. The calculation procedure for absorption of oil flow rate is described, together with equipment description, in Chapter 15.

14.2.7 Processing Soybean Flakes for Edible Applications

Spent flakes of dehulled soybean may be processed either for animal feed or for human food. The DT is the equipment used to produce meal which is used in animal feed. If flakes are to be desolventised to produce edible flour, the desolventising process must be much milder. Specifically, the flakes are not to be 'toasted', so that the native form of protein is maintained as much as possible. The native form of protein is of critical importance, since the flakes, or the flour, have to retain 'water absorption' quality, which is the foremost essential parameter of a food component. In industry parlance, the product coming out of such a desolventiser is called 'untoasted flakes', or 'white flakes'. The protein quality is measured in terms of 'Protein Dispersibility Index', the PDI.

It may be noted that the Trypsin Inhibitor does not get activated in the process. In fact, the measured Urease Activity should be close (2.0–2.2) to the original activity in soybean (2.3–2.4).

Since toasting is to be avoided, live steam, which is the mainstay of the desolventising process in a Schumacher-type DT, cannot be used as the primary means for solvent evaporation. Also, the temperature should be 90 C max.

Mainly three types of equipment are used for the production of 'untoasted flakes'. The first and oldest type is the classical 'Dryer' that was used for desolventising of soybean meal until about the 1940s. These are essentially horizontal paddle screw mixer-conveyor of large diameter. The second type is the classical DT, with vertical multi-stage design, using mainly the indirect steam as the heating medium. This may be referred to as the DS (for desolventiser), and not DT, since toasting is not a part of the process. A small amount of live steam is injected only for contact in the last stage, to reduce residual solvent.

The third, and the most elaborate, is the 'Flash Desolventiser System', the FDS. The FDS uses superheated solvent vapours to transfer heat to the flakes and evaporate solvent. Extra solvent vapour is bled out, and the remaining solvent vapour is re-heated and passed through the bed of flakes in a closed circuit. A large fan keeps the vapours, with the flakes, in circulation (Figure 14.12). The solvent vapours are heated to 125–150 C, and vapour bleed quantity is regulated by controlling the pressure at fan outlet. The higher the temperature of the solvent vapours, the better it is for desolventising. The temperature reduces to 85–90 C, as heat is transferred to meal for evaporation of solvent. The desolventised flakes, also at 85–90 C, are separated in a cyclone, discharged by means of an rotary air-lock, and are subjected to final stripping with live steam in a 'Dryer' or a steaming 'pan' which is like a small DS. In some designs, the Dryer or the DS may be operated under vacuum.

Comparison of the product characteristics obtained from, and energy input in, the three types of desolventisers is given in Table 14.2.

It may be noted that, for all the three processes, a finishing stage is added where small quantity of live steam is injected to reduce hexane content to less than 1,500 ppm. This is still much higher than the residual solvent in meal processed in the DT. The residual hexane is then removed by air purging during cooling operation, prior to bagging.

For capacities above 150 Tpd, the surface area available in screw dryers becomes a limiting factor, and often this is made up with more live steam; this is the reason for lower PDI in white flakes processed thru screw dryers. The PDI of 75%, as achieved with vertical desolventisers (DS) suffices for most food applications. However, for special products, such as soy protein isolates, higher PDI is necessary, and for such application, the FDS is the preferred option.

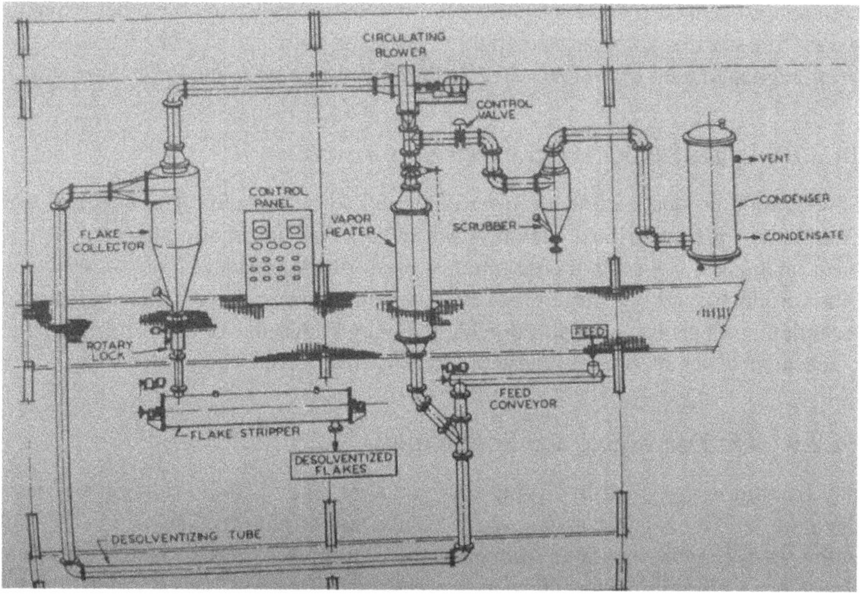

FIGURE 14.12 Flow Sheet for Flash Desolventisation.

TABLE 14.2
Product Characteristics and Energy Requirement of Untoasted Flakes Obtained from Various Desolventisers

	Screw Dryer	Vertical DS	FDS
Protein Dispersibility Index (PDI)	60–65	74–78	82–86
Electric Energy, kWh/T (basis: 200 Tpd white flakes)	2.5–3.0	5–6	9–10

It may be noted, that the FDS is an elaborate system, with much higher initial investment, and higher energy requirement. Also, in view of the presence of superheated solvent vapours, the process must be well-regulated, with good auto-control systems. These factors make the other alternatives, especially the vertical DS, attractive even with a lower PDI product.

One operational issue faced for processing of white flakes is on account of the fines getting into the DT gas. As the fines are scrubbed in Dust Scrubber, these get into the scrubbing water. Since the flakes are not toasted, the fines absorb too much water and create foaming in the water heater, which is used to heat the scrubbing water. The foam may hold solvent, and cause 'pressure bursts', causing pressure disturbance in the DT-Economiser-Condensor system. To minimise this problem, fines entrainment must be minimised, by the use of low gas velocities, with large-sized ducts.

White flakes are a special product, and typical plant size may be between 100 and 300 Tpd. Some plants of large soybean crush capacities are equipped with two parallel desolventising sections, one for normal toasted meal, and the other for white flakes. The streams are separated after the extractor, with feed to white flakes section being regulated with a variable speed feeder, the rest going to the DTDC.

14.2.8 Edible Products from Untoasted Flakes

Several food products are made from untoasted soybean flakes. These include soy flour, used for protein fortification of wheat or other flours; soy nuggets, used as meat analogues; soy protein concentrates, used for the production of soy milk and other products; and soy protein isolates, used in pharmaceutical and other special formulations. This is a specialised field, involving knowledge of food chemistry and biochemistry. An overview of these products is covered in Chapter 16.

14.2.9 Oil Degumming and Soy Lecithin

Soy lecithin is produced by drying the 'gums' separated from crude oil by water degumming. Gums is the name used for a mixture of phosphatides and neutral oil with water, which is separated from oil at the degumming process, which is the first operation of oil refining. Lecithins are good emulsifiers and are used widely in food applications. The phosphatides (PHT) content of soybean oil is high, nearly 2–2.6%, compared to most other oils which contain less than 1% PHT.

The degumming operation may either be conducted in an oil refinery or in the solvent extraction plant. The practice of degumming in extraction plant is prevalent in large extraction capacity plants located near ports, without an associated refinery, where the crude oil is to be shipped out. It is difficult to ship out crude soybean oil without degumming, as the gums tend to settle during transportation.

When degumming is carried out in extraction plant, the degumming process is incorporated with the distillation section (Figure 14.13). Miscella from the Stripper, prior to the Oil Dryer, is taken to a hydration tank for mixing with water. Slow mixing for nearly 15–20 minutes helps form agglomerates of water-PHS, known as 'micelles', which remain suspended in oil. These are then separated with high-speed centrifugation, taking advantage of the higher density of the water phase. The degummed oil is sent back to the oil dryer for the final drying operation. The gums are sent to the lecithin section.

Gums separated at the centrifuge typically contain around 30% phosphatides, 20% neutral oil, and 50% water. When this water content is reduced to below 1%, the product is called lecithin. The process consists of mainly three steps: heating, flashing under vacuum, and cooling.

Gums are a highly viscous material, viscosity in the range of 500 Poise. These should preferably be heated in scraped surface heat exchangers (SSHE) to avoid localised over-heating. Another design of gums heater incorporates rotating coils in a tank of gums. Temperature may be limited to 90C for good quality lecithin, and heating must be done under vacuum. Rising film SSHEs are quite suitable, as the water vapour pushes the film upwards. As the gums are flashed under vacuum, water evaporates instantly. Vacuum @ 20 TorrA ensures moisture below 0.5% at 90C. Should the vacuum be lower, say the water film vacuum of 60–70 TorrA, higher temperature, nearly

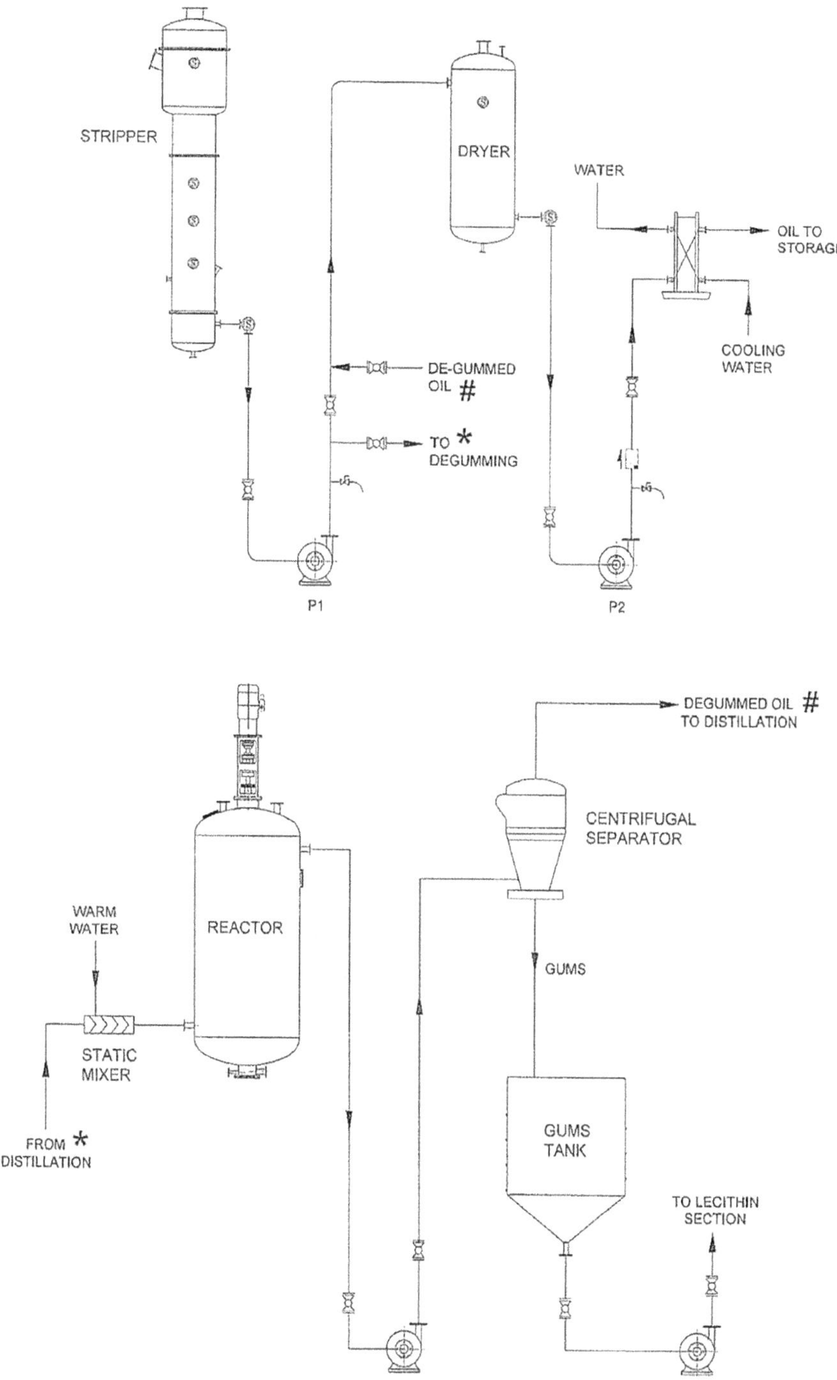

FIGURE 14.13 Oil Degumming in Solvent Extraction Plant.

100C, is required to achieve moisture lower than 1%. Cooling is also conducted in scraped surface heat exchangers, using cooling water as the cooling medium.

One important quality parameter of lecithin is the percentage of phosphatides in lecithin, which should preferably be between 60% and 62%; this is a function primarily of the centrifuge operation in the degumming process.

Another parameter, the most important one, for lecithin quality is the presence of hexane-insoluble fines. Commercial standard for insolubles is mostly 0.3% max. To achieve this, the miscella must be filtered well prior to distillation. A good miscella filtration should limit the insolubles in oil to 0.03%. Beyond this level, the insolubles in lecithin are bound to be in excess of 0.3%. Higher insolubles, and higher heating temperatures, also affect the colour of lecithin. Lecithins with colour units less than 12 on Gardner scale are accepted as light coloured.

14.2.10 Utilities Requirement for Soybean Extraction Plant

The main utilities, which contribute to processing cost, are electricity, steam and solvent. The electricity requirement can vary significantly with plant capacity; larger capacity plants consume significantly lower electricity per ton of soybean. The steam requirement and solvent losses also do change with plant capacity, but to a much lesser extent. Typical utilities requirement for a modern 1,000 Tpd soybean crushing plant are tabulated below; section-wise figures would help plant engineers have a better understanding of, and hence control on, processing costs.

As may be seen, the desolventising-drying-cooling section is the major consumer of electricity and steam within the SEP. The major loss of solvent from DT is also the maximum from among the processing sections. The other major contributor to solvent loss is the leakages from large flanges, as also loss during plant start-stops. The former may vary on (a) plant size, larger plants incurring lower loss per ton; and the latter would depend on (b) mechanical integrity and continuous operation of plant (Table 14.3).

TABLE 14.3
Utilities Requirement (Section-Wise) for a Soybean Crushing Plant (Plant Capacity 1,000 Tpd, without Dehulling)

Plant Section	Electricity, kWh/T	Steam, kg/T	Solvent Loss, kg/T
Soy Preparatory[a]	14	80	–
Solvent Extraction Plant (SEP)			
Extraction	1	15	–
Desolventising-Drying-Cooling	7	140	0.4
Distillation	3 incl. cooling water pump	32	0.1
Vent Recovery	0.1	3	0.1
Leakages & Maintenance			0.2–0.4
Total for SEP	**11**	**190**	**0.8–1.0**
Total Prep + SEP	**25**	**270**	**0.8–1.0**

[a] flakes to SEP, no expander

14.3 RICE BRAN PROCESSING

14.3.1 Extraction

As noted before, rice bran (RB) is a low-protein, high-carbohydrates material. As a result, the pellets do not have a mass of coagulated protein to hold it together tightly, hence the pellets tend to crumble under solvent spray. It is common to see a layer of fines on top of bed, after the first two sprays in the extractor. This results in low percolation rates of solvent. Therefore, longer extraction times, nearly 90–100 minutes, are required to achieve ROC of 0.5%. Since RB oil is low on phosphatides, it is common to see plants operate with ROC 0.4%, with even longer extraction times. If expanded collets are used, in place of pellets, extraction time may be reduced to 75–80 minutes. Solvent percolation rates may be just about 12–15 $m^3/m^2/h$ for pellets, but nearly 20–25 $m^3/m^2/h$ for collets. With lower percolation rates, even after drain time of 20 min, solvent in spent pellets may be nearly 32–35% on wet basis. Expander collets may drain to lower solvent content, @ 28–30%.

It may be noted that the low level of ROC is possible, in spite of low percolation rates, because the MD for RB pellets is very low, 0.1% max.

The low protein does offer a small operational advantage. Pellets do not stick to the extractor screen (Linear Extractor). Thus, the screen rinsing requirement is not rigorous, and the extractor may be operated for months on end, without the fear of screen fouling.

There is no DT Jump in RB meal. Thus, an ROC of 0.4%, analysed on sample at bagging station, would mean actual ROC at extractor discharge to be also 0.4%.

14.3.2 Desolventisation

Desolventisation is accomplished in the DT; the name is somewhat misnomer, since no 'toasting' is required, but is the common name in industry parlance. The DT is either a classical design, with live steam injected through sweep arms, or through perforated pipes; or a cross between the Schumacher design and the classical. A full Schumacher design, for total counter-current configuration, with holes in bottom of all stages, cannot be operated with rice bran. The spent RB coming to the DT consists of a lot of fines. These fines tend to create high pressure drop for passage of steam and vapour through the bed in each stage. If all the vapour has to pass through all the double bottoms and all the beds in a counter-current fashion, the total pressure drop can be very high. This would create high pressure in the lower stages, high enough to obstruct the meal flow from upper stages. For this reason, an intermediate design, with holes in the lower one or two bottom only, is conveniently used for RB. Vapour has free passage in upper stages, either through central vapour drums, or via annular space along the shell wall (Figure 14.14). Steam injection in multiple stages result in higher temperature of DT gases, around 75 C.

As discussed before, a classical DT, with steam injection through sweep arms in multiple stages, requires more desolventisation time, compared to the Schumacher DT with counter-current design. Typically, desolventising time of 35 min may be considered. Since there is no requirement of toasting, moisture levels are maintained

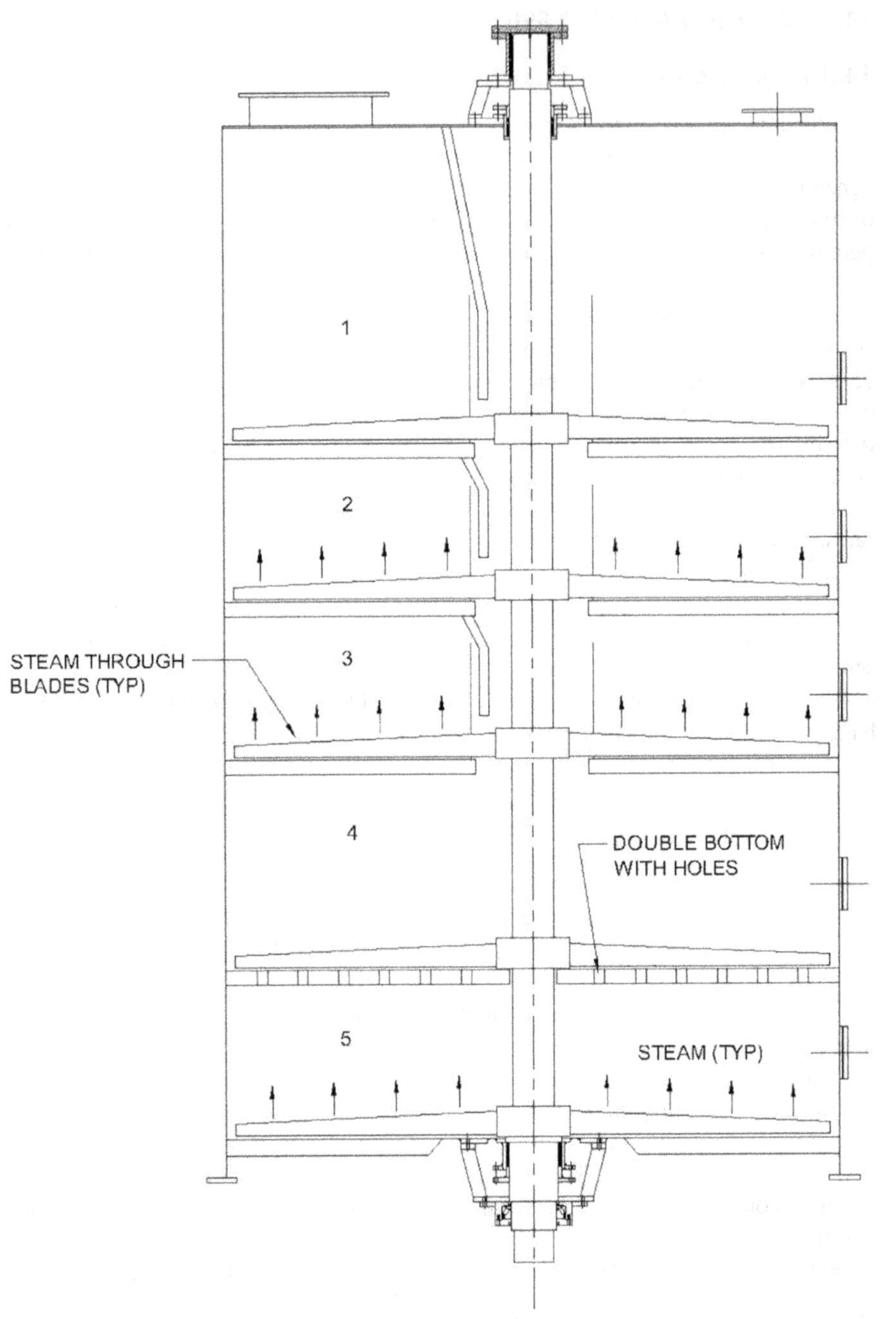

FIGURE 14.14 Desolventiser for Rice Bran.

low. This is accomplished by more indirect heating. So, a DT for rice bran is bigger in size compared to that for soybean of comparable tonnage. Bed heights then may be maintained much lower, to save on electric energy for sweep agitation.

Typically, RB meal is discharged at around 13–15% moisture content. Hence, drying with hot air is not necessary. Just a single stage of cooling, with ambient air, is sufficient to reduce the moisture to below 12%, while cooling the meal. For highly humid ambient conditions, the air may be heated to 50C. This may also be done when meal moisture is more than 14%.

14.3.3 Distillation

Miscella concentration, when pellets are extracted, may be low, within 15–18%; or be 22–24% in case of collets. Because of lower miscella concentrations with pellets, the Economiser may only help enrich the miscella to 40–50%. This means the steam consumption for evaporation of solvent is significantly more than for soybean, and so is the size of second Evaporator. Typically, steam requirement for RB extraction is 30–40 kg/T more than that for soybean. And, evaporator size is two to three times that for soybean, for similar plant capacity.

The stripping stages are very similar as those for soybean, since the oil quantity is nearly the same.

The absorption section is designed for similar load as in case of soybean.

Miscella filtration is somewhat difficult, due to the presence of fibres coming from rice bran. Hence only a pre-filtration is done, using a hydro-cyclone. Resultant insolubles content in oil is always more than 0.2%. This puts a limitation on the production of lecithin from RBO. Whatever small quantities of lecithin are produced, these are used mostly in animal feed at the present time.

For the production of food grade lecithin, improved versions of the wedge-wire screen filters would be required. Recently, filters with plunger mechanism, to force-remove all the fines deposited on the screen, have become available; results are awaited from a plant in Nigeria.

14.3.4 Colour of RBO – A Problem

A major issue with RB oil, apart from high FFA, is its dark colour (Vadke, 2013). The Red in the crude oil usually ranges between 3 and 5 units (1/4 inch cell, Lovibond). It is difficult to remove this red colour during refining. Even after double-bleach, with high dosages of bleaching earth, causing high oil losses, the refined oil is still reddish. It should be noted that the red colour is not native to RBO, unlike palm oil where the red is due to native carotenoids. The red in RBO is created during the preparation of RB, specifically in the pelleting step. As the material is subjected to high temperature and shear, there are fast reactions between starch and protein, called 'browning' reactions. Some of the coloured products produced by these reactions are oil soluble. These molecules combine with oil molecules when miscella is subjected to temperatures above 80C. This is considered as 'colour fixation'. It is very difficult to remove this 'fixed' colour by bleaching.

Miscella refining, as practised on cottonseed oil for removal of red pigment gossypol (along with FFA), is not a practical solution for RBO, for two reasons; one, the higher phosphatide (gums) content and high wax content in RBO, and second, the presence of Oryzanol. The 'gums' and waxes form emulsion when alkali is mixed with the miscella; soap separation becomes very difficult, resulting in high oil losses. Also, the alkali used for miscella refining would remove the beneficial component, 'Oryzanol', from RBO. The present method of refining of RBO, 'Physical Refining', wherein no alkali is used, helps retain much of the Oryzanol.

Thus, the only control at the present time is to keep distillation temperature low as possible, say below 80C, to minimise the colour fixation. But, more importantly, novel preparation steps need to be developed for RB, to avoid the red colour development in the first place. Alternatively, new extraction methods may be adopted, wherein such 'preparation' may not be necessary. These issues are discussed later in this chapter.

14.4 COTTONSEED MEAT PROCESSING

14.4.1 Extraction

As discussed in Chapter 13, cottonseed meats, obtained from delinting and dehulling, are prepared either as whole-fat collets or partially de-oiled collets, with oil content 22–24%. The first type is produced with an undrained expander, and the latter from a drained expander, using more electric energy. The whole-fat collets are soft and tend to pack to some extent in the extractor, especially in beds deeper than 1 m. The partially de-oiled collets (PD collets) are firm, and when properly hardened with slow cooling, retain their shape in extractor.

Since the PD collets retain their shape better, solvent percolation rates are superior, which results in superior extraction. The relative performance figures are given in Table 14.4.

The PD collets not only are extracted somewhat faster, they are also extracted to lower ROC, offering a significant benefit. This is thanks mainly to a lower MD, achieved by more rigorous work on the mass in drained expander, apart from the higher percolation rates.

TABLE 14.4
Extraction Performance of Whole-Fat and Partially Defatted Collets of Cottonseed Meat

Parameter	Whole Fat Collets	Partially Defatted Collets
Solvent percolation rate, $m^3/m^2/h$	25–30	32–34
Extraction time	80 min	75 min
ROC	0.9–1.0%	0.6–0.7%
Miscella Concentration	22–24%	24–26%
Solvent carryover in spent collets	28–30%	26–28%

Since the solvent carryover is less, the steam consumption in the DT is lower than for whole fat collets. Thus, the extra energy input on the drained expander is more than offset with benefits in the extraction plant. As shown in Chapter 12, the drained expander operation offers the most optimum option for processing of cottonseed meats, compared to undrained expander or the prepress.

Another advantage associated with a lower ROC of cottonseed meal is the lower level of 'free gossypol'. Typically, while meal from undrained expander process may contain up to 500–700 ppm free gossypol, that from drained expander process may have less than 400 ppm. This difference becomes important when cottonseed meal is to be used up to 30% in ruminant feed, where the quality standard is 200 ppm max.

The 'DT Jump' in cotton seed meal is around 0.1%. Thus, when a meal sample collected at bagging station is analysed to have ROC of 0.7%, corresponding sample at extractor discharge would only show ROC of 0.6%.

14.4.2 Desolventising – Cooling

The spent PD collets may be desolventised in Schumacher-type DT. Traditionally, however, classical DTs are used on cottonseed. This is due to the fear, from past experience, that too much live steam may cause formation of large lumps. Such lumps formation would be a result of inadequate cooking. This was so because cottonseed extraction industry was not technically well-organised as the soybean industry; the preparation prior to pressing, especially cooking, used to be inadequate. With better preparatory operations of conditioning and cooking, and with additional cooking in the drained expander, the risk of large lumps formation is eliminated.

In many plants, soap obtained from miscella refining operation, is mixed with meal in DT. This may conveniently be done by spraying the soap in second-last stage of DT. This stage must be lined with SS 304 plates, to prevent corrosion. The additional heat load of evaporation of moisture in soap has to be considered in DT design.

The meal discharged from the DT has moisture less than 14% and does not require hot air for drying. Blowing of ambient air through the meal for cooling purpose can reduce the moisture content to less than 12%.

14.4.3 Distillation

Extracted cottonseed oil contains pigments of 'gossypol', which give a dark red colour. The colour must be removed prior to subjecting the miscella to high temperature. Hence, the miscella is treated with 'Miscella Refining' process, prior to final distillation, wherein the colour pigments are removed with 'soap'. Benefits of miscella refining are discussed in the next section. Miscella concentrated to 60–65% is taken for miscella refining (MR), and the refined miscella is taken back to complete the distillation process. Should cottonseed flakes, or whole-fat collets be extracted, the miscella concentration out of the Economiser is just about in the required range. So, miscella discharged from the Economiser is sent to the MR section.

However, when PD collets are extracted, the miscella concentration out of extractor is higher, and oil content is less; thus Economiser may enrich the miscella to 75%. When the expander oil is mixed with the miscella, the concentration would further increase to 80%. If this miscella were subjected to refining, some of the advantages of miscella refining process would be lost, as explained in the next section. Hence, two Economisers may be utilised, in place of one, the first to enrich to 50–55%, and the second to further concentrate the refined miscella to 75%. The partially enriched miscella from first Economiser is mixed with the oil from expander, resultant concentration being in the range of 60–65%. The miscella, after refining, is sent to the second Economiser, where the remaining heat in DT gases is used to further enrich the miscella to 75–80%.

The quantity of additional oil, coming from the expander, has to be considered while designing the stripping equipment.

Absorption system parameters for the vent are similar to those for soybean.

14.4.4 Miscella Refining

This process is unique to cottonseed processing. Cottonseed contains a red pigment, gossypol, which gives the extracted oil a dark red colour. Gossypol is also a toxin that affects fertility in mammals. When the oil is subjected to high temperature during distillation, the pigment partly attaches with molecules of oil. The darkened oils are difficult to bleach, during refining. This phenomenon is known as 'colour fixation'. Hence, the pigment must be removed before miscella is subjected to temperatures above 60C. Bulk of it may readily be removed by caustic wash, as used in normal neutralising, also called 'refining', process. In this case, however, the caustic mixing, and separation of resultant soap, happens in the miscella phase, as against in oil phase during normal refining.

Miscella refining also offers the added benefit of **lower refining losses** of neutral oil, compared to normal oil-phase refining. Comparative data on miscella refining at different concentrations and normal oil phase refining is presented in Table 14.5.

TABLE 14.5
Comparative Data on Miscella Refining and Oil-phase Refining of Cotton Seed Oil

Miscella Concentration	Viscosity, cP	Specific Gravity (S.G.)	S.G. Difference with Water	Refining Loss Factor[a]	Residual Soap in Oil, ppm
100% (Oil) (at 80C)	20	0.88	0.12	2.2	600–800
80% (at 60C)[b]	8	0.83	0.17	1.5	300–400
70%	3	0.80	0.20	1.4	200–300
60%	1.5	0.77	0.23	1.35	150–200
50%	1	0.74	0.25	1.3	120–150

[a] Refining loss = FFA × Loss Factor (Figures as achieved with self-cleaning separators; comparative figures with manual separators may be higher).

[b] Oil temperature, at the centrifugal separator, is 80C during normal oil refining; however, miscella temperature must be maintained lower at 55–60C for miscella refining.

As can be seen from the data, the refining loss reduces significantly in miscella phase, thanks to the larger density difference as also the lower viscosity. At lower miscella concentrations, losses are even lower. However, with lower concentrations, it is also difficult to mix the alkali with oil, due to the presence of a large volume of solvent (At 50% w/w concentration, volume of solvent is nearly 58%). Therefore, the optimum range for miscella refining is considered as 60–65% w/w. AT 80% concentration, while refining loss may not be much higher, the residual soap in refined oil is high, which necessitates a water-washing step during oil refining.

Miscella from first Economiser is taken to an 'equilibration' tank, where oil from expander is mixed. Multiple equilibration tanks are used, so that concentration in each tank may be checked, and maintained in the desired range by the addition of expander oil or solvent as required. Once a tank is filled till 3/4th level, miscella is directed to another tank. Expander oil is added, concentration checked and adjusted by solvent addition if required. This miscella is then ready for miscella refining.

The miscella from equilibration tank is pumped to an alkali reaction tank. Upstream of the tank, alkali is dosed online into a two-stage 'static mixer'. The mixture is stirred in the reaction tank for nearly 15 minutes, to form soapstock (sodium salt of fatty acids). The mixture is then pumped into a high-speed centrifuge for the separation of soapstock. A heater may be placed online, upstream of the separator, to maintain the temperature at 60 °C. The process flow of miscella refining is shown in Figure 14.15.

The centrifuges utilised for miscella refining are blanketed with nitrogen, to prevent the escape of solvent. In most cases, self-cleaning centrifuges are utilised. These are quite expensive compared to manual machines, but offer multiple advantages of lower refining loss, higher fatty matter in soap (TFM) meaning less water content, zero down time on account of bowl cleaning, apart from prevention of solvent loss at bowl cleaning in case of miscella refining.

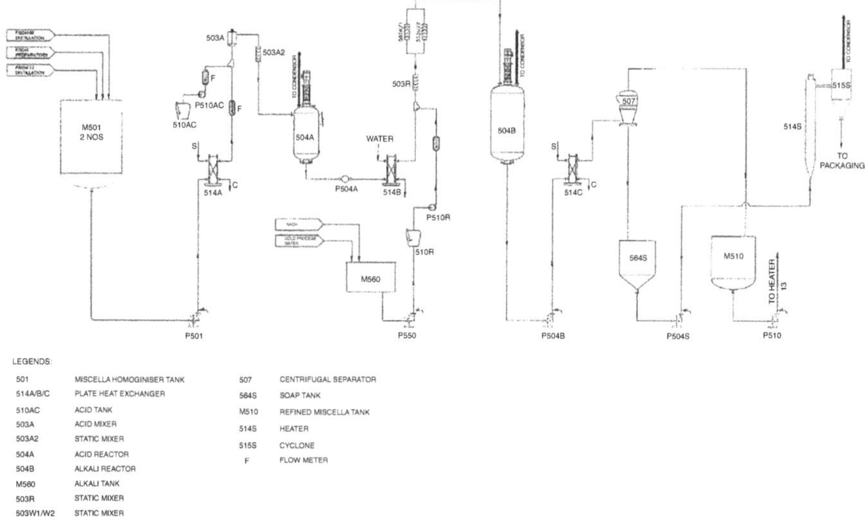

FIGURE 14.15 Flow Sheet for Miscella Refining of Cottonseed.

During off-season, when oil quality may be poor, a prior reaction with a mild acid, such as citric or phosphoric acid, for 30 minutes, can help improve the quality, by converting non-hydratable phosphatides (NHP) into hydratable and making them reactive with alkali to form soap. Quantity of alkali may be computed as 10–15% excess of the stoichiometric amount. Should acid be used, alkali dosage be increased by an equivalent amount. Acid dosage may be 10% in excess of the stoichiometric quantity required to react with the manganese or calcium moiety in NHP.

The soapstock is sent either to DT for mixing with meal; or for drying, where hexane is recovered, and the desolventised soap is stored prior to sale. The miscella containing refined oil and solvent, loosely called 'refined miscella', is sent back to distillation section. The soap drying process is discussed in the next section.

Thanks to the larger difference in S.G. of water and miscella phases, as also to the low viscosity of miscella, the separation of phases during miscella refining is almost complete. This means, while on one hand the loss of oil in water phase is minimal, resulting in the low refining loss factor; on the other hand, the retention of soap in the oil phase is also very low. This level of soap, up to 150–200 ppm, is readily handled during bleaching operation in oil refinery. Hence, additional water wash, which is a must for oil-phase refining, is not required. This is a major benefit, as the wash water generation, and the subsequent effluent treatment, is totally avoided. This is the third major benefit of miscella refining, in addition to light-colour oil and low oil loss.

14.4.5 Drying of Soapstock

Soapstock coming from the centrifugal separator (self-cleaning) contains nearly 38–40% TFM, 6–8% solvent, rest being water. It is a thick viscous material and is maintained warm in jacketed 'soap tank'.

The viscous soap is pumped, preferably by a progressive-cavity (screw) pump, to a scraped surface heat exchanger, jacketed for steam heating. Temperature is raised to 85–90 C and is flashed in a vessel under vacuum, to evaporate the solvent. Vacuum is maintained low, just about 100–150 Torr (absolute pressure 600–650 TorrA), to avoid too much foaming. Under these conditions of temperature and vacuum, the solvent content in discharge soap reduces to below 2%. For further reduction in residual solvent, to 0.1%, the soap may be flashed into second flasher under higher vacuum and higher temperature @ 100–105 C. The vapour line from flasher is connected to the Distillation Condensor in SEP. Since the condensor is usually at a higher vacuum (300–350 TorrA), a throttling valve is provided in the vapour line. Care must be taken to install the valve at least a meter away from vapour outlet of the Flasher, so as not to increase the escape velocity of vapour from flasher; a high escape velocity would suck the foam into the vapour line. A foam-break steam connection should be provided on top of the flasher for safety.

The soapstock composition mentioned above relates to a self-cleaning centrifugal separator. For 'manual' separators, water flush line is provided to help remove sediments, hence TFM is usually somewhat lower, between 30% and 35%. Also, solvent content tends to be higher, around 10–12%. For such a soap, the entry nozzle to flasher be such that the feed enters tangentially at high velocity and goes round the vessel wall. This is required to suppress foam, which may be excessive in view of the higher solvent content.

The 'dry' soap, low in solvent but containing 50–55% water, is much safer to handle and may be pumped to storage tank, even without cooling. However, in view of residual solvent, a vent line may be connected to DT condensor in SEP.

14.5 PRESSCAKES OF RAPESEED, SUNFLOWER SEED AND OTHER OILSEEDS

Next to soybean, the solvent extraction industry worldwide processes presscakes of rapeseed and sunflower seed. There are also many plants which process cakes of palm kernel and groundnut. Cottonseed is now processed more with expanders; this has been discussed in the previous section. A few sheanut cake plants have started mainly during the last decade, and more are expected to come up.

14.5.1 Cakes of Rapeseed and Sunflower Seed

As discussed in chapter on cakes preparation, all these cakes are broken into small size, prior to extraction. The performance, and process conditions, for presscakes of rapeseed and sunflower seed, either dehulled or whole, are quite similar. The cakes hold reasonably well under solvent spray. Extraction times are between 90 and 110 minutes, to reach ROC of 0.7%, provided the milling defect of cakes is not more than 0.3%. Solvent percolation rates are around 25 $m^3/m^2/h$, and outlet miscella concentration may be maintained around 20%.

The 'DT Jump' in ROC for rape seed and sunflower seed is around 0.1%.

The cakes of rapeseed and dehulled sunflower seed, being high in protein content, tend to stick to the screen, in the Linear extractor. Rinsing arrangement must be provided on the return path of screen conveyor.

Solvent carryover in spent cakes is nearly 28–30%. These are conveniently desolventised in Schumacher-type DT, to residual solvent less than 300 ppm.

The desolventised meal may contain up to 14–15% moisture, especially in sunflower meal, depending on moisture in presscakes and may require warm air, at say 80C, for reduction of moisture content to less than 12%. For rapeseed meal, cooling with ambient air should suffice. The cooling stage may conveniently be built below the DT.

Distillation process is similar to that for soybean miscella, but more solvent has to be handled because of lower miscella concentration. This means the Evaporators are bigger, although Strippers are the same size as for comparable capacity soybean plant. Foaming may be encountered if fines are not filtered out of the miscella. Absorption system performance is similar to that for soybean.

In general, the performance of shallow-bed extractors is better on most presscakes, as these are prone to make fines, less or more depending on cake quality, its protein content, etc. Especially the solvent drainage is superior, with less carryover of solvent in spent cakes.

Many sunflower cake extraction plants are now equipped with a degumming section within the SEP, as is the case for many soybean plants. Lecithin produced from sunflower oil is of similar quality and is gaining market. The trend is slowly starting with rapeseed oil as well.

14.5.2 Groundnut Cake

Groundnut cakes are very well formed, usually contain less oil than the norm (of 18%), just about 14–16%, and may be extracted in shorter time, 60–70 minutes. The other parameters, percolation rate and miscella concentration, are quite similar to those of rapeseed and sunflower seed. In Linear extractors, cakes tend to stick to the screen, hence effective screen rinsing arrangement should be provided.

Although the cake oil content is low, milling defect of the cake is often high, nearly 0.4–0.5%. GN is so soft and easy to press, that quite often the cooking operation is ignored. Also, since the 'soft' mass is compacted too fast in the press, uncrushed particles tend to get trapped inside the compact mass. If the cutting and mixing devices within the press are not adequately maintained, the uncrushed particles tend to pass in the cake. These result in high MD and high ROC after solvent extraction.

The well-formed cake has superior drainage property, hence solvent carry over to DT is less than that with cakes of rapeseed and sunflower seed. However, desolventisation may not be done in Schumacher-type DT. As discussed above, the seed preparation, especially cooking, is ignored in many plants. This results in cakes having uncooked protein. Spent cake of such nature is prone to protein coagulation in DT and formation of big lumps. For this reason, classical DTs, with less live steam injection, and more indirect heating, are commonly used.

Groundnut is an edible product, and the meal does not contain any native antinutritional components. However, groundnut meal is only used for animal feed at the present time. The main difficulty in its use for food is the presence of 'aflatoxins'; toxins coming from fungus, which may infect groundnut during harvesting or during storage. When groundnut is processed for 'table nut', the infected nuts are separated first with colour-sorters and then manually; identifying the infected nuts by their slight discolouration. When colour sorter machines are used, 85–90% of the discoloured nuts are removed, the balance 10–15% are separated manually. The manual intervention adds to the process cost, making it too expensive for the low-margin operation of GN crushing for oil. So, we have to await the progress on colour sorting technology. Better farm practices may also reduce the aflatoxins infection in future.

Once the aflatoxins issue is taken care of, three major process changes will be required for the production of edible flakes. One, the thin skin on the nut may have to be removed; second, the cooking and pre-pressing operations will have to be tuned to effect minimal changes in protein structure; and third, desolventising method will have to change to that used for soybean (edible 'white' flakes).

14.5.3 Cottonseed Cake

Some cottonseed oil mills still work with prepress plus solvent extraction process. Cottonseed cakes are also well-formed, and perform similar to the groundnut cake, with extraction time 60–65 minutes. If the meats were well-prepared, ROC may be reduced to 0.5%.

At such low ROC, free gossypol levels are low, usually less than 300 ppm. This is a distinct advantage when premium quality meal is desired.

The miscella percolation rates, as also solvent carry over to DT is similar to that of GN cake. Desolventisation and cooling may be accomplished in a combined DTC, as for other cakes. For cottonseed cakes, again, as for groundnut, traditionally classical DTs have been preferred. But with better seed preparation, the use of Schumacher DT has started.

Miscella concentration at extractor outlet is higher, more like that for soybean. Foaming is not a usual phenomenon for cottonseed miscella distillation, since the phosphatides content is very low.

Plants that process cottonseed cake are mostly old. New plants are equipped with expander technology instead. Most of the old plants are not equipped with Miscella refining. The dark oil produced in distillation then has to be double-refined, with incurrent high refining losses. Also, the reddish wash water poses a major issue for effluent treatment.

14.5.4 Corn Germ Cake

Corn germ cakes tend to crumble to a degree, being in the medium protein category. This causes somewhat higher ROC, close to 1%, with extraction time of nearly 90–110 min. The solvent carryover is also higher @ 30–33%, leading to higher steam requirement in DT.

14.5.5 Castor Bean Cake

These are well-formed cakes, thanks to the high protein content. Performance is similar to those of sunflower seed cake. Can be extracted to low ROC, with less than 90 min extraction time, and desolventised to low residual solvent. However, since the core of the beans is soft, there is a risk of high milling defect in press cake, if proper attention is not paid to multiple cake-cutting devices. This is especially true for large presses, with deep worm channels.

14.5.6 Palm Kernel Cake

Cakes of palm kernel are not as strong as the cakes discussed above, mainly due to lower protein content (14–15%), and contain fines. This results in somewhat lower percolation rates of solvent, and higher ROC, nearly 1.2%. The solvent carryover is also higher, almost 35–38%, which causes higher steam requirement in DT; also, the residual solvent in meal tends to be high, with resultant higher solvent losses from the overall process, often as high as 3–4 kg/T cake, compared to just about 1–2 kg/T for most other cakes.

The main problem with processing of PK cake is the difficulty in desolventising. Despite long desolventising time, residual solvent tends to be high, resulting in multiple 'flames' in the meal sample. The main culprit here is inadequate steam conditioning of the kernel. An adequate Conditioner prior to flaking, with at least 30 min residence, and minimum 7% moisture in PK, can help produce a far better press cake, which is more amenable to desolventising.

14.5.7 Shea Nut Cake

Sheanut cakes are the most fragile, thanks to very low protein content, just about 10%. The cakes crumble under solvent spray, which affects the solvent percolation rates drastically. Percolation rates may vary between 8 and 12 $m^3/m^2/h$, depending on the quantity of filter cake mixed with the Press cake. Extraction time may be as high as 3–4 hours. Even with such long extraction time, ROC may be as high as 1.5%.

The fines also affect the solvent drainage, and carryover of solvent in spent cake is high @ 33–35%.

In most plants, as the filter cake is mixed with Press cake manually, the quantity of filter cake (known in industry as the 'filter mud') to Extractor may vary significantly. A higher dosage of filter mud may severely affect percolation rates, which then affects the extraction rate. At times, the extractor may have to be stopped for some time, until the last miscella spray becomes 'clear', i.e. oil in the spray is reduced to low value, indicating completion of extraction. This is why the filter mud should preferably be pressed separately in a 'Foots Press', and then the formed cake may be sent to Extractor with the Press cake.

Higher solvent carryover to DT means high steam requirement for desolventising, as also a DT with larger heating surface area. Classical DTs are preferred, in view of the high fines content in spent cake. However, it is possible to desolventise the spent cake to low residual solvent level, unlike with cake of palm kernel.

Fines also get into miscella going out of extractor. These must be filtered out, using a screen of 30 or 50 micron, to prevent fouling of tubes in evaporators, and darkening of oil. Distillation of miscella is a simple affair, without any foaming.

As discussed in the chapter on seed preparation prior to mechanical expression, better preparation, suitable for the large-sized seed, is expected to improve the performance of both, mechanical expression and solvent extraction. Results from a plant in Nigeria are awaited.

14.6 SUNFLOWER SEED – DIRECT EXTRACTION

As discussed before, on the chapter on seed preparation for solvent extraction, sunflower seed may be flaked and sent for direct solvent extraction. Such plants were not uncommon until the 1980s. However, with the advent of large capacity presses, for prepress duty, most of these plants have now employed presses. Nevertheless, a discussion on direct extraction of sunflower seed may still be of interest to those operators who might mostly crush soybean, but may want to process sunseed on and off. It is the whole seed, without dehulling, that is subjected to direct extraction.

The prepared flakes, from whole sunseed, do contain some fines. Hence percolation rates are just about 20 $m^3/m^2/h$, and miscella concentration of 22–24% may be maintained. Final ROC will be higher compared to that obtained from prepressed cakes; even ROC of 1% may require a long extraction time of almost 2–2½ hours.

Thus for a given Extractor, capacity may reduce to just about one-third of that on soybean.

Solvent carryover in spent flakes is also higher @ nearly 35–40%. Thus, steam consumption in DT is high, but that may be utilised beneficially to distill solvent in Economiser. For a given DT, the capacity may just be half of that achieved with soybean.

As discussed before, steam requirement for solvent evaporation is high, due to the nearly 3½–4 times the oil compared to that in prepressed cake.

Thus, overall, when a plant designed for soybean is operated with sunflower seed flakes, plant capacity may reduce to just about 1/3rd.

14.7 PALM KERNEL

The performance of palm kernel flakes in solvent extraction is very similar to that of whole sunseed flakes. However, should the PK grits, obtained from grinder, not be adequately pre-conditioned, the flaking operation may produce a lot of fines, which would disturb the entire operation. The ROC may then be higher than 1.2%, miscella concentration may be lower at around 20–22%.

Desolventising also may be difficult due to the presence of too much fines; Schumacher-type DT may not be used, as explained for rice bran desolventising, thus steam consumption would be higher, together with high residual solvent in meal. Thus, the importance of adequate pre-conditioning of PK grits should always be remembered. In any case, the solvent losses tend to be much higher than that for sunflower flakes.

14.8 CORN GERM

The flakes of corn germ, with very low protein content, tend to crumble under solvent spray. Extraction is difficult, leading to high ROC, nearly 1.8–2.0%, even with extraction time over 3 hours. Solvent carryover is also high, nearly 40–45%, resulting in high steam requirement in DT. Thankfully, all this extra heat may be beneficially utilised for distillation because of high oil content in germ.

If the germ is further prepared with second cooking and light-pressed in an expander with drain cage, like that for cottonseed, the extraction performance may be significantly improved. The collets, containing nearly 28–30% oil, maybe extracted to ROC of 1.2–1.5%, with extraction time of 2 hours. Solvent percolation and drainage are also improved, resulting in less solvent carryover to the DT.

14.9 MINOR SEEDS – SAL SEED, MANGO KERNEL

14.9.1 Sal Seed

Sal seed flakes crumble under solvent spray, causing percolation difficulties. Percolation rates are just about 12–15 $m^3/m^2/h$. This is typical of low-protein materials, as discussed before. Expanded collets perform better, with good percolation, and ROC of 0.7% may be achieved within 60 min (Tiwari, 2019). Miscella concentration can be maintained at 25%. Solvent carryover to DT is high @ 40%.

While spent flakes cannot be desolventised in a Schumacher-type DT, due to the presence of too much fines, and require a classical DT; spent collets may safely be processed in the former design, to less than 500 ppm residual solvent.

14.9.2 Mango Kernel

In small factories, mango grits are extracted directly, without cooking and flaking. While it is possible to reduce ROC to less than 1%, it comes only after a long extraction time of 2 hours, for extraction of only 12–15% oil. On the other hand, if the grits were conditioned and flaked, as discussed in the chapter on preparation, the extraction time may be reduced to just 1 hour, for ROC of 0.7%. Solvent carryover to DT is high @ 40%, similar with flakes of sal seed.

Mango fat has a high pour point and solidifies at ambient temperature (30 °C). Hence, all the oil lines, as also lines in stripping section, should be traced with steam tubing.

15 SEP Equipment
Operation, Construction and Design Aspects

15.1 EXTRACTOR SECTION

15.1.1 Feed Conveyor

The first equipment in a solvent extraction plant (SEP) is the Feed Conveyor, which collects flakes (collets/cake) from the preparation plant (or Oil Mill). The conveyor is typically a drag-chain type, with chain movement speed of 20–25 m/min. The bottom and sides of the carrying section are preferably lined with friction-reducing wear-resistant plastics such as UHMWPE (ultra high molecular weight polyethylene). The lining not only improves the life of the chain and the conveyor casing, but also reduces load on the conveyor drive.

Usually the conveyor has an initial horizontal portion, as it collects flakes from multiple flakers or collets/cake from cake cooler. It then rises, at the boundary of the preparation plant, at an angle, to reach the top of extractor in the SEP. Typically, the angle may be between 30 and 40 deg. The design of the 'bend' segment must be such as to prevent material accumulation. First, the curvature has to be smooth and long; second, a narrow wheel to keep the chain lightly pressed to the bottom helps. Alternatively, in some plants, two conveyors are used: one for the horizontal portion and the other for the straight incline portion.

A **safety fan** should be provided at the top of the discharge end to ventilate the conveyor. In case of accidental escape from extractor, this safety helps ensure that any solvent vapours would be sucked out and would not travel to the preparation plant through the conveyor.

Size of the casing is determined by carrying capacity and the bulk density. For example, bulk density of soybean flakes is nearly 0.35 kg/L, but that of soy collets is nearly 0.6 kg/L. Thus, for a capacity of 1,000 Tpd (42 Tph), considering excess capacity of 20% and a filling factor of 0.65 for the inclined conveyor, and linear speed of 25 m/min, carrying cross-section should be 14.8 dm^2.

$$\text{Casing cross-section} = \text{Capacity Tph/Bulk density}\left(\text{T/m}^3\right)\text{/Filling factor}$$

$$\text{*Excess factor/60}\,(\text{min/h})\text{/Linear Velocity}\,(\text{m/min})$$

For a width-to-height ratio of 1:0.5 (for taper conveyor), the conveyor size becomes 543 mm wide. So, we may opt for a 550 wide conveyor. On the other hand, for the

DOI: 10.1201/9781003309475-17

same capacity on collets, a much smaller conveyor, 400 mm wide, would do the job. For conveying of most cakes, bulk density between 0.4 and 0.45 kg/L may be considered.

15.1.2 Rotary Air Lock

An effective air lock between the feed conveyor and the extractor is required to act as a seal for solvent vapours. Rotary airlock (RAL) is a simplest such device. This is a rotary valve, machined fine to minimise the clearance between the vanes and the casing, as also between side flanges and the vanes. At times, replaceable soft wear strips may be fitted on vanes, which would maintain contact with casing during rotation. It is common to see RALs with four vanes. However, for wide inlet and outlet, five vanes are preferred to maintain vapour seal. Also, a semi-circular lining at the root of vanes helps prevent material accumulation (Figure 15.1).

For small- and medium-sized extractors, with bed width up to 2.5 m, square RALs, whose length is equal to the dia, are used. For large extractors, with bed width 3 m or more, however, long RALs may be used for better distribution of material over the width.

An ideal airlock, one which would maintain air leakage at zero, in fact would act to pump air out of the extractor. Since that would not be desirable, as then some solvent vapours might escape out, minor leakage through the vanes is allowed, so that the net transfer of air is close to zero. For this reason, it is not a good idea to fix wear strips on the vanes to maintain contact with casing.

RALs may be operated at slow speeds, say 25–30 rpm, to prevent fines generation. The size may be determined considering a filling factor between 0.65 and 0.8, depending on size (higher filling factor for bigger sizes).

$$\begin{aligned}&\text{RAL Capacity}\,(\text{Tph})\\&= \text{D}^2\,\left(\text{m}^2\right) * \pi / 4 * \text{Length}(\text{m}) * \text{Filling factor} * \text{Bulk density}\left(\text{T/m}^3\right)\\&* \text{rpm} * 60\,(\text{min/h})\end{aligned}$$

So, for a 1,000 Tpd soybean flakes capacity, considering excess capacity of 20%, a square RAL D 550×L550, would suffice, at a speed of 26 rpm. For 1,000 Tpd soy collets, the size could just be D450×L450 at 28 rpm.

Some plant suppliers install a 'sluice valve' in place of an RAL. It acts as an auto-seal when feed to extractor is stopped. During normal operation, it remains open. It is better to use the sluice valve in addition to an RAL, as an additional **safety device**.

15.1.3 Extractor Feed Hopper

The extractor may be seen to have three main sections: feed hopper, extractor proper, and discharge hopper. Operation of the two hoppers, and their regulation, is closely related to the extractor.

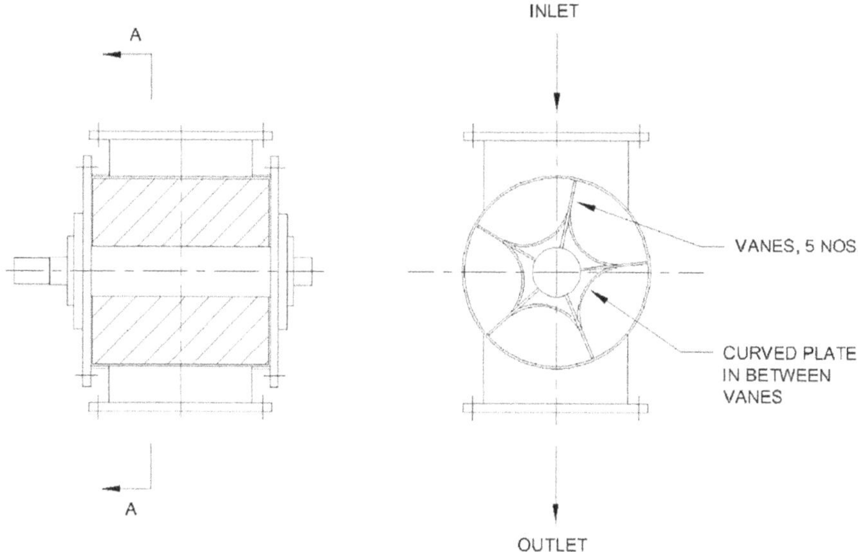

FIGURE 15.1 Rotary Air Lock.

The feed hopper acts as a buffer to hold flakes (to be read as flakes/collets/cakes) coming from the preparation plant (Prep), so that the extractor is fed continuously and uniformly, in spite of any flow variations from the Prep. The hold-up also acts as a barrier between the extractor, whose vapour space is full of solvent vapours, and the RAL.

In most extractors, the feed hopper comes fitted with an electronic level sensor together with a level control system. The level is maintained close to a set point, so that the barrier is not broken. This is achieved by varying the speed of extractor bed just so slightly, as not to affect the overall extraction time to any major degree. That much variation is enough to balance out the variations in feed rate from the Prep. Safeties are provided in the form of high-level and low-level switching actions. In case the control system fails to maintain the level, maybe due to high variations in flow from Prep, and if low-level be reached, the extractor bed movement is stopped. If high-level is reached, feed is cut off, until the level is restored. It may be noted that frequent start-stops of the heavy machinery in the Prep are not desirable. Hence, a better option is to sound a warning signal to Prep at an HL, so the Prep operator may reduce the output; should it fail, then to stop the machinery only when an HHL is reached.

For soybean flakes, a slow agitator may be fitted in the feed hopper, to avoid bridging. This is sometimes necessary, especially for flakes with high moisture, close to 11%. Volume of feed hopper is normally determined to hold flakes for nearly 5–6 minutes. The feed hopper is tapered on all four sides (Figure 15.2). Along the extractor width, the hopper dimension has to match the RAL width at top and extractor width at the bottom. A taper of nearly 12–15 degrees is maintained for the purpose. Along the extractor length, the hopper is tapered to facilitate the flow of flakes

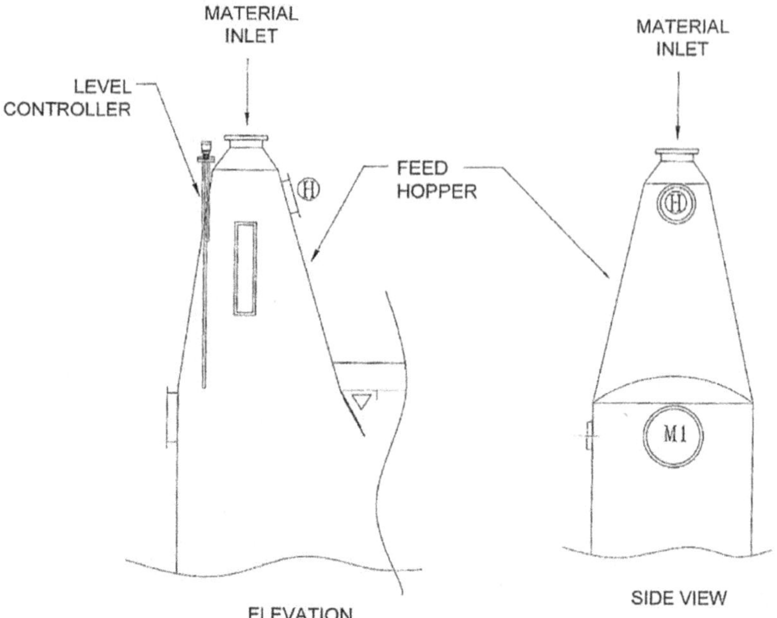

FIGURE 15.2 Extractor Feed Hopper.

smoothly into the extractor. Should the slope be less than 15 deg, the flakes bed drops considerably at the entry line.

15.1.4 The Extractor

As mentioned in Chapter 11, there are mainly three types of extractors in use worldwide at the present time. The fourth, the Lurgi design, is in use in some European countries. Other types, including the version of linear moving screen with screen segments opening up to discharge the bed onto screen below, so that bed moves in a loop; have been used in some plants, but are not much visible at the present time. So, we will mainly discuss the three main types, with a brief discussion of the Lurgi design.

15.1.4.1 The Linear Extractor

15.1.4.1.1 Construction Features

The Linear Extractor (Figure 15.3) was invented by DeSmet[1] in Europe towards the end of the 1940s. This was around the same time that the Rotary design was invented in USA. In the linear extractor, the screen moves, on two chains, and the bed moves with the screen. The bed depth is usually between 2.0 and 2.5 m; however, for materials that contain much fines, bed depth may be reduced down to 1.5 m or even less. The bed depth may either be adjusted by an adjustable taper plate in between the feed hopper and the extraction zone; or with a damper plate that may be adjusted

[1] Linear Extractor was invented by Mr. DeSmet, who formed the company Extraction De Smet at Antwerp, Belgium, to market the same.

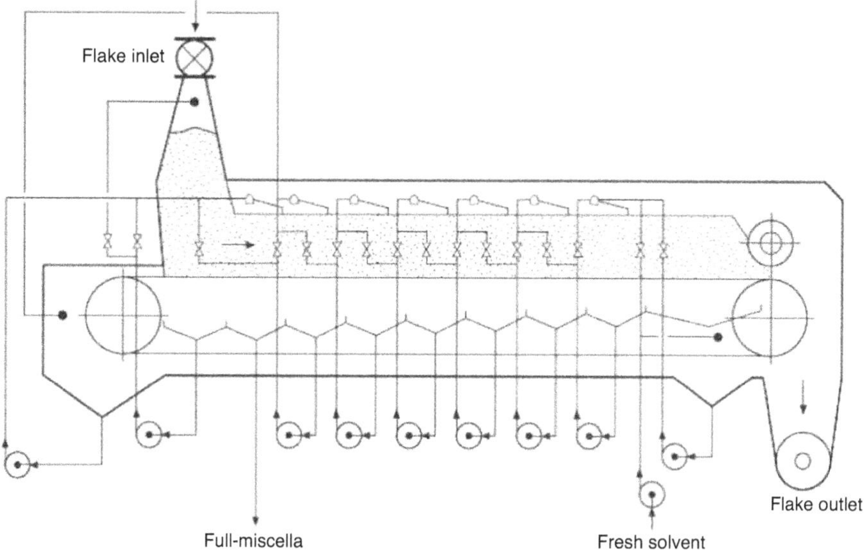

FIGURE 15.3 Linear Extractor – Miscella Hoppers and Pumps.

from outside. The internal adjustable taper plate may only help very the bed height by about 200 mm; and may not be adjusted from outside. Width of the bed may vary from 1 m, for small extractors, to as much as 5 m for very large ones.

Solvent and miscella sprays are fitted about 300–400 mm above the bed, and spacing between adjacent sprays is 1.5–2.0 m. While there is no partition between sprays, a small separation is created by a small heap of material collected near bed Scraper Arm. This full-width scraper, with teeth embedded in the bed, is hung on top of the bed (Figure 15.4). The main job of the scraper is to scrape off the fines, which get circulated with miscella, and get deposited on top of bed. The teeth create a minor obstruction to movement at the top of bed and create a heap, which acts as a barrier between two spray zones, so that the two sprays do not mix on top of bed. This barrier also works to create a small 'drain' zone between two extraction segments.

The chains are fitted with rollers that roll over rails, to minimise the frictional force, thus minimizing the load on the extractor drive. The return chains move on rails supported on the bottom casing, below the miscella hopers. Chain pitch is between 300 and 800 mm, depending on the extractor width. Discrete screen segments are supported on each pitch by 'cadder frames'. The gap between two adjacent screen segments is covered either by flexible cloth strips or by overlapping SS plates. Until the turn of the century, the screen used to be of wire mesh, between 24 and 30 mesh, supported by a perforated plate. Now the mesh and the plate are replaced by segments of wedge-wire screen.

A vertical side-sealing plate prevents material from going over to the chain, on either side (Figure 15.5). A square rod of a soft material fitted behind the plate keeps contact with the moving chain, so as to maintain the seal. The rods used to be of

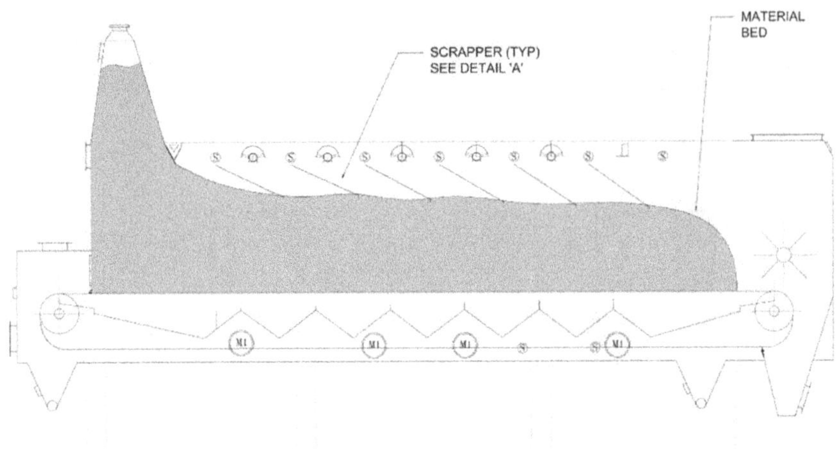

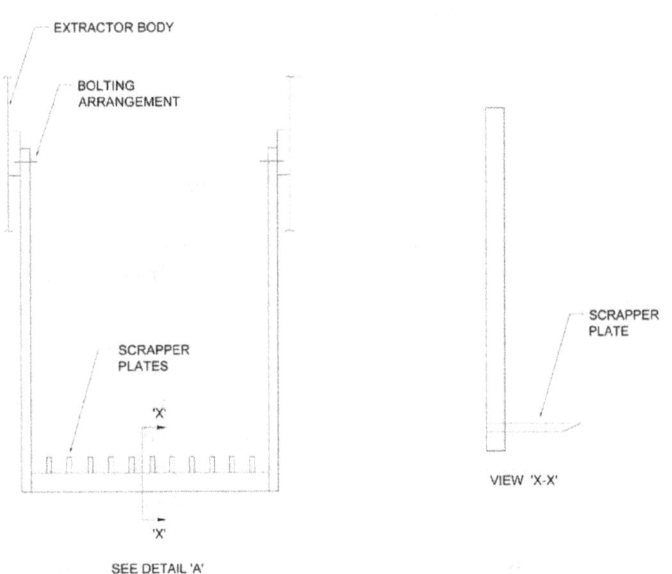

FIGURE 15.4 Bed Scraper in Extractor.

brass before, but are mostly of UHMWPE at the present time; the latter material offers multiple advantages, less cost and wear-resistance.

Miscella lines, from the circulation pumps to the spray, are jacketed with steam, to allow heating of miscella, to maintain temperature around 55–58 C. In fact, there is value in maintaining a higher temperature of 58–60 C; but that would require a close control, to avoid any higher temperature which would cause pressurisation

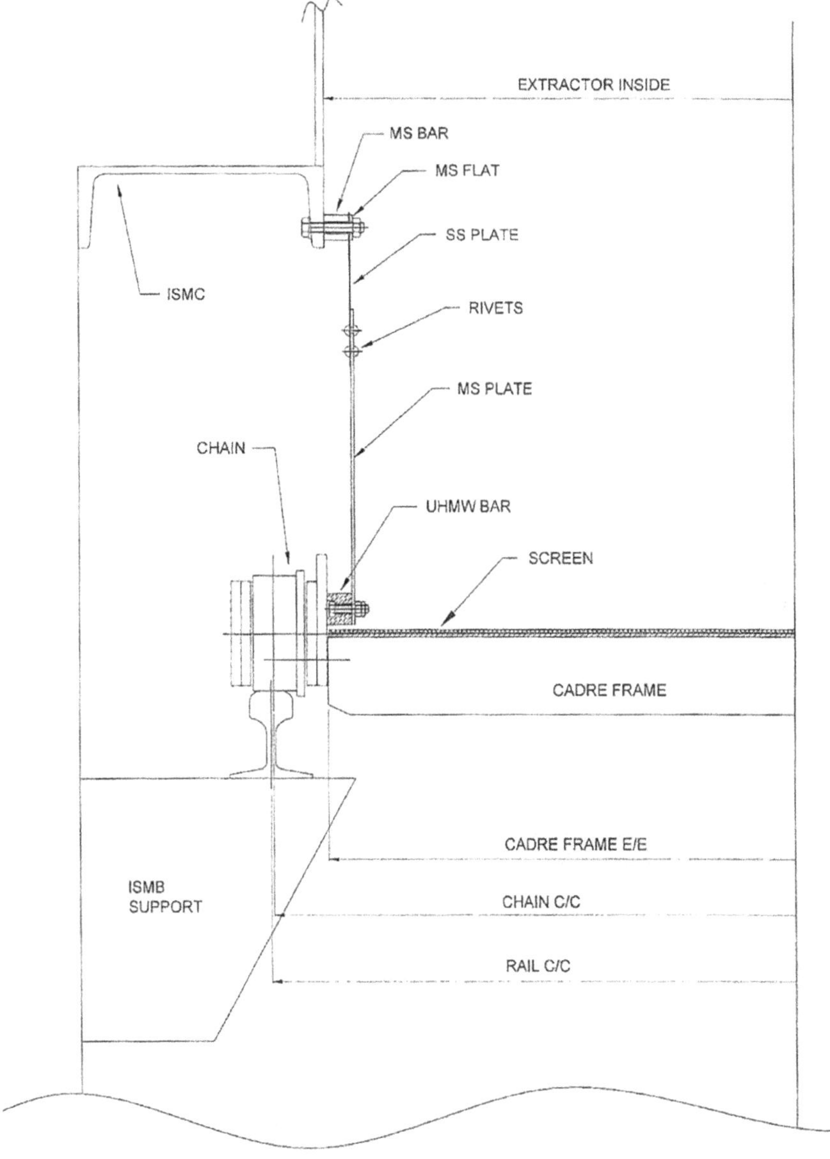

FIGURE 15.5 Side Sealing Arrangement in Extractor.

of extractor. Auto-control of miscella temperature would be advisable in the latter case.

Extractor is continuously vented, so as to maintain a slight suction, just about 1–3 mmWc (mm water column), to prevent any leakage of solvent vapour out of extractor. A safety pressure release valve is provided on top cover, to prevent accidental pressure build-up.

Since there is no relative movement between the material and the screen, some material tends to stick to the screen surface. This is especially true for high-protein oilseeds such as soybean, rapeseed, sunseed, cottonseed, castor, etc. The screen is rinsed with high-velocity spray of solvent on the return path, just after material discharge. Special spray nozzles, with the linear spray pattern, are used for the purpose. A scraper plate may also be used to scrape off large lumps sticking to the screen, just above discharge hopper.

15.1.4.1.2 Counter-Current Flow

As the miscella percolates through the bed and collects in the hopper below, it is pumped up in the same extraction segment. The miscella is not forwarded, towards the feed end, by circulation pumps, but overflows from one hopper to the hopper before. This is how the miscella flows in direction opposite to the flow of material bed. The last miscella spray to contact the bed, just prior to the fresh solvent spray, is a control point for extraction operation. Miscella concentration in the last miscella spray below 0.2% is a major indication of completion of extraction, and of low ROC in spent flakes.

For materials which allow very high percolation rates, like well-cooled collets of soybean obtained from a die-plate of expander, or tight pellets of rice bran which have been dried with hot air, a peculiar issue may be encountered. At the first spray, on fresh collets, since there are no fines to obstruct the flow of miscella, the bulk of the miscella may quickly flow towards the dry zone below the feed hopper and not collect into the hopper below the bed. This would cause the hopper, below the bed, to starve of miscella, and the pump would not have enough miscella to pump. This problem may be faced for the first two or three hoppers. Resultant extraction would be poor, with high ROC. One way to overcome the problem is to advance the miscella towards discharge end, by successive pumps. Another way, of course, would be to not make the pellets so tight and allow creation of some fines, as is the practice for rice bran.

The spray mechanism in the Linear extractor consists of a large dia full-width pipe, which gets miscella from circulation pump at one end, with holes (or slots) on the top; as the miscella jets out through the slots, it hits a full-width umbrella, and falls along both the sides of the umbrella. Thus there are two spray curtains from each spray. Since the spray openings are on the top side of the pipe, fines tend to get collected on the pipe bottom. To prevent fines accumulation, a small outlet line is provided on the other end, flush with the bottom of spray pipe; fines keep moving to the other end and are discharged back on top of the bed.

The bed of flakes (collets/cakes) at the beginning of the extraction zone is un-compacted by solvent, and has sufficient gaps for fine solid particles to travel through with miscella. When presscakes are fed to extractor, the quantity of fines can be high. Some of these solids pass the screen and get collected in the miscella hopper below. This is the hopper from which the rich miscella is to be pumped out to distillation. To avoid too much solids going to distillation, the miscella is filtered, solids discharged on to the bed, and the filtered miscella sent to distillation.

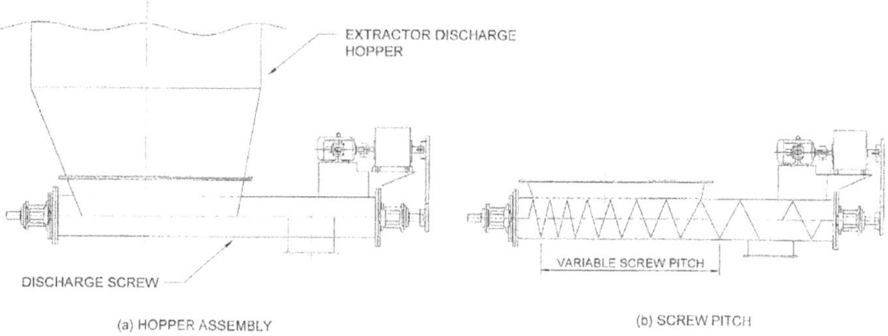

FIGURE 15.6 Extractor Discharge Screw.

15.1.4.1.3 Fully Wet Bed in Percolation Extractor

Since there are no partitions in between the sprays, miscella layer above the bed is not permitted, and sprays have to be adjusted by operator, to avoid inter-mixing. Too much throttling will, while preventing inter-mixing, make the beds dry which would affect the extraction adversely. The only correct way to ensure fully wet bed, without seeing the miscella layer on top, is to observe a good spread of the spray as it hits the bed, a spread that may extend up to 150–200 mm before disappearing into the bed.

15.1.4.1.4 Discharge Hopper

The spent flakes are discharged into a discharge hopper. A slow-rotating scraper drum is fitted above the hopper, with multiple scraper flats to scrape the bed, to ensure a continuous discharge.

Discharge hopper is fitted with a screw conveyor at the bottom. It should be designed with increasing pitch, so as to be able to evacuate flakes from the entire width uniformly (Figure 15.6). A constant pitch screw would get filled with material at one end, and would not retrieve material from parts towards discharge side. The screw is equipped with variable speed drive, and the speed is adjusted so that the flakes removal rate is uniform, and not affected by variations in discharge from the scraper drum. In fact, the speed regulation may be automated, in tandem with the speed of the extractor screen.

15.1.4.1.5 Design Considerations

The theoretical models for the design of extraction process have been discussed in Section 14.1. In present practice, the sizing of extractor is determined from the extraction time required, and the bulk density of material bed. Extraction times must be determined by laboratory experimentation, with a safety factor (say 25% extra) considered for inefficiencies of an industrial extractor. As discussed before, the method proposed by Karnofsky (1987) under-predicts the extraction time.

It may be noted that the **bulk density** of materials in the bed is always higher than one measured in air. It may also change with bed depth. For example, bulk density of soybean flakes may be measured outside the extractor as between 0.32 and 0.35 kg/L.

However, under the solvent spray, the flakes get compacted to 0.5–0.55 kg/L depending on bed depth. This bulk density may be determined in a laboratory by repeated solvent spray on flakes in a vertical cylinder, with an outlet nozzle at the bottom below a screen. The effect of bed depth may be simulated by the addition of an equivalent weight on top.

Thus, for 1,000 Tpd Soybean flakes, the volume of flakes under the extraction zone may be calculated, for **extraction time** of 50 min and bulk density of 0.55, as 63.2 m^3. For bed width 2 m, and bed height 2 m, extraction length would be 15.8 m. This length may be covered with eight miscella sprays, spaced 2 m apart, and a fresh hexane spray. The length of the drainage zone may be calculated, for 15 min, as 4.75 m. The length may be shortened by nearly 10%, and 1 spray saved, if bed height were increased to 2.2 m instead. It may be noted that the same extractor could extract almost 1,250 Tpd soy collets, thanks to higher bulk density of 0.62 (under solvent) and shorter extraction time of 45 min. Thus, the addition of an **expander** in the Soy Prep may help increase extractor capacity by 25%.

Percolation rates may also be measured with the same apparatus as used to determine bulk density, the vertical cylinder. This helps determine the capacity of circulation pumps. The first circulation pump, which circulates rich miscella on the fresh flakes, has to be of much higher capacity than other pumps, as the flakes are not yet compact and also as it may take the additional flow from the discharge of miscella filter.

Size of the spray pipe should be determined such that the flow from all the slots is uniform. This may be achieved when pressure drop in the pipe is negligible, say less than 5%, compared to that for flow out of a slot.

While matters of the **mechanical design** of equipment are generally out of the scope of this book, a brief discussion on design of moving parts of extractor may not be out of place. At times, it is seen that the main shaft, that moves the chains, is undersized; this has led to **shaft failures**; and at times it is oversized. The discrepancies appear to arise from a miscalculation of consumed power. The power should be calculated by determining the force on the chains, and multiplying it with the chain speed. The force consists of three factors, the load of flakes on the chains, weight of the chains themselves, and the friction factor between the rollers of chains and the rails. The friction factor may be considered as 0.2 for good rolling movement. However, during operation, it is likely that some rollers just skid on the rails. Hence, for safety, the **friction factor** may be considered as 0.3. If forces, power and torque, are calculated with this friction factor, one would arrive at safe yet realistic sizes of shaft and chain components.

15.1.4.2 Rotary Extractor

Out of the three main extractor designs, this design has the simplest drive mechanism and least moving parts. The central shaft rotates, and the compartments rotate with it. The screen is stationary, and there is no movement of carrying chains. Compartments are created by vertical partition plates welded to the shaft on the inside and the circular casing on the outside (Figure 15.7). The compartments are open at top for miscella sprays, and move with small clearance on screen at the bottom. The entire mechanism, as also the hoppers below, are covered with outer casing.

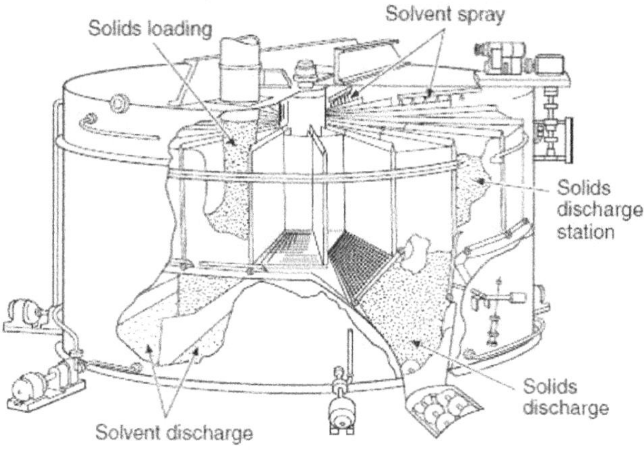

FIGURE 15.7 Rotary Extractor (Inside View).

This is a 'deep bed' extractor design, with bed depth ranging from 2 to 3.5 m. For materials which contain fines, such as most presscakes, percolation rates are limited, not only due to the bed depth, but also because there are no bed scrapers available, as with Linear extractor, to scrape the top layer of fines. Scrapers cannot be fitted, as these would interfere with the moving partition plates.

The biggest advantage of this design is that alternative compartments are devoid of sprays for most of the time, except for the short periods when the sprays are just above the partition plates. Thus, dry drain zones alternate with spray zones, and intermixing of sprays is mostly avoided. Therefore, a solvent layer may safely be maintained on the bed, which means the regulation of miscella circulation is easier, and the bed is fully flooded in extraction zones.

A small Feed Hopper feeds flakes continuously into a moving compartment below, and spent flakes are discharged after circular movement through more than 300 degrees. There is no screen below the discharge section, and the bed falls in a big heap. Thus, the discharge hopper is of bigger size, with a screw fitted below, equipped with variable speed drive, to remove the flakes in uniform flow to the DT.

The volume calculation for this extractor is similar as for the Linear Extractor. However, intermediate drain zones have to be added. So, the overall volume of bed is greater in this case. Drain volume is also required to be more, in view of longer drainage time required for the deeper beds.

15.1.4.3 Loop Extractor

The Loop extractor is a proprietary design of Crown Iron Works, USA. It consists of a shallow material bed pulled en-masse by a slow-moving drag chain conveyor, over fixed screens in two horizontal segments of a loop (Figure 15.8). Material is fed at one end of the upper segment. After the bed covers the first segment, it collapses in the curved casing between the two segments, before it is formed again on the second screen segment. Spent flakes are discharged at the end of the second segment. The second curved casing covers the return path for the chain, after material discharge.

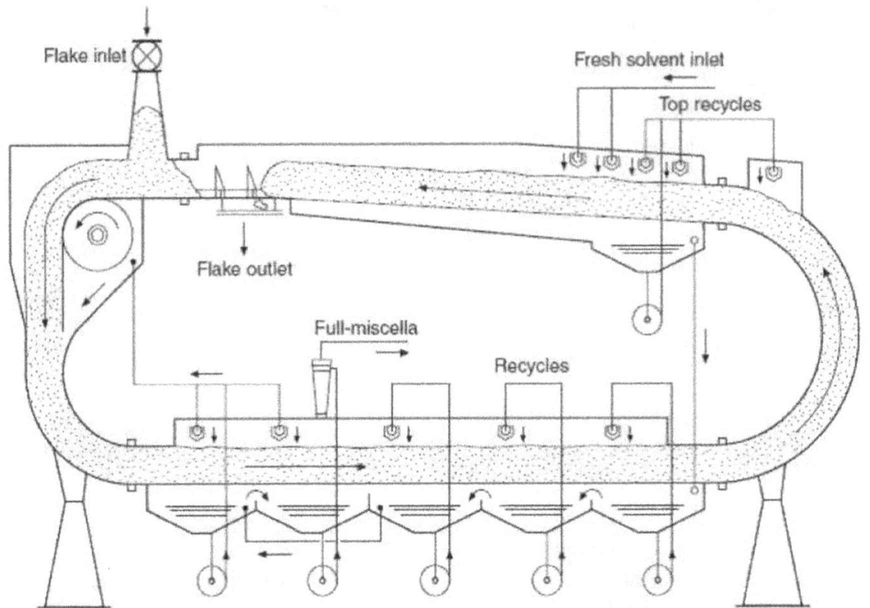

FIGURE 15.8 Loop Extractor (Crown Model III).

Bed depth is maintained at less than 1 m. Due to shallow bed, material compaction at the bottom layer is minimum, ensuring uniform extraction and higher percolation rates. Also, as the bed collapses in the first loop, and re-forms on the second screen, any channeling of solvent is disturbed, and all parts of bed are contacted. Of course, it is best to have flooded beds, so that there is no possibility of any channeling. The technique to have flooded beds is discussed in section on Linear extractor.

This design offers a major advantage for materials with much fines, such as most presscakes. The shallow bed ensures reasonable percolation rates. For this reason, the extraction time requirement is somewhat lower. Certainly, the drainage is much faster, hence the drainage times are significantly lower, say just about 5–6 min for soybean collets, compared to 15–20 min for the Rotary type, and 12–15 min for Linear type.

The flow patterns and miscella advancement is very similar to that in Linear design. Even though the bed is shallow, the width and length of this extractor are similar to those of Linear design, for comparable processing capacity, thanks to the two legs of extraction, one above the other. The overall height of the extractor is, of course, much higher, and requires two operation floors.

The discharge of spent flakes is continuous, as in, and more uniform than, Linear design. Thus, the discharge hopper can be much smaller than with the other two designs.

Mechanically, this design is simpler than the Linear extractor, since it involves a single moving chain, in place of two chains that require close parallel alignment in

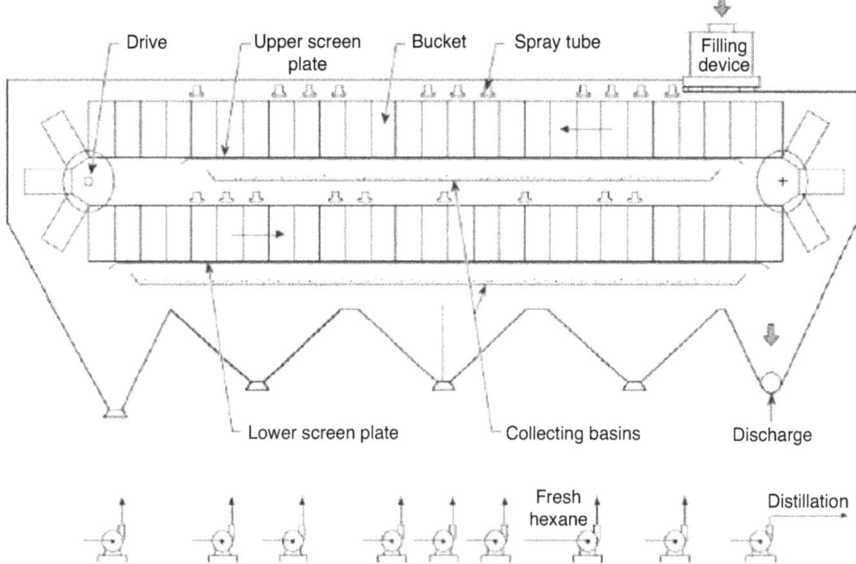

FIGURE 15.9 Flow Patterns in a Sliding Cell Extractor.

the Linear design. Even the single chain does not require multiple rollers, as it simply slides over the screens. Also, no side-sealing is required to protect the chains. Thus the maintenance of the Loop extractor is really a simple matter.

15.1.4.4 Sliding Cell Extractor

Sliding cell extractor is a patented design of Lurgi. The basic design is similar to the Loop extractor, except that dual-chains drag the baskets (cells) over the upper and lower screens (Figure 15.9). As a cell travels over the edge of upper screen, the material is dumped into cell below on the lower screen.

This design offers most of the advantages of a Loop extractor. But construction is somewhat complicated, since the moving element involves the dual-chains, as with the Linear extractor.

15.1.4.5 Materials of Construction, Drive and Insulation, of Extractor

For large capacity plants, larger than say 1,000 Tpd, There is a special requirement of minimum downtime. Thus, even a long 'annual maintenance' may be avoided, if possible. Towards that end, extractors constructed fully in stainless steel are preferred. Extra cost, due to the expensive material of construction (MoC) hardly matters in overall project cost for large capacity plants, which have to invest heavily in storage and logistics infrastructure.

For most small and medium-sized plants, however, extractor casings are mostly built in mild steel. The inner surfaces of casings are protected from corrosion by a coating of epoxy paint, given every year. Internal fittings, except the moving parts, are preferably constructed of SS. These include mainly the spray assemblies and miscella hoppers. In linear extractors, the chain links are built from toughened steel, whereas the rollers are hardened and ground. A soft, non-wearing plastic sleeve is preferably fitted between the moving rollers and stationary pins. The sleeve serves the purpose of a self-lubricating bush bearing. Extractor shafts are made of high-tensile steel, such as EN 19, sprockets being made of cast steel, with teeth usually flame-hardened.

Extractor drive is an elaborate affair, in view of the very slow speed of the screen, or the central shaft, or the chain, for the three types of extractors. The motor itself may be very small, just about 3 hp for a 1,000 Tpd extractor, but the speed reduction is a major multi-stage assembly. There used to be three or four speed reduction devices until about two decades back. These have now been reduced to just two in modern plants. The first reduction, coupled with motor, is a large reduction gearbox, ratio up to 2000:1. The gearbox may either be a five-stage reduction helical type, or a planetary gearbox. The second stage reduction is a very high torque zone, and usually comprises a pinion and gear wheel combination, with a reduction ratio of around 10:1. In recent years, there is a trend towards mounting a single gearbox, of very high-speed reduction ratio, directly on the extractor shaft. Finally, the motor is driven through a variable speed drive, usually a VFD, for regulation of extraction time and for the purpose of process control.

At times, the extractor is left un-insulated, especially in warm countries, in view of the moderate temperature of the bed. This is basically to avoid the cost of insulation of the large equipment. However, it should be noted that, even in warm conditions, the heat loss from the surface could lead to increase in steam consumption to the tune of 4–8 kg/T soybean, depending on the size of plant. The second problem is the condensation on top cover and upper side casing when left un-insulated, which is a sure cause of corrosion.

15.2 DESOLVENTISATION SECTION

15.2.1 Feed Mechanism

A solvent-tight conveyor (STC), of drag chain type, carries the spent flakes to the DT. Since it must collect flakes from the screw discharger at bottom of extractor and convey these to the top of DT, it has a bend portion and a steep vertical portion at 80–85 degrees angle (Figure 15.10). Only in the case of Loop Extractor type III, where spent flakes are discharged at a height, from the upper part of the loop, the degree may be much smaller. The bend is usually short, because of the limitation of distance between the extractor and the DT. Each segment is of welded construction, to minimise the risk of solvent leakage; the segments are bolted to each other and sealed with rubber gaskets.

Because of the vertical conveying requirement, the STC is fitted with full length central partition plate. Both the plates, the bottom plate on which material is conveyed, and the central partition plate which supports the return chain, are made of wear-resistant material, typically Manganese-steel, which is a work-hardening steel. A critical detail is a diversion plate at the discharge section, which prevents material

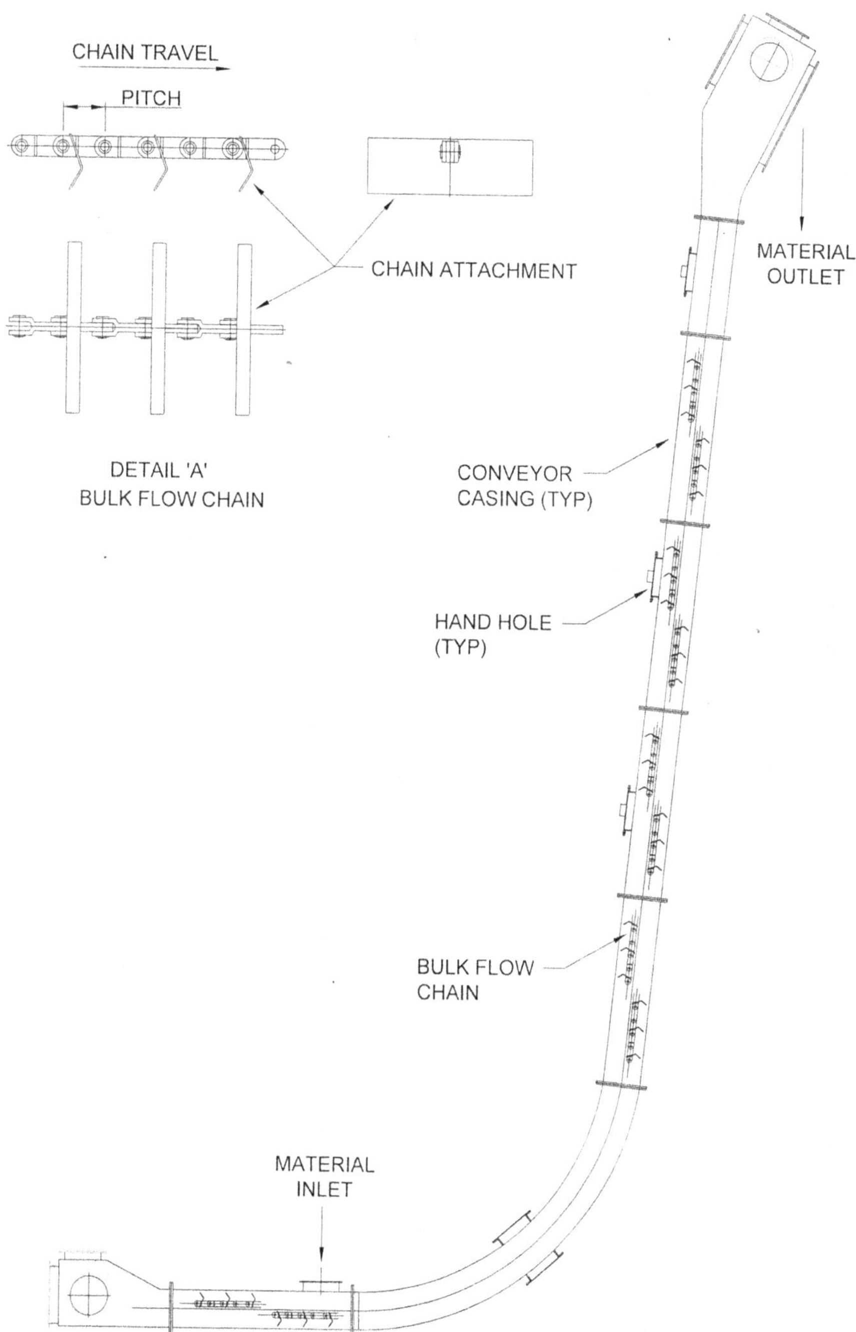

FIGURE 15.10 Solvent-Tight Conveyor.

drop into the return chain. This detail is sometimes missed, causing reduction in the carrying capacity of the conveyor. Because of the vertical take-up, the load on this conveyor is much more than a normal horizontal conveyor. The load should be calculated carefully, for the adequate design of the chain and the drive.

Quite often, STC is left un-insulated, in view of moderate temperature of the marc. However, insulation should be provided, as its absence leads to condensation of vapours, which may cause material deposition, which then would lead to load 'kicks' on the conveyor. The 'kicks' may be minimised by fitting a guide strip below the partition plate of bend portion, to keep the chain lightly pressed to the bottom plate.

The chain is heavy duty, preferably of forged links. In some designs, inverted buckets are fitted as attachments to the chain (bulk flow). This construction makes the chain heavy; such a chain may not be stretched tight; a slightly loose chain scrapes the bottom of the bend portion effectively, and prevents load 'kicks'.

A rotary airlock is fitted in between the STC and the DT, to isolate the latter from the vapour atmosphere of the extractor. The design is similar to the RAL above extractor, except that this one is always square, as there is no requirement to spread the material over a width. The semi-circular lining at the root of vanes becomes critical for this RAL as the material is wet and prone to deposition. A steam jacket should be provided on this RAL, just to keep the casing warm, to prevent deposition of the wet material, especially during the cold season.

In some plants, the RAL is driven by drive of the STC. But, this is not a good idea, since the STC is prone to load 'kicks', which may only become worse with a combined drive.

A sluice valve may be provided additionally, between the RAL and DT, as a safety device for plant maintenance.

15.2.2 DT

As mentioned before, the desolventiser is commonly known as the DT, thanks to the domination of soybean in solvent extraction industry, since the spent flakes of soybean require to undergo toasting along with desolventising. For soybean, adaptations of the Schumacher DT are used worldwide. So, we will discuss this design first, and the 'classical' DT, whose usage is now limited to only a few materials, will be discussed after.

As discussed before, the DT is the most important equipment in the plant for operation costs. The DT is also mainly responsible for quality of meal, which tends to be the main product of soybean crushing, ahead of soybean oil. The DT, therefore, is the focus of attention of the plant operator, and is also subjected to maximum auto-controls.

15.2.2.1 Schumacher Type DT

15.2.2.1.1 The Schumacher Principle

The Schumacher DT, invented by Dr. Heinz Schumacher, was introduced during the early 1980s, and ushered in a revolution in the solvent extraction industry. Thanks to the counter-current stripping principle, the DT became compact, steam usage

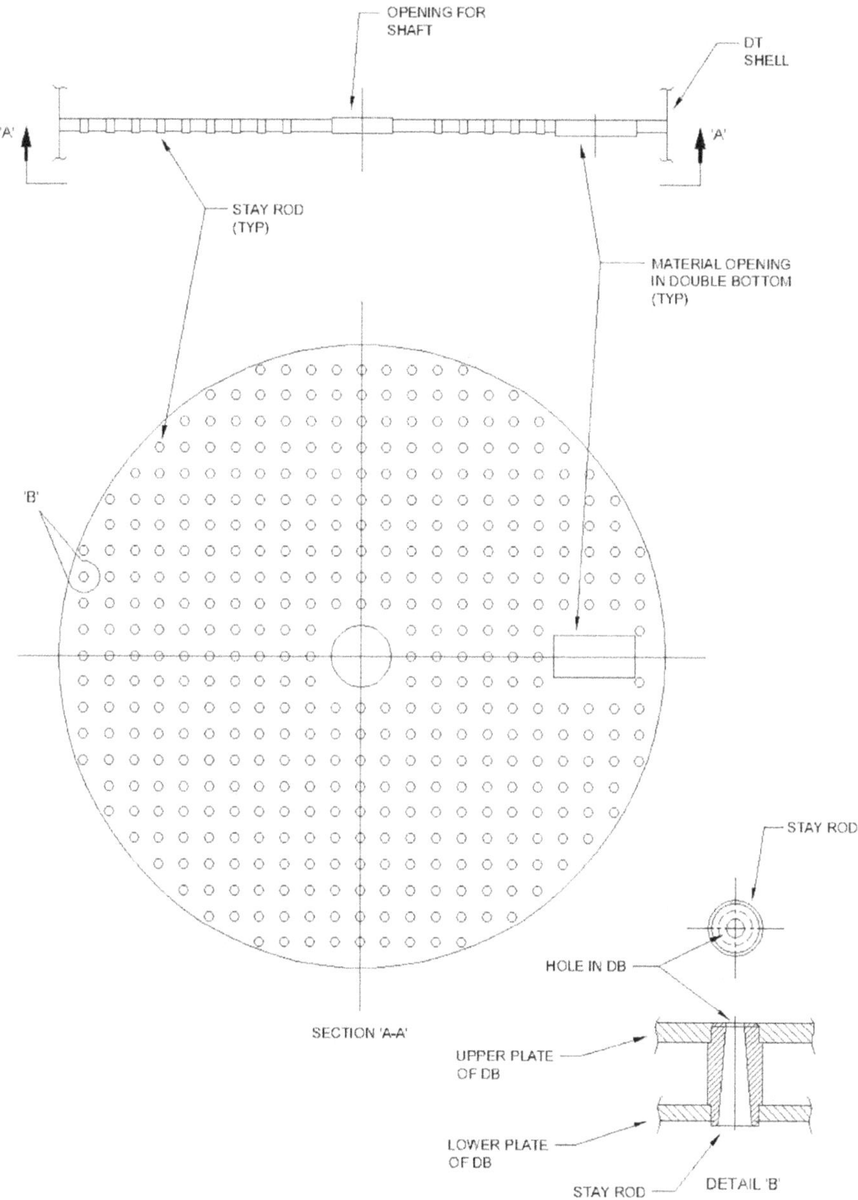

FIGURE 15.11 Schumacher Principle – Holes in Double Bottom.

reduced, and it was possible to produce meal with low residual solvent. The numerous holes on the double-bottom (DB) helped distribute live steam throughout the bed of spent flakes (Figure 15.11). The holes in DBs of upper stages allowed passage of vapour consisting of steam and solvent through the bed in those stages.

15.2.2.1.2 Pre-Desolventising Stages

As the DT became compact, the number of stages reduced. This meant the heating surface reduced significantly. As more live steam was used to strip off the solvent, it raised the moisture level higher than that required for toasting. Thus the load on dryer increased, and so did the steam consumption in dryer. A solution was found by the addition of pre-desolventising (PD) stages, where only indirect heating steam was used to evaporate the first part of the solvent. A typical Schumacher-type DT, consisting of the PD stages is shown below (Figure 15.12).

Since the PDs are in top portion of the DT, the vapours coming from desolventising stages have to be provided a path around the PD stages. This is accomplished by either providing annular space around the stages, or central vapour drums (Figure 15.13). The central drums are prone to deposition of solids and must be fitted with an effective scraper.

Bed height, in PD stages, is typically maintained between 300 and 350mm. Deeper beds offer no benefits since the objective is to transfer heat from the DB only. Evaporation of solvent is a function of temperature and is not time-dependent; hence longer dwell time does not help. Vapour space, as also space for level float mechanism, of say 300–400mm is provided on top of the bed in each stage. So, typically, the clear stage height of a PD stage is 700–800mm. However, much larger vapour space is provided on the topmost stage. This is to avoid carryover of much fines into the outlet gas, as surface solvent flashes off as the marc enters the hot environment of DT. Typically, the vapour space above the bed of flakes in the top stage is as high as 1.2–1.4m.

A sight-glass fitted to the DT wall, at the desired bed height, helps the plant operator have a continuous view of the material bed. Additionally, a level float mechanism helps indicate the level with a pointer outside the DT.

15.2.2.1.3 Desolventising Stages

In desolventising stages (DS), heat is transferred by two means, one by indirect heating from the DBs, and the other by direct steam injection. The live steam injection is the distinguishing feature of the DS stages. The steam as also the solvent vapours pass through numerous holes in DBs of all DS stages. Thus, perforated DBs are the specialty of all the DS stages. As discussed before, the steam is injected below the bottom stage. It rises through the holes in the DB. The live steam does not condense in the bottom stage, since the temperature of the flakes is more than 100C, and strips the last residual solvent from the flakes. As the steam rises from the bed in bottom stage, carrying a small quantity of solvent vapours, the combined vapours pass through the holes in the immediate upper DB and contact the flakes in the upper stage. Here again, some more solvent is stripped off. Thus the gas slowly gets richer in solvent vapours as it rises from bottom to top. The temperature of the gas reduces correspondingly, as the solvent fraction increases.

Most of the steam condenses in the top two stages of desolventising section, where the flakes reach a temperature of 100C. Here the moisture content of flakes rises by nearly 6–8%, depending on the amount of solvent evaporated. In fact, nearly 98% of the solvent is evaporated in the upper two desolventising stages, most of it in the top

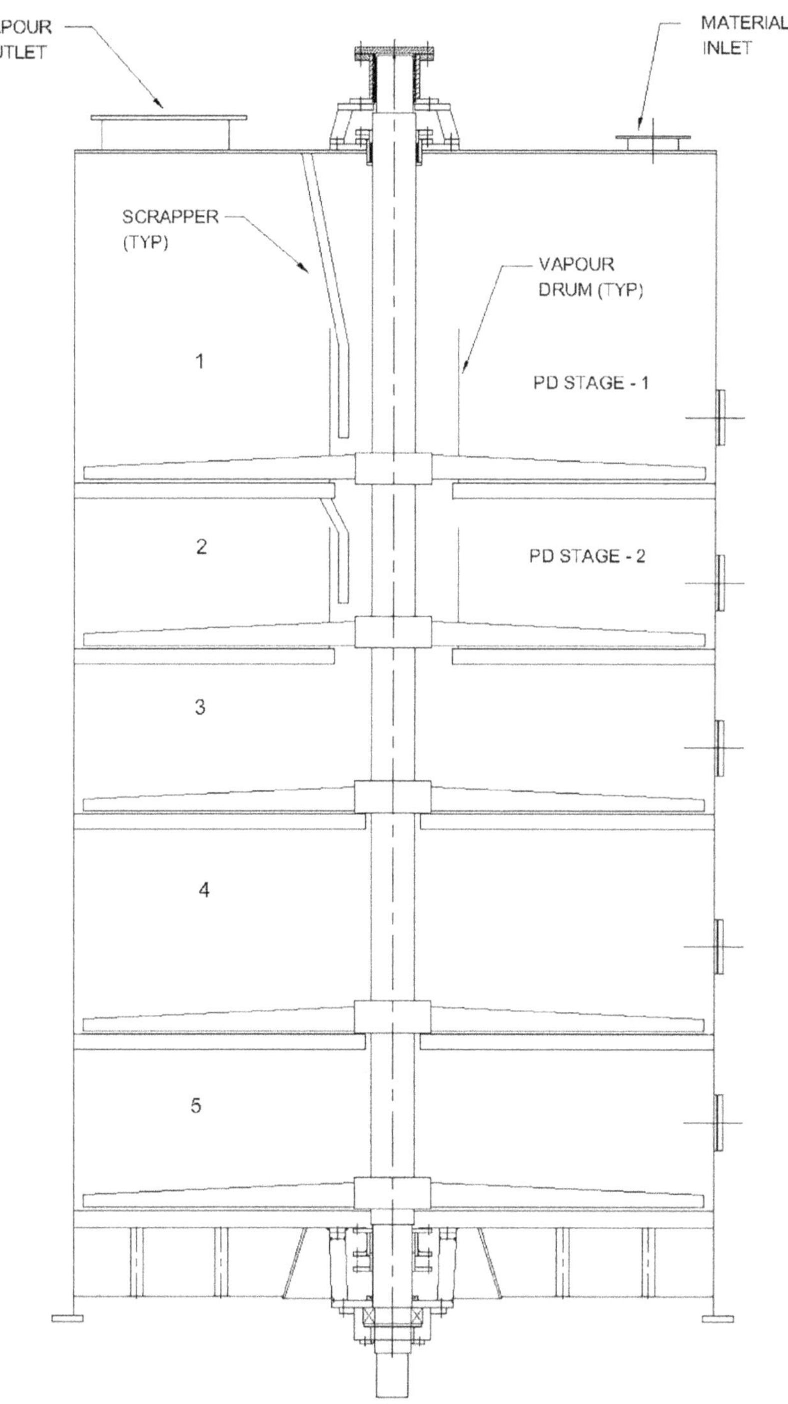

FIGURE 15.12 Schumacher DT with 2 PD Stages.

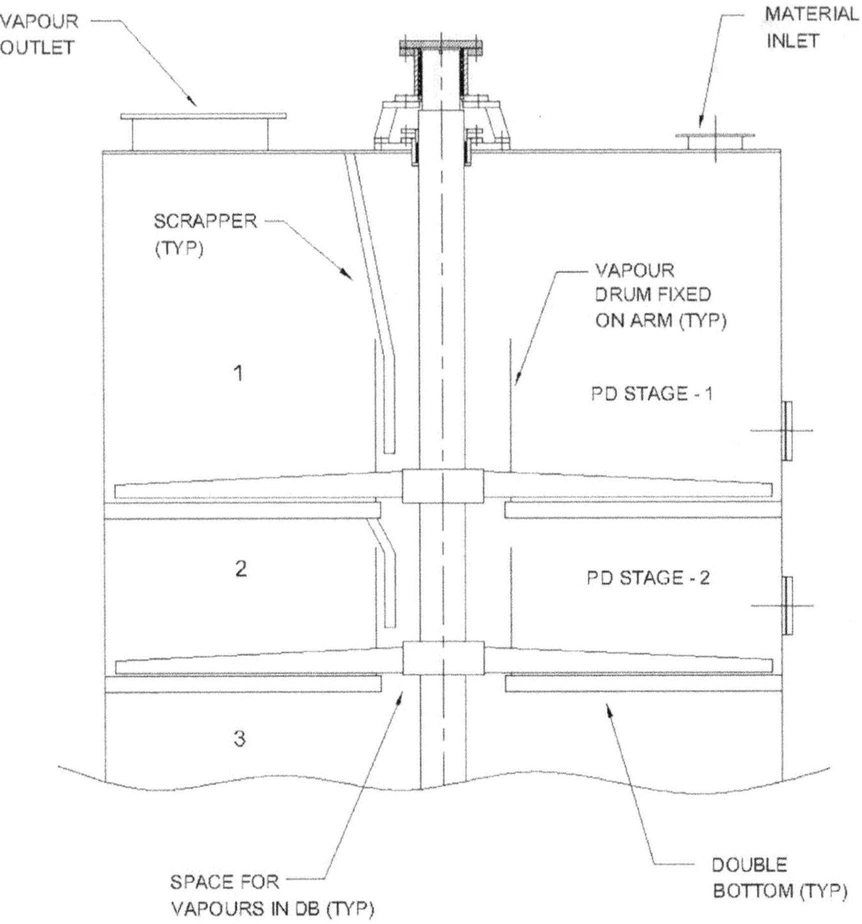

FIGURE 15.13 Central Vapour Drum in a PD Stage.

stage itself. The temperature rises to around 105 C towards the bottom stages, and moisture in meal reduces by about 1%.

Deeper beds are maintained in DS stages, compared to those in PD stages. Deeper beds help improve the contact between the passing steam and the flakes. The deeper the bed, the better the contact. However, deeper bed also reduces the agitation of flakes in the upper portion. Hence, an optimum height, to allow sufficient contact with steam, while ensuring adequate agitation, of around 600–700 mm is usually set in DS stages. If for a given DT, if the dwell time in DS stages is a bit short of the required toasting time, bed height in one of the DS stages may be increased up to 1,000 mm; but, this requires an additional sweep arm fitted about 300 mm below the top of bed, to ensure sufficient stirring of the top layer.

As with the PD stages, vapour space is provided above the bed, which also facilitates the movement of level float, as also of the door mechanism for stage above.

Typically, for bed h of 700 mm, clear height 300 mm may be provided above. Here again, since more vapours are produced in the first DS stage, the vapour space there may be as much as 400–500 mm.

As for the PD stages, a sight glass, fitted at the desired bed height, on each of the DS stages, allows the plant operator to monitor the bed readily. This monitoring is in addition to the level indication by means of a float resting on the bed. The level indication also is utilised in auto-control of bed levels; this is discussed further in the section on DT controls.

A steam chest is provided below the bottom-most DS stage. The height of the chest is usually maintained minimum 350 mm, to allow entry of a workman for any repairs. Steam pressure within the chest is low, 0.5–1.0 BarG, depending on the number and size of holes in the DB above.

15.2.2.1.4 Double-Bottoms (Trays) and Sweep Arms

Double bottoms, as the name suggests, are made of two plates with a spacing to hold the heating steam. As the latent heat of steam is transferred to flakes, steam is condensed, and condensate is removed continuously through steam traps. Float-type traps are preferred, to minimise the loss of steam. Spacing between the plates may be maintained up to 40 mm for DT up to diameter 3 m and 50 mm for larger dia DTs. Should low-pressure steam (pressure below 4 BarG) be used for heating, in place of the normal M.P. steam (pressure around 10 BarG), spacing should be increased to accommodate the higher volume of steam and maintain the velocity low enough as not to disturb the evacuation of condensate. For steam pressure of 3 Barg, spacing of 60 mm should suffice. When M.P. steam is used, the DBs are preferably made from boiler-quality carbon steel plates. For L.P. steam usage, however, lower-grade M.S. plates may be used.

While machined hollow studs are commonly used for DBs of desolventising stages, to accommodate the holes for vapour passage, these also serve as DB stiffeners. However, the DBs of PD stages may simply be stiffened with pipe pieces. In some recent designs of desolventising stage trays, the machined hollow studs are replaced with pie-shaped enclosures capped with slotted screens (Kemper, 2005). The latter design offers much more open area (more than 5%) compared to the design with hollow studs (just 1–2%). The screens may in fact be made from wedge wires.

Among the major oilseeds listed above, rapeseed is the most difficult to desolventise, thanks to the high compaction of the soft core in press, which may cause entrapment of semi-crushed portions in the cake. These semi-crushed portions contain more oil and hold the solvent tightly. Extra effort is, therefore, needed to reduce residual solvent in rapeseed meal. This requires more number of holes in DBs; for example, if 60–70 holes per sq. m. may suffice for soybean and sunflower seed, 100 holes may be required for rapeseed.

It is critical to maintain the straightness of the upper plate of DB to a minimum possible, to avoid material accumulation in 'low' pockets. For a fabrication shop with reasonable facilities, it should be possible to maintain the straightness within 6 mm. This then allows to maintain the clearance between DB and sweep arms to within 15 mm. With this clearance, it is possible to ensure complete sweep of material on the DB, thereby not only maintaining good heat transfer, but also to avoid overheating of any portion of flakes.

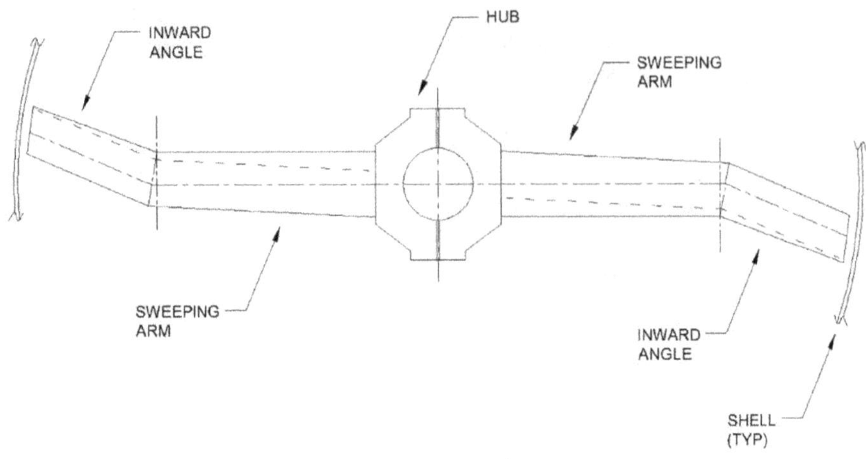

FIGURE 15.14 Sweep Arms with Inward Angle.

Two sweep arms, at 180 degrees apart, are preferred over a single arm per stage. Each sweep arm is attached to a boss, which is made with ID a bit less than the shaft OD, so that the bosses of the two arms, when bolted tightly, have a firm grip on the shaft. The arms may be made simply with a thick straight taper plate, for small DTs. However, for DTs of dia larger than 3 m, an angled arm is more suitable, to avoid higher bed height at wall due to centrifugal flow (Figure 15.14). For cases where live steam is to be injected through the arms, box construction may be used; this is discussed in the section on 'Classical DT'. A box construction also works as a suitable stiffener for long sweep arms.

Angle of sweep arms, to the horizontal, has direct impact on power consumption in DT. Higher the angle, higher is the power consumption. But too low an angle would not create enough stirring of the meal. The optimum range, for adequate stirring with low power requirement, is 15–20 degrees.

Moderate angle of the sweep arm also has another beneficial effect. Since the stirring on the surface is mild, the level float also rises mildly with each sweep. Too much of a 'jump' of the float can cause high wear of shaft bearing, which then may cause leakage of solvent vapours. To reduce the jump, the float also should be designed in a manner that the resting surface is not horizontal, but makes a slight angle, say 10 degrees, with the bed surface. This ensures that the impact of the bed stirring is not taken by the float in an instant, but is spread over longer duration. The float will then rise slowly, instead of jumping with each sweep.

As discussed before, bored studs are provided on all DS stage DBs. As the steam-solvent vapour passes through the studs, it incurs a pressure loss due to friction in the bore as also the entry-loss. The entry-loss may be reduced by providing a smooth chamfer at the entry to the bore in each stud. Reduction in the pressure drop at the entry to studs helps reduce the overall pressure drop, which in turn helps maintain low pressure in the last stage of DS. Such a low pressure directly helps remove last residual solvent in the final DS stage.

15.2.2.1.5 Discharge Mechanism

Just as an effective seal is necessary at inlet of extractor, a good vapour lock is required at discharge of DT, to prevent the escape of solvent vapours to the surroundings. Typical locks are a Rotary Airlock or a Plug Screw. Long RAL, with a length much more than the dia, to ensure proper uniform discharge, is used commonly. The length would depend on the DT diameter. For example, for a DT of dia 3,000 mm, the length of the RAL might be 800, the dia being 300–350 mm, depending on the throughput capacity of the plant.

For a plug screw, one intermediate flight may be cut, to ensure a plug of meal below the discharge gate. The cut flight should be at least one and half pitch away from the gate. Above the screw, a slide gate should be installed for a total seal when DT is not in operation. Such a slide gate also prevents packing of meal in the chute above the screw, when the screw is stopped, say because of low level in the last stage of DT, or for any other reason. The RAL is normally flush with the bottom plate of the last DT stage, hence a sliding gate is not required. The RAL or the screw is driven by a variable speed drive, to help maintain a constant level of bed in the last stage. This is discussed further in the section on control systems.

15.2.2.1.6 Operating Parameters for Oilseeds

This type of DT is used extensively not only for soybean but also for other major oilseeds including rapeseed and sunflower seed. In most cases, saturated M.P steam is used for indirect heating in DBs, typically at 10 Barg. Saturation temperature at this pressure is 185 C; thus a large temperature difference of nearly 90 C average is achieved for heat transfer. The meal is mostly discharged at around 102–104 C, to allow effective stripping of the last part of solvent. The temperature above 100 C also helps prevent condensation of steam on the discharging meal. The temperature of spent flakes coming from extractor is normally 55–60 C. The temperature in PD stages rises slowly, around 5 C in each stage. However, at the first desolventising stage, where part of steam coming from bottom condenses, the temperature rises to almost 85–90 C. By the second stage of desolventising, the temperature is already 100 C. In lower stages, all the heat of live & indirect steam is utilised mostly for the evaporation of solvent, with only a small fraction of heat used for a slight temperature rise of 2–4 C.

Live steam is adjusted so that outlet vapour temperature is 70–72 C for soybean and sunflower seed. For rapeseed cakes, a slightly higher temperature, 72–74 C may be required for complete desolventisation.

The desolventising time, in DS section, of 25 minutes suffices for soybean flakes. However, higher time of at least 30 min is required for cakes of sunflower seed and rape seed. For soybean collets, the time may be even shorter @22 min. The shorter time, combined with higher bulk density of the spent collets, mean capacity of the DT on collets can be higher by nearly 30% over flakes. This is of special interest, since extractor capacity may also increase by 25%, as seen earlier.

The DT is operated at atmospheric pressure. A slight suction is maintained so as to prevent vapour leakage to surrounding. The suction is maintained by a ventilation fan, or a steam ejector, which maintains suction in the entire vapour passage through the DT – Dust Scrubber – Economiser – Condensor – Absorber. The fan speed, or steam pressure of the ejector, is adjusted to have 10–20 mmWc suction (below atmosphere) in the DT duct. The pressure rises in successive DS stages, and the pressure in the bottom DS stage may be in the range of 300–500 mmWc, depending mainly on the number of stages, as also on velocity of gases through the holes in DBs. The pressure may be even higher if cakes, especially of sunflower seed, contain much fines.

15.2.2.1.7 Control Systems

Two sets of control systems are applied to the DT, namely temperature control and level control. Temperature control scheme may consist of two control points: the temperature of DT gases and the temperature of meal at outlet. For the first control, temperature in the vapour duct is sensed, processed in PID controller, and signal sent to control valve for live steam going to the steam chest. It must be noted that the response time of this parameter is long; hence, the PID settings should be set for slow response. It may be noted that the live steam usage may be at least three times the indirect heating steam; hence, this control is critical not only for the effective stripping of solvent from flakes but also for the optimum usage of steam.

The second control is optional and involves sensing the temperature of outlet meal, and adjusting the header valve which regulates the pressure of indirect heating steam in all DBs. Manual valves are provided on all DBs; these should preferably be kept fully open. Here again the response is slow. A pressure control scheme works better, with pressure sensing in the steam line going to DB of the last stage. As the temperature of the outlet meal varies, the meal would absorb more or less heat from the steam-heated DB; the resultant change in steam pressure in the DB can be sensed quickly, and suitable action is taken to restore the pressure.

Levels of all stages are closely controlled, as this alone ensures the desolventising time as also adequate contact of steam with flakes. Level control in the PD stages tends to be simple, in that bed level in the stage determines the opening of the gate below. Gates are usually the rotary cam type. A float rests on the bed, connected to a shaft which extends outside the DT wall. As the level changes, the float moves and the shaft rotates. The rotation may be sensed either by a pressure valve or by an electronic 'rotation sensor'. The pressure signal is conditioned, and an output signal is sent to a pneumatic proportional valve, which opens or closes the gate proportionally. Where an electronic sensor is used, it sends a 4–20 mA signal to the PID controller in the plant control panel, a conditioned signal comes back to the pneumatic valve through an I/P convertor.

Levels in desolventising (DS) stages also may be controlled in the same manner as for PD stages. Alternatively, in some installations, the level in the top DS stage regulates the gate to the bottom DS stage. For the intermediate stages, either fixed chutes are used to maintain maximum height, or reverse-acting cylinders are used to maintain the levels. In this system, level in, say, stage 5 is used to reverse-regulate the damper in the chute above, i.e. chute below stage 4. If the level of stage 5 increases, the damper would close a bit. The advantage of this system is that the DS stages, except the top one, are maintained at full bed height. Then, the live steam cannot by-pass through the gate and is forced to travel through the holes in DBs. However, this control system is more intricate, difficult to tune and requires special expertise.

The last stage is equipped with a rotating discharge device, either an RAL or a plug screw, with a variable speed drive. The control signal changes the speed to maintain the bed height in the stage above, or in the first DS stage, as the case may be.

One peculiar difficulty in level control of various DT stages must be understood. As noted before, the level sensing is from the float which rests on the material bed. As the sweep arm moves below the float, part of the bed below the float moves, the float is temporarily lifted before it rests on the bed again. This movement of float creates a false signal, which must be ignored for control action. When pneumatic proportional valves are used for gate actuation, the air buffer in the valve absorbs the temporary pressure fluctuation. Alternatively, a digital filter may be used to filter out the high-frequency signal. This is a very important aspect, the absence of which may cause continuous 'hunting' of the actuator, which would disturb the control action, apart from damaging the mechanical parts of the actuator, or the drive.

Apart from the above two main control systems, there is another control applied to DT operation. The suction (negative pressure, below the atmosphere) in the duct is controlled by either varying the speed of Vent Fan, or by varying the motive steam pressure of Vent Ejector. The response of suction to motive steam pressure is not linear. Therefore, fan with variable speed drive is preferred for the purpose.

15.2.2.1.8 Recent Innovations

In recent years, a few further improvements have been effected in the design of Schumacher-type DT. First, in the bottom stage, where live steam is injected, the perforated double bottom plate, has been replaced with segments of wedge-wire screen (Figure 15.15) by some machinery manufacturers. This has further improved the contact between steam and flakes and has helped reach even lower residual solvent levels.

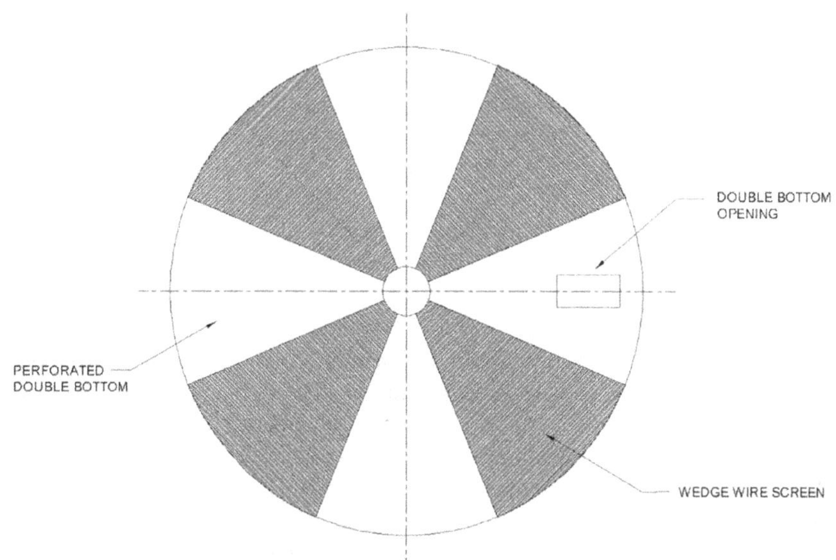

FIGURE 15.15 Last DT Stage with Wedge Wire Screen Bottom.

The second improvement is the reduction in pressure drops through the stages. In most DTs, the pressure drop in each stage was almost 80–100 mmWc (8–10 mBar); thus, for a DT with six desolventising stages, the pressure in the last stage was almost 400–500 mmWc. Since this was the last stripping stage, the pressure affected the stripping efficiency. With recent corrections, the pressure in last stage has been reduced below 200 mmWc (Extech, 2016). This reduction has been achieved mainly with rationalising the vapour velocities through the holes in DBs of the DS stages, and by increasing the open area, as by replacing the hollow studs with pie-shaped enclosures with wedge-wire screens.

The third improvement has come into practice during the last decade and has been patented by Crown (Anderson, 2001) and DeSmet (Van Damme, 2009). The innovation relates to a 'Flash' stage, below the last stripping stage, where the meal is exposed to slight vacuum, to flash off a part of the moisture. This arrangement may also act as a safety for the removal of excess solvent in case of mis-operation of the DT. The stage has a heating DB and provides the additional heat for moisture evaporation. The moisture removed in the stage, along with the steam of vacuum ejector, is used as stripping steam for desolventising. Thus, the net steam consumption for heating and vacuum is zero. Load on the dryer is reduced, resulting in the reduction of overall steam consumption in DTDC. Should there be any fluctuation in temperature control, or level controls, leading to excess residual solvent escaping the stripping stages, the flash stage would act to suck back the excess solvent.

15.2.2.1.9 Design Considerations Heat and Mass Balance Model:

The size of the desolventising section is determined by the time required for desolventising and toasting. For soybean, for example. With good contact between live steam and flakes, the desolventising may be accomplished within 25 min. That time period also suffices for toasting, should moisture content be 17.5% or more. Thus, the size is first determined by the time requirement, and moisture content is computed for the configuration. If the moisture content is much higher than that required for toasting, pre-desolventising stages may be added as necessary. Let us take an example for a soybean plant of 600 Tpd capacity.

Daily capacity of 600 T means hourly 25 T, which after de-oiling becomes 20.5 Tph. For 25 min, with average bulk density of 0.5, the volume requirement becomes 17.1 m^3. Let us select a DT of dia 3 m. Then total bed height requirement, with a filling factor of 0.95, works out to be 2.54 m. This can be achieved in four stages, with bed height of 650 mm in each stage. For safety, add 10%, we may take bed height as 700 mm. Should we consider collets, in place of flakes, the bulk density may be taken as 0.6. Also, desolventising time may just be 22 min. In that case, for same dia, height requirement reduces to 1.86 m. This may be achieved, including 10% safety, with only three stages, with bed height of 700 mm each.

The moisture-in-meal calculation, for the above configuration is shown in Appendix 2.2.

It is seen that with 4 DS stages, the moisture-in-meal will be 18.6%, which is high, slightly above the required range of 17.5–18.0% for toasting, and would add unnecessarily to the drying load.

As we repeat the calculations with an additional PD stage, the moisture-in-meal is seen to reduce to 18.0%, which is within the range. Thus, the configuration selected, for the 600 Tpd soybean (flakes) capacity, is Dia 3,000×(1 PD+4 DS stages). We may note that the meal moisture may further reduce to 17.4% if two PD stages were added instead. However, that may not be desirable, since the toasting time at 17.4% moisture would need to be slightly longer.

The calculation also yields the quantity of live steam to be injected in DT. Whereas, in the first case, with only four DS stages, it was 2,413 kg/h; with the addition of a PD stage, the requirement reduces to 2,260 kg/h. Thus, the additional heating steam input to the PD stage (176 kg/h) is nearly balanced with less live steam (153 kg/h). The difference in the quantities is due to the higher heat utilisation from live steam (550 kCal/kg) against only the latent heat from the heating steam (475 kCal/kg).

Similar calculation may be conducted for the case of 600 Tpd Soy collets.

Readers will be able to perform these calculations for any processing capacity for soybean, and indeed for any other oilseed. Correct bulk density, and moisture in cake, and the desolventising time, should be used for practical results.

It may be noted that, in the above calculation procedure, we presumed the desolventising time, and the DT vapour temperature, based on experience. Let us now examine if these could be determined with the help of theoretical models of the desolventising process.

Theoretical Models: Cardarelli et al. (2002) have presented a mathematical model, based on solvent diffusion from pores of flakes, which relates desolventising time and steam velocity to residual solvent. The 'effective diffusivity' value is assumed for the calculations, based on a plant operation data. However, the model predicts high residual solvent values even with high steam injection rates, and also with low starting solvent values. The assumed value of effective diffusivity of $4.0\times10^{-10}m^2/s$ may be increased to get practical values. In any case, a method has to be found, either experimental or theoretical, to determine the diffusivity; without this, the model will be of little value to predict the performance of new oilseed materials.

Another model has been presented based on stage-wise material and heat balance (Paraiso et al., 2008). An ideal stage was considered as equivalent to change of flakes temperature by 1 C. Predicted values of meal moisture and residual solvent matched well with those observed in a running plant, capacity 1,200 Tpd soybean. The main limitation of this model is the lack of relation of ideal stages with time or mass flow of flakes. Without such a correlation, it is not possible to calculate the total residence time required for desolventising, and hence the DT size. Also, the solvent sorption data are taken from the plant. That again makes the model applicable only to soybean of certain quality.

So, studies are required to determine the diffusivity of solvent from flakes based on experimental procedure. Only then will it be possible to predict the performance of DT, and predict desolventising time with a theoretical model. The desolventising time is the most important parameter for design of a DT.

Cardarelli and Crapiste (1996) have presented sorption/desorption isotherms for soybean and sunflower meals, using nitrogen as the stripping vapour medium. As expected, the 'activity of hexane is lower at lower temperature and at lower hexane content in meal. They have presented results from another study (Roques et al., 1984) conducted on rapeseed meal, and compared the desorption data of the three meals. The 'activity' of hexane (akin to 'volatility' in a liquid mixture) from rapeseed meal is seen to be much lower than the other two meals. This agrees well with the industry experience that it is much more difficult to reduce the residual solvent in rapeseed meal, compared with other meals. However, it is interesting to note that the activity of sunflower meal is higher than that of soybean meal. This, in principle, should allow to reach the required residual solvent level in sunflower meal with somewhat less live steam, compared to that required for soybean meal. In effect, it should be possible to achieve complete desolventisation of sunflower meal with a lower exhaust vapour temperature from DT. This, however, does not match with the industry experience, where nearly the same proportion of live steam (at equal DT vapour temperature) is required for both the meals.

Power calculation: Absorbed power in each stage may be calculated by the following empirical formula:

$$P_{stage} = f_p * \text{no. of sweep arms} * D^2 \tag{15.1}$$

where, f_p is the power factor and D is the stage diameter (m). Power is then estimated in kW.

The power factor, when angle of sweep arms is 15–20 degree, and shaft speed is around 15 rpm, may be considered as follows:

PD stage (bed h 300–400): 0.4
DS stage (bed h 600–700): 0.6
DS stage (bed h 1,000): 0.75
DC stages (bed h 500–600): 0.5

Absorbed Power for the DT may be computed by summation of absorbed power for individual stages. Thus, for the example of DT for 600 Tpd soybean flakes, with two arms per stage, the absorbed power (for 1 PD stage and 4 DS stages) may be computed as:

$$P_{DT} = (0.4 * 2 * 1 + 0.6 * 2 * 4) * 3^2 = 50.4$$

For this much absorbed power, a motor of 55 kW may be too tight, considering fluctuations in load on DT. Hence, motor of 75 kW, the next higher standard rating, may be selected for the drive. Gearbox may be selected with a safety power factor of 1.8 on the absorbed power.

15.2.2.2 Classical DT

While most of the major oilseeds are processed through the Schumacher type DT, some others, either those which contain fines such as palm kernel and rice bran, or those which are prone to lumps formation with too much live steam, such as groundnut, are still processed through versions of classical DT.

The main difference between the two designs is in system of live steam injection, and vapour flow within the DT. While in Schumacher type, all the live steam is injected below the last desolventising stage, and vapours pass through successive beds in upper stages in counter-current fashion, the steam in classical design is of cross-flow pattern, and so is the vapour flow. Thus, in the latter design, live steam is injected in multiple stages, and vapour from lower stages does not contact the flakes in upper stages (Figure 15.16). The vapours from all the stages travel either through

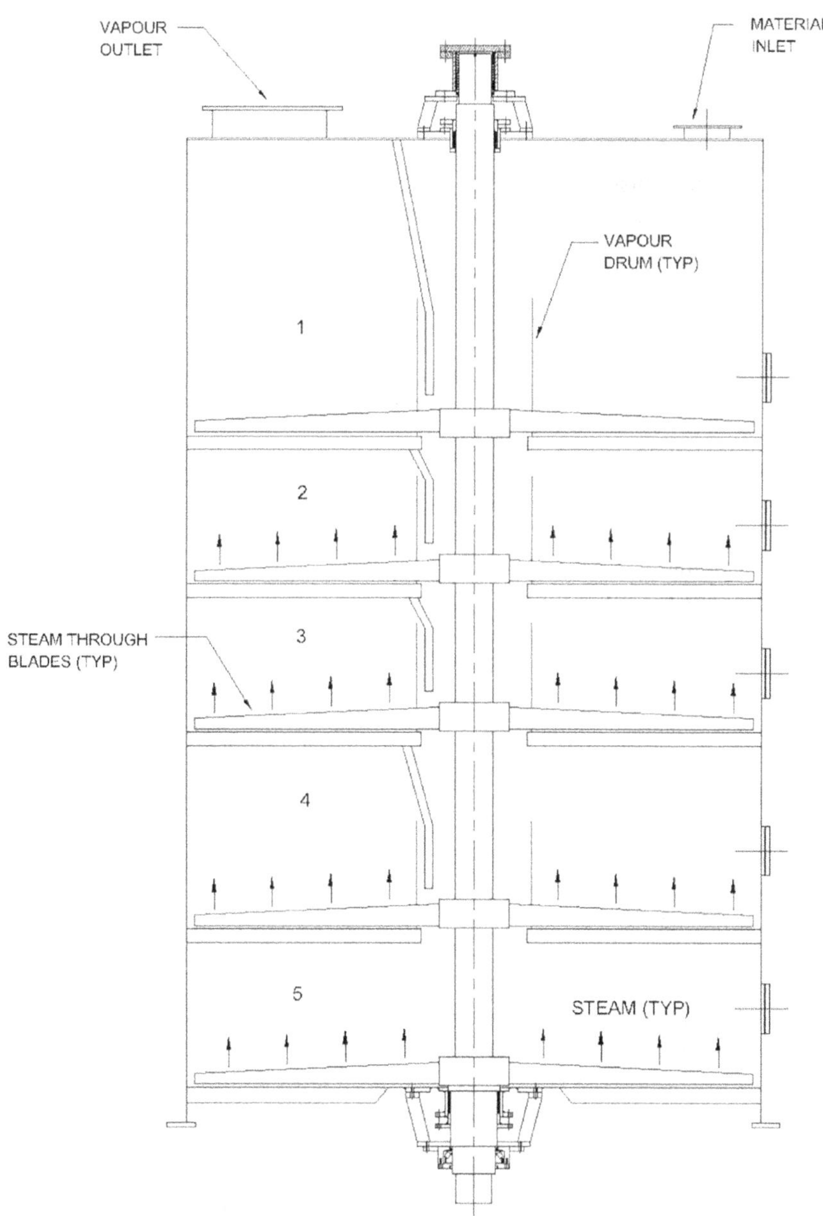

FIGURE 15.16 Classical DT – Steam Injection in Multiple Stages.

central vapour drums in, or annular spaces around, all upper stages. Thus, the stages are similar to the PD stages of a Schumacher DT, but with additional provision for steam injection. Obviously, there are no perforated DBs in the classical DT.

For rice bran, a modification in the classical configuration has been adopted (Modi, 2012), wherein counter-current flow of vapour has been adopted in the last two stages, as shown in Figure 14.14. Such a flow in more number of stages is not permitted, since the pressure drop in each stage is high, due to the fines in spent RB. This adaptation has reduced the overall steam consumption in the DT. Such a cross-design has not been successful for spent flakes of palm kernel, which packs the bed even more compact causing even higher pressure drops.

Since steam is injected in almost all stages, and vapours taken out from individual stages, the steam consumption is much higher, as is reflected in higher vapour duct temperature, often around 78–80C. It may be noted that fraction of water vapour in DT gases at 80C is as high as 18.4%, compared to just 9.4% at 70C. Thus, excess steam, as much as 9% of w/w solvent, is 'wasted' in this design. Rather, this is the kind of steam saving in Schumacher design. It is another matter that some of this excess steam may be used beneficially in distillation section; but that may be possible only if the miscella concentration at distillation feed is lower than 25%.

Material bed heights in the stages where live steam is injected may be maintained around 400–500mm. Deeper beds, as the DS stages in Schumacher DT, do not help, since the small quantity of steam injected in each stage rises slowly and gets sufficient contact time in moderate bed height. It may be noted that less bed height means less load on drive motor. Should live steam be not injected in upper stages, the bed height should be maintained lower @ 300–350mm, as with PD stages of Schumacher DT.

The mechanisms for feed of spent flakes, as also for discharge of meal, are similar to those used in the Schumacher DT. The level control mechanisms are simpler, since there aren't multiple desolventising stages with counter-current vapour flow. All the stages, except the last, are subjected to direct proportional control, mostly with pneumatic control valves. Level control on the last stage is similar to that in Schumacher DT, with speed regulation of the discharge device. In rice bran industry, many plants still adopt the mechanical lever mechanisms of yore. However, these are prone to mechanical failures, and result in higher solvent losses, due to leakages through seals of the shafts of level float as also of gate mechanism.

The construction of double-bottoms for Classical DT (CDT) is simpler than those in Schumacher DT (SDT). Since there are no holes in bottom, specially machined studs are not necessary, and simple pipe stiffeners, spaced wide, are sufficient. In short, the construction of DBs is like those of the PD stages in the SDT. This means that the cost of a CDT is much lower than a SDT of similar size. The requirement of straightness of the upper plate of DB, and the clearance under sweep arm, is the same as in SDT.

Live steam injection in each stage is preferably done through the sweep arms, which are fed from the hollow central shaft. For this purpose, sweep arms may be made of a box design (Figure 15.17). Steam nozzles may be fitted on the rear face of the arm. Alternatively, a pipe may be run along the length of the arm, fitted with steam nozzles. Locations and spacing of nozzles should be decided so that the flakes get uniform steam contact across the radius. Since the sweep perimeter increases along the radius away from shaft and towards the DT wall, more nozzles are to be

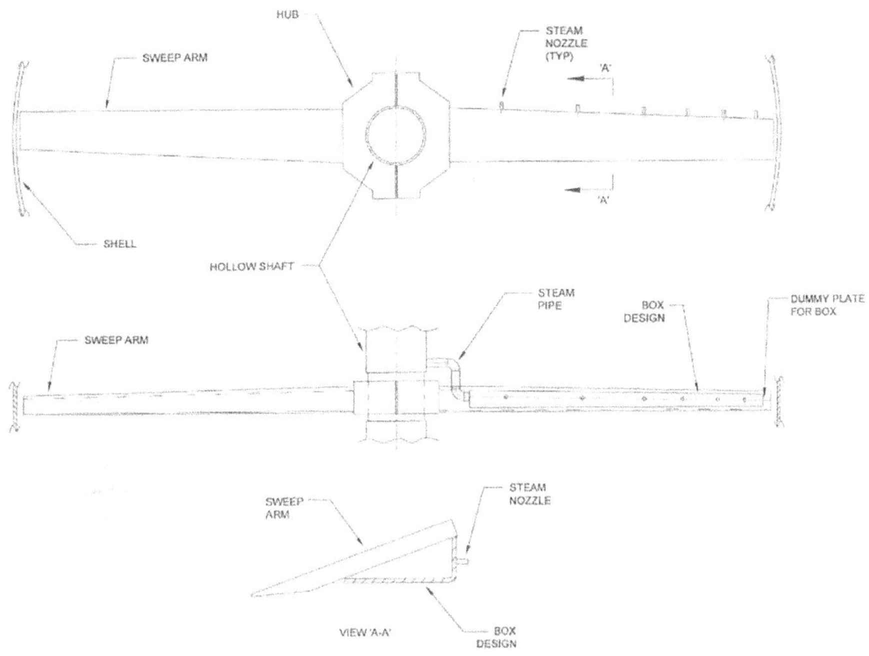

FIGURE 15.17 Steam Injection through Sweep Arms.

located towards the wall than near the shaft. The spacing between nozzles would, therefore, reduce progressively towards the wall.

While these DTs are in wide use for groundnut cake at present, a shift to the Schumacher DT may be expected, once preparation of GN improves, with adequate cooking of flakes.

Out of the three recent innovations listed for the Schumacher-type DT, the first two are specific to that type. However, the third one, the Flash Chamber, may readily be used for classical DT, provided a good seal gate is fitted on the last DT stage, such as an RAL or a Plug Screw. This would help reduce moisture content of meal and serve as a safety against accidental high solvent residual in meal.

In the absence of counter-current flow, the time of desolventisation is longer, compared to the Schumacher DT. Thus, there are more stages, hence more heat surface area. As a result, the moisture in meal is always lower than 14%. This means, the meal may only be cooled with ambient air (or mildly warm air, in case of high humidity), which would reduce the moisture below 12%. No drying operation, with hot air, is necessary.

Design methods are similar to those for Schumacher DT. Higher desolventising time, and lower bed heights mean more number of stages than an SDT. The steam requirement, and the meal moisture, may be calculated by the method described for SDT. Power requirement may be calculated using Eq. (15.1), with f_p of 0.4 for upper stages where live steam may not be injected, and 0.5 for lower stages with live steam injection.

15.2.2.3 Material of Construction, and Drive, of DT

As in case of extractors, for large capacity plants, there is a tendency towards building entire DT casing in stainless steel. However, this is limited only to a few plants, and in most other cases, the casing is constructed of mild steel. The top cover and the casing of top stage(s) may be built of SS, or these may be lined with SS sheets, since it is the top stages which are subjected to most corrosion due to the condensation of volatile gases of sulfur compounds at low stage temperature. For corrosive gases coming from rapeseed and palm kernel, casing of the upper stages, including all PD stages and the top DS stage, may preferably be lined with SS sheet.

Inside fittings, such as central vapour drums and vapour baffle, should be built in SS, to prevent corrosion and need for maintenance. The level floats, as also the gates, either cam gates or chutes with dampers, are also built in SS for trouble-free service.

The double bottoms should be constructed with boiler quality plates, when M.P. steam is used for heating. For L.P. steam, which may be used when the crushing complex has a co-generation facility, lower grade mild steel plates may be used.

The central shaft is usually made of carbon steel hollow pipe, and end pieces are machined out of high-tensile steels. The sweep arms are preferably built of wear-resistant steels. A simple option is work-hardening steels, such as Manganese-steel. The boss of sweep arms is made of cast steel, and these are held tightly on the shaft with high-tensile bolts. Other fasteners inside the DT should all be of SS.

The DT shell temperature is mostly above 100C, hence the DT should always be well-insulated.

Drive of a DT is always mounted below, supported on the ground floor of the extraction plant. The motor is the biggest in the plant, and so is the gearbox, thanks also due to large reduction ratio. Apart from the motor and the gearbox, the couplings also are important components. The gearbox is preferably a helical, three or four-stage reduction, with the last reduction component being a bevel gear for 90 degree change in direction, from a horizontal input shaft to a vertically upwards output shaft. The shaft speed being in the range of 12–20 rpm, with a motor of 1,500 rpm, the reduction ratio is in the range of 75:1 to 125:1. The gearbox torque rating may be selected for a safety factor of at least 1.65 (or, 1.8 on absorbed power), in view of possibility of sudden rise in torque requirement if bed height was to rise in any stage, in case of failure of a level control.

A good flexible coupling should be provided between the motor and gearbox, which should not only provide a flexibility of alignment, and safety for motor, but also provide a low-torque start of DT when it is full of material. These conditions are met by a fluid coupling. Alternatively, a soft-starter may be provided in the control panel, but that is an expensive solution. The coupling between the gearbox and the DT shaft has to have high torque rating, and moderate flexibility for alignment; a ‘gear’ coupling is a good choice for the purpose.

In large DTs, a small **ancillary drive** may be provided to allow emptying of the DT in case of power failure. The power rating of the drive may typically be 10% of the main drive, with a higher gear ratio to run the DT sweep arms at a very low speed (nearly 1/10th of nominal speed), just to empty out the DT within a few minutes. Such an arrangement prevents overheating, and charring, of material in the DT in case of power failure. Such a drive, a gearmotor of output speed around 150 rpm, is connected to the main gearbox, through a clutch which may be put on when the drive is to be used, using a stand-by generator.

15.2.3 Dryer Cooler

As mentioned before, meal drying and cooling used to be carried out separately from the DT, usually in the meal bagging building. But since the 1980s, the DC started to be combined with the DT, in what is known as the DTDC. The advantages of a combined DTDC over the separate DT and DC have been listed in the section on Schumacher DT. We will discuss the combined design first and then describe the separate DCs.

15.2.3.1 DC in a DTDC

The DC is usually built as a two-stage section below the DT, in a single vessel, with a combined agitator shaft and combined drive, as shown in Figure 14.10. In some cases, if DT meal moisture is higher than 18%, usually only for soybean, two dryer stages may be used. For most cakes, such as those of rapeseed and sunflower seed, the meal moisture is lower than 14%, and a dryer stage may not be required. Thus, only a DTC would suffice in such cases.

The dryer stage receives hot and moist meal from the DT through the discharge RAL. Meal level is maintained between 600 and 800 mm, sufficient for dwell time of 6–8 minutes. However, the dwell time may be higher depending on quantity of hot air required for the drying purpose; this is discussed later. Hot air is blown through the meal through holes in upper plate of air chest below. Warm moist air, with moisture from meal, is let out at the top of stage. Hole size and spread is determined so that the pressure drop through the holes and through the meal together does not exceed 400 mmWc. Typically, holes dia 5–6 mm, at triangular pitch of 50 mm may suffice for the purpose.

Since the dryer stage is fitted just below the last desolventising stage, it is important to have minimum air pressure above the bed, so as to minimise the risk of air entry into the DT. For this purpose, two balancing fans are provided, one FD fan and an ID fan. The ID fan is designed to only suck the air out through the outlet duct and downstream air cyclone, so that the pressure above the bed be maintained close to zero.

Process control involves control of air temperature, level control and pressure in outlet duct. Temperature control is a simple feedback loop, sensing the temperature and adjusting the opening of steam control valve to the air heater (radiator). Level control is similar as for the last stage of DT. The level float position is sensed as the rotation of float shaft, by a rotation sensor giving out 4–20 mA signal to the controller. The conditioned signal adjusts the speed of discharge RAL to maintain the level. Air pressure in the outlet duct is sensed with a pressure transmitter, conditioned in a controller, and the speed of ID fan regulated to maintain a slight suction (of few mmWc) in the duct.

For incoming meal moisture range 16% or more, as for soybean, hot air at 130–140C is utilised. The quantity of air required to reduce moisture below 12.5% is calculated using the air-water psychrometric chart. For incoming moisture range of 14–16%, air temperature of 80–100 C should suffice. It should be remembered that higher the air temperature, lower is the air quantity required. For incoming moisture below 14%, the moisture removal may be accomplished by cooling air alone, utilising the heat of meal for evaporation; in such a case, a Dryer stage may be avoided. However, should ambient air be humid, with relative humidity at 60% or more, then moisture removal from meal becomes difficult; and it would require very large quantities of air, which means high fan power. Then, it is better to slightly heat up the incoming air to say 50–60 C.

Once the quantity of hot air is determined, the diameter of the dryer stage may be determined for a nominal air velocity of 1,500 $m^3/h/m^2$. At this velocity, there is sufficient fluidisation of meal for effective contact. At higher velocities, carryover of fines with outgoing air would increase. If the determined diameter of the dryer stage is within the dia of DT, or slightly bigger, one dryer stage would be ok. Should the required dia be much bigger than the DT dia, two dryer stages are preferred. Normally, for incoming meal moisture up to 18%, a single dryer stage would suffice, subject to low ambient air humidity.

The Cooler stage operates, and is constructed, similarly as the Dryer stage, except that usually it is fitted with only an FD fan. Part cooling is accomplished as the evaporating moisture, 1–2%, takes the latent heat from the meal. The balance cooling is accomplished by transfer of the sensitive heat to air. Since the air flow is somewhat counter-current to the flow of meal, air outlet temperature may be much higher than the temperature of outgoing meal. With higher temperature gain, air requirement reduces. The gain increases with longer contact time with meal.

Power requirement for the DC stages may be computed using Eq. (15.1):

$$P_{stage} = f_p * \text{no. of sweep arms} * D^2$$

where, P is in kW, f_p is the power factor and D is the stage diameter in metre.

As discussed in section on DT design, f_p for DC stages may be considered as 0.5. So, for the two stages, with dia 3 m,

$$P_{DC} = (0.5 * 2 * 2) * 3^2 = 18$$

Thus, the DC stages would absorb additional 18 kW power. Together with the power absorbed in DT stages (see section on DT), the total power consumption is 68.4 kW. Motor of 75 kW may not be sufficient to take load fluctuations, hence, motor power may be selected at the next higher standard rating, i.e. 90 kW.

15.2.3.2 A Separate DC

When the DC is built as a separate unit, it may either be designed as a vertical unit, similar to the DC part of the DTDC, or as a horizontal slow moving screen unit, as in a cake cooler. A vertical DC works exactly as described above, except that a separate feed valve is required, to receive the hot moist meal from meal conveyor. The latter should be built similar to the RAL at DT feed.

The horizontal Dryer-Cooler is described in the chapter on the preparation of soybean collets (Chapter 12). The difference here is much more moisture is to be removed. So, the air temperature is maintained higher, up to 140C, as also more air quantity would be required. Thus more Dryer modules would be required in the assembly of the modular DC. Drying time up to 10min would be required, somewhat higher than that in a vertical DC, because of shallow bed of 200–250mm.

Because of lower heat transfer efficiency with shallow beds, more air flow is required, hence the steam requirement for air heating is higher than for a vertical dryer. However, electric energy required for air blowing is nearly the same, even with somewhat more air quantity, because of lower head required for fans for the shallow bed. The power for bed movement is much less than for the vertical DT. However, the power saving does not balance the extra steam requirement. Hence, the vertical DC is the preferred equipment.

A separate DC may either be installed within the SEP or in another building, together with meal bagging section.

15.2.3.3 Air Cyclones, Bag Filter and Air Heater

The outgoing air from the dryer and cooler stages contains some amount of fines. Air cyclones are used for the separation of the fines. The cyclone for the dryer stage must be steam-traced and insulated to prevent deposition of the wet solids on cyclone wall. This cyclone is preferably constructed in SS, to prevent corrosion with the wet air. Air duct from dryer stage to the cyclone, as also out of cyclone, should also be in SS. The duct up to cyclone must be insulated.

Cyclone on the cooler stage may be constructed of carbon steel. In fact, this cyclone is preferably replaced with a bag filtration unit, so that air pollution is eliminated. Air bag filters cannot be used for dryer stage, as the wet solids tend to stick to the filter bags.

The commonly used air cyclones are of intermediate pressure design and are designed to remove particles up to 15–20 micron size. Smaller fines escape to the atmosphere and cause air pollution. To minimise the pollution, from the dryer stage, high-pressure (high efficiency) cyclones, with smaller diameter and longer taper portions, should be used. These are designed to remove particles up to 10 micron.

Fines from cyclones or bag filters are discharged with RAL fitted at the bottom. Some cyclones come fitted with a buffer hold-up above the RAL, to ensure that the RAL does not run empty, and the air seal is maintained.

The **air heater** is usually a steam radiator type, with external fins on tubes to increase the heating surface area as also to improve the contact of air with hot surface. These should be designed with a condensate heat recovery module, with the steam condensate heating the incoming air, to extract maximum heat from steam. Air heaters should be constructed fully in SS. Care must be taken to have clean air coming into the heater. Should the incoming air come from a dusty atmosphere, fines would deposit on the fins, and over time, as these get heated, could catch fire. Square fins, instead of helical ones, are preferred for ease of cleaning.

15.2.4 Dust Scrubber

DT vapours always carry some fines. The fines content may be reduced by means of suitable diverters in the vapour passage inside DT, and by lower exit velocity with bigger vapour nozzle, but cannot be eliminated. As the gases are sent to the shell side of the 1st Evaporator, the Economiser, in distillation section, the fines may stick to the outside surfaces of tubes, and reduce the heat transfer capacity. The fines, therefore, must be scrubbed off upstream of the Economiser.

The fines are scrubbed either by solvent or by water. In the first case, in order to minimise the solvent requirement, fines are first separated with a centrifugal action in a cyclone, and the outgoing vapour is contacted with a solvent spray. The solvent, with scrubbed fines, is sent back to the DT. For ease of access, the cyclone is mounted on top of DT. This system increases the heat load on DT and hence is not practised commonly. The other system, water scrubbing is more commonly used worldwide.

A typical Dust Scrubber, using water, is of vertical cylindrical design, where vapours may travel co-current or counter-current to the spray of water (Figure 15.18). The co-current configuration entails lower pressure drop in the gas flow. Multiple spray nozzles may be installed in the vapour path to ensure complete contact. Water flow should be such that it is able to scrub all the fines. At the minimum, two nozzles should be used, one a full-cone type, and the other a hollow-cone type. A full cone nozzle covers entire area below, but the flow is sparsely distributed. On the other hand, hollow cone nozzles only cover the periphery, like an umbrella, but with a thick water curtain (Figure 15.19). Thus, a full-cone nozzle is useful for catching more quantity of fines, while a hollow-cone nozzle is useful as a safety last contact to ensure that the small remaining quantity of fines do not escape un-contacted. To benefit from the qualities of both the types of nozzles, the first contact with vapour is made with a full-cone nozzle, and the later contact with a hollow-cone. Should the fines quantity be high, more than one full cone nozzles may be used. Sight-glasses fitted at the level of the spray nozzles help monitor the uniformity of the sprays.

The diameter of a Dust Scrubber may safely be determined for vapour velocity around 1.5–1.8 m/s. It may be noted that the lower the velocity, the longer the dwell time, and the better the chance of fines removal. However, too low a velocity can only be achieved with a very large diameter, with increasing equipment cost. Also, the larger the vessel diameter, the higher the water flow requirement. Hence, an optimum vapour velocity is chosen.

Dust Scrubber, also referred to as Dust Catcher in industry parlance, should be made in SS grade 304 or higher, to prevent corrosion on contact with hot water.

The hot water is pumped from a tank, known commonly as the **Dirty Water Heater**, and goes back to the tank from the Scrubber, in a closed loop. Water temperature in the tank is maintained around 80C by steam heating, to remove solvent associated with the fines. The solvent vapours pipe is connected to the DT condensor. As fines accumulate in the water over time, a small portion of water is bled off, and fresh water taken in, to maintain fines at low concentration in water. Water pumping rate is maintained at 1.5–3 times the DT vapour flow by weight, with lower ratio for higher capacity plants. The small quantity of bleed water may be drained to a sand pit, and left for drying. Multiple pits allow scraping off the fines deposited on the

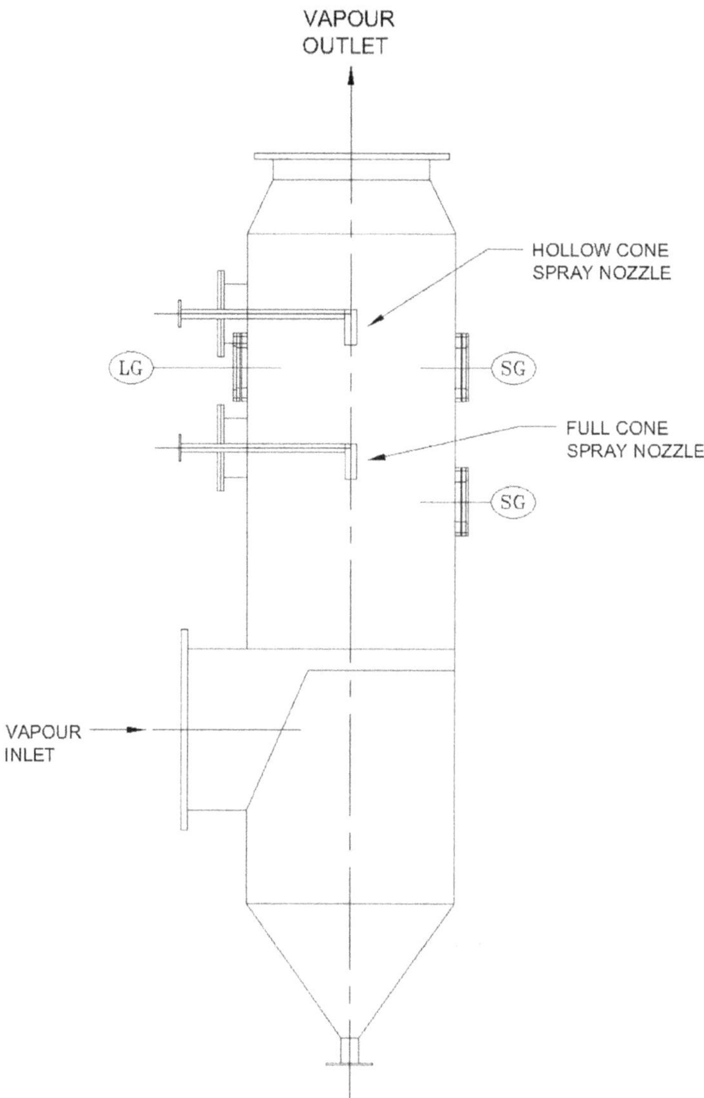

FIGURE 15.18 Dust Scrubber (Counter-Current Flow).

sand. Bleed water quantity may be nearly 0.5 L/T. Thus, for a 600 Tpd plant, daily bleed-off may just be 300 L.

15.2.5 Steam Economiser

This equipment may either be considered a part of the desolventiser section, or a part of distillation section. This is so because while it is rightly the first evaporator, it uses

FIGURE 15.19 (a) Full Cone Spray. (b) Hollow Cone Spray.

the DT vapours as the heating medium. Thus, it is a part of the path of DT vapours, through to the DT condensor; and hence, is discussed in this section.

An Economiser is typically a shell-n-tube equipment. The miscella flows inside the tubes, and the DT vapours are in the shell side. The tube side is under vacuum, whereas the shell side vapours are at atmospheric pressure. Recently plate type exchangers have been tried for this application, but these have not been accepted widely.

Economiser tends to be the largest equipment in distillation line, since it is responsible for up to 90% solvent evaporation. Also, the area of this exchanger, per unit kCal exchanged, is large compared to the second evaporator, because of lower temperature difference between the hot and cold streams. In the second evaporator, the temp difference is larger, thanks to the higher temperature of steam, the heating medium, usually at 140–150C, compared to the temp of DT gases @ 70–75C.

Economisers may be designed as either 'rising film' type, or as 'falling film' type. The rising film design is more common, mainly due to the 'natural' flow pattern on both sides; the miscella rising, as it gains temperature, with the rising solvent vapours; and the DT vapour flowing downwards as they are cooled and condensed. The following discussion refers to the rising film design, although it is equally applicable to the falling film type.

For the rising film Economisers, a flasher is mounted on top, to separate the solvent vapours from concentrated miscella (Figure 15.20). The flasher, when mounted on top, is fitted with suitable baffles to prevent liquid entrainment with the vapours. Alternatively, the flasher may be mounted on the side, with tangential entry of the miscella-vapour mixture, for effective separation without baffles. However, the flasher mounted on top saves floor space, hence is preferred over the side flasher.

It may be noted that the components of DT vapour, namely, hexane and water vapours, form an azeotrope (heterogeneous azeotrope) at 61.6C at atmospheric pressure. Since food grade hexane also contains higher molecular boiling components such as cyclohexane and heptane, we may consider the azeotropic temperature as 62C. At this temperature, the vapour components condense at a constant weight

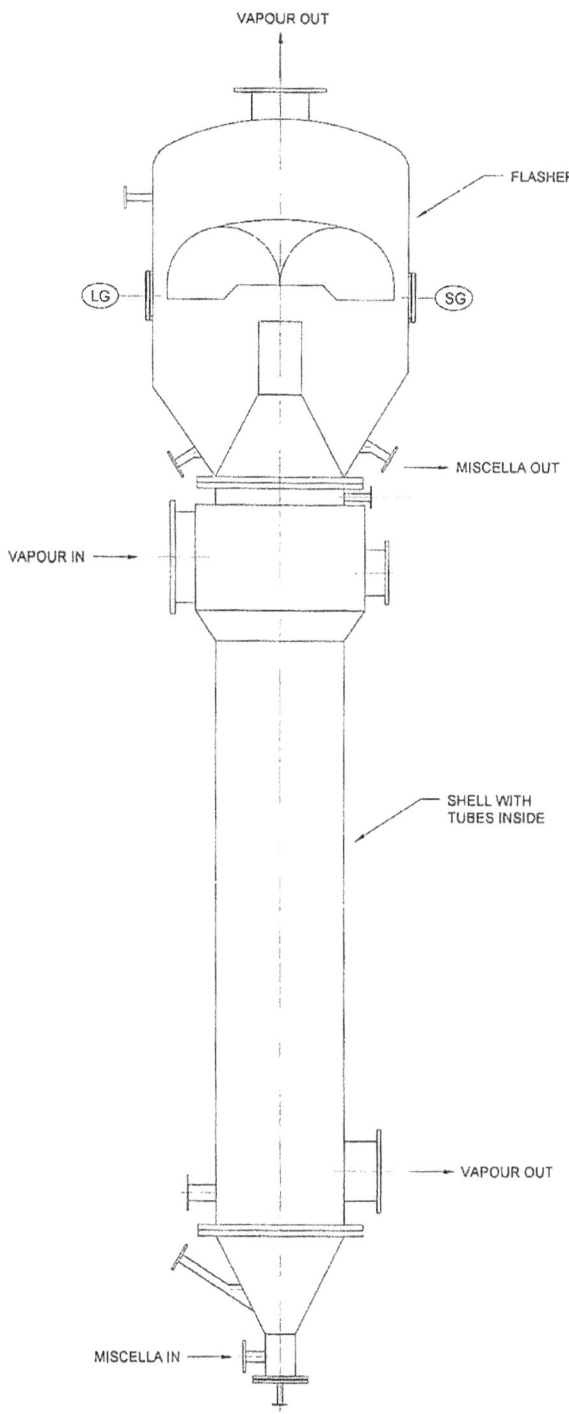

FIGURE 15.20 Economiser (Rising Film) with Flasher on Top.

fractions. For our azeotrope, the respective concentrations of hexane and water are 94.6% and 5.4%. Since the DT vapours contain nearly 9%, or more, water, only water vapour condenses initially in Economiser, on the top portion of tubes. The temperature of vapour reduces from 70–72 C gradually to 62 C, when the water fraction reduces to 5.4%. Hereafter, the condensation occurs at a constant temperature and constant concentration. Thus, shell temperature in the lower zone is constant at 62 C.

Thus, most of the heat transfer takes place when vapour temperature is 62 C. Therefore, targeted temperature of miscella should be not more than 57–58 C. If miscella is exposed to vacuum of 350 TorrA, temperature of 57 C is achieved when miscella concentration rises to 75%. Thus, this is about the maximum concentration which should be targeted in the main distillation zone. Of course, as this miscella, at 75%, contacts the condensing water in the top zone, some more solvent may evaporate, taking the outlet concentration up to 80–82%. These are the usual parameters for design of Economiser for soybean collets, where miscella concentration may rise from around 28% to 80% in Economiser. In case of soybean flakes, since the starting concentration is just about 25%, only 75% outlet conc. may be targeted. For other seeds and cakes, the miscella enrichment may be the same or lower, depending on the miscella concentration and also depending on the heat content of DT vapours. The latter depends on quantity of solvent vapours and the vapour temperature.

High heat transfer rates may be achieved in Economiser, thanks to the change of phase on both sides of the tubes. The overall coefficient can be as high as 300 kCal/m^2/C/h, especially in the middle portions of tubes where miscella concentration is low. In the lower tube portions, the miscella is still in the liquid phase, and the coefficient may be low. Hence, the overall coefficient of 250 kCal/m^2/C/h may be safely used for design of heating surface.

In some plants, the Economiser is not insulated. Justification given is two-fold; one, the temperature of the shell is not high, and two, the DT vapours are to be subsequently condensed, with water cooling, anyway. The first argument has been examined in case of extractor, and found not correct; in this case, the heat loss per unit surface area will be even more than that from the extractor. Second, while it is true that the balance DT vapours out of Economiser have to be condensed, maximum possible heat must be transferred to the miscella in Economiser; thus heat cannot be allowed to dissipate from the shell surface. Thus, an Economiser must always be insulated.

There is always a risk of deposition of fines on Economiser tubes, with accidental escape of fines from the Dust Catcher. One or two hand-holes should be provided for inspection, and for water wash with high-pressure jets.

In most soybean plants, the shell of Economiser is made of mild steel. The tubes are always made in SS 304. In case of rapeseed and palm kernel, the DT vapours contain corrosive sulfur gases, hence even the shell must be constructed in SS grade 304 or higher.

15.2.6 Solvent Heater

Nearly 80–90% of the transferrable heat available in DT vapours is utilised in the Economiser. The balance may be utilised to heat the solvent which is to be sent to the extractor. The solvent condensing temperature is around 37–39 C. The DT vapours exiting from the Economiser are at 62 C (constant Azeotropic temperature), as explained in the previous section. The solvent is to be heated up to nearly 58 C. For a shell-n-tube design, with solvent in the tubes, the overall heat transfer coefficient may be considered as 250 kCal/m^2/h/C. The surface area requirement works out to just about 5–8% of the Economiser.

While the Economiser is designed with a tube length of between 5 and 8 m, depending on size, the Solvent Heater should be designed with much shorter tubes, so as to have a reasonable cross-section area for flow of DT vapours. Small cross-section would entail high pressure drop in the vapour path. The tube length may thus be selected between 2 and 3 m.

Unlike the Economiser, there is no flasher on this equipment, as we do not seek to vapourise the solvent. While the tubes should be made of SS, the shell and the inlet/outlet dishes, or cones, may be of mild steel.

It may be noted that the Solvent Heater also works as a condensor, just like the Economiser, and helps reduce the heat load on the DT Condensor.

Alternatively, the solvent heating may be accomplished by gaining heat from condensing vapours from distillation section, but this is not much beneficial, since the condensing vapours are under vacuum, causing lower temperature of condensation, thereby reducing the temperature difference between the hot and cold streams.

15.2.7 DT Condensor

Most of the heat available in DT vapours is utilised in the Economiser, thus most vapours are condensed there. Only a small portion of vapours, if at all, flow to the DT condensor. However, the condensor is designed for a higher heat load, as a safety against higher-than-normal solvent carryover in DT. Even then, this condensor is relatively much smaller than the Distillation Condensor where all the vapours from distillation section are condensed.

An exception to the above rule may be when full-press cakes are solvent-extracted. In this case, the oil quantity is far less, and spent cake quantity more, than those for prepress cakes. Therefore, the heat load in DT vapours may be high, but requirement for heat in Economiser would be low, in view of less solvent going with less oil. Since, a major portion of DT vapours would not be condensed in Economiser, these would pass over to the DT condensor, which then would have to be bigger in size.

Condensors are mostly shell-n-tube design, although plate type units have been installed in a few plants. The shell-n-tube condensor may be installed vertically or in horizontal position. A vertical installation may save floor space, but horizontal installations are preferred, for the ease of maintenance of the tube bundle.

A horizontal shell-n-tube condensor is usually fitted with vertical baffles on shell side, to improve the contact of vapours with the tubes. Cooling water flows in the tubes, usually in multiple-pass configuration so as to achieve a reasonably high velocity of water in tubes. Higher velocities not only improve heat transfer coefficients on the tube side but also help prevent deposition of fines on tube walls. However, high velocities also entail higher pressure drop, hence higher water pumping energy.

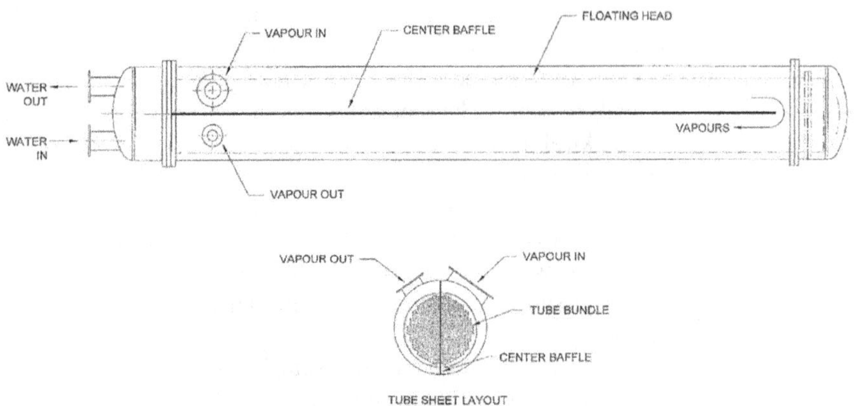

FIGURE 15.21 DT Condenser with Central Baffle.

A moderate range of water velocity, around 2 m/s, meets the earlier two objectives, without requiring high pumping power.

Because of several baffles, perpendicular to the axis, and with multiple passes of water, the flow configuration is a mixture of counter-current, co-current, and cross-current patterns. The effective overall heat transfer rate, per unit surface area, with this pattern is almost 15–20% lower than that for a counter-current configuration; thus the surface area requirement is higher in that proportion. If, in place of several baffles perpendicular to the axis of condensor, a single vertical baffle is placed along the axis, it may then be possible to have a perfect counter-current flow pattern, with two water passes on tube side (Figure 15.21). However, special care must be taken, in this case, to minimise the clearance between the central baffle and condensor shell. Failure on this count may lead to significant by-pass of the vapour from vapour inlet straight to the outlet. Also, in case of floating tube bundle design, as the space between outer tubes and shell may be large, peripheral baffles would be required to prevent significant vapour flow in the open space by-passing the tube bundle.

Until recently, multiple condensors used to be provided for the duty. In fact, two condensors were the norm, often the first being placed vertically next to the Economiser. A times, a third condensor was provided, using water at lower temperature. Advantages of a single condensor are multiple: one, less water pumping costs (this is discussed in later section); two, ease of cleaning and maintenance; and three, lower equipment cost.

The DT Condensor should be built with a floating tube bundle, to enable easy inspection, and complete cleaning, of outer surfaces of tubes. This is a safety for clearing of any fines which might escape accidentally from the Dust Scrubber and pass through the tube bundle of Economiser; or for fines coming with vapour from the Dirty Water Heater. As discussed above, a floating tube bundle entails more empty space between outer tubes and shell (Figure 15.22); thus, peripheral baffles are necessary to prevent major vapour flow in the empty space, by-passing the tube bundle.

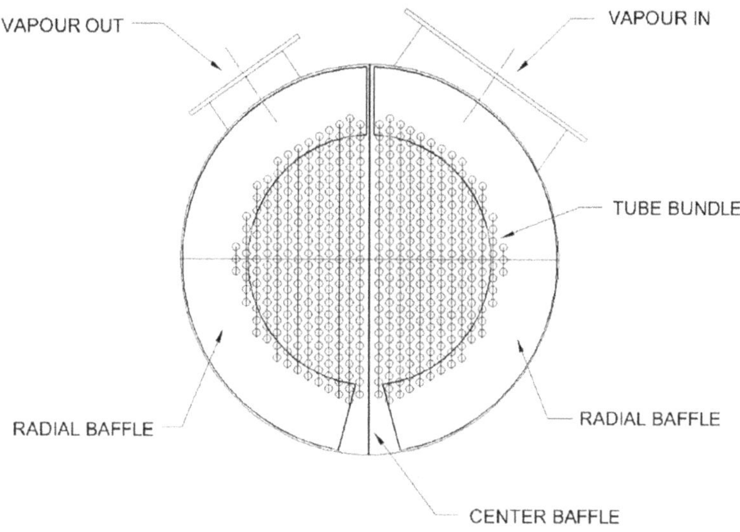

FIGURE 15.22 Central & Radial Baffles in DT Condenser.

The material of construction is similar to the Economiser. For corrosive vapours of rapeseed and palm kernel, the entire condensor should be built in SS. For other oilseeds, the shell and the tube flanges may be made in carbon steel.

The uncondensed vapours, mainly air with small fraction of solvent vapours, flow to the Vent Absorption section. The amount of solvent vapours in the air depends on the outlet air temperature.

15.2.8 Ducts and Pressure Drop

The vapour ducts on the path of DT vapours are all made of stainless steel. The ducts downstream of the DT Condensor, through the Vent Recovery system, may be of carbon steel, as these vapours are at low temperature. For sizing of the ducts, the following vapour velocities are recommended.

DT duct: At the DT outlet, vapour velocity should not exceed 3 m/s, so to minimise entrainment of fines. DT duct is built with a sudden direction change, so that part of fines fall back into the DT. After the change of direction, the duct dia may be reduced to let velocity be 3–5 m/s.

Duct between Dust Scrubber and Economiser: velocity 4–6 m/s.

Duct between Economiser and DT condensor: velocity 4–6 m/s.

The velocity may be higher for large-capacity plants, and lower for small and medium plants. These are calculated to ensure that the pressure drop through all the ducts does not exceed 20 mmWc.

The pressure drop in individual equipment in the vapour path, namely, Dust Scrubber, Economiser and DT condensor, may be calculated as a sum of entry frictional loss, drop of pressure through the water spray or through tube bundle, and the frictional loss at exit. The total pressure drop through the system is expected to be less than 50–60 mmWc. This is to have a moderate pressure head on the ventilation fan, which has to overcome additional pressure drops through Vent Absorption system. During plant operation, the fan speed is regulated to maintain a steady suction of 10–20 mmWc in the DT duct.

15.3 DISTILLATION SECTION

The main equipment in distillation section are the Evaporator(s), the Stripper and the Distillation Condensor. These are supported by various other equipment. All these are discussed below. The first Evaporator, namely the Economiser, has been covered in the Desolventisation section.

15.3.1 Miscella Filtration

Rich miscella from extractor contains fines, which should be removed before the miscella is fed to distillation. Fines tend to deposit on inner surfaces of tubes in evaporators, get charred over time, and impart red colour to oil. Also, the fines in oil will get concentrated in gums, during refining of oil, and then in lecithin, spoiling the lecithin quality. To be able to meet lecithin quality parameters, fines (insolubles) in crude soybean oil should be below 0.04%.

Many plants use hydro cyclones for the purpose of fines removal. However, these would remove only the bigger particles and not meet the standard mentioned above. Miscella filters described below offer a better option to meet the requirement. It may be noted that even the ‘miscella filter’ is basically a screener, and not a true filter. A true Filter is not self-cleaning and may need occasional opening. This is not desirable for the miscella phase, with so much volatile solvent in it.

15.3.1.1 Hydro Cyclone

Hydro cyclones, or hydroclones as these may be called, are of similar shape as air cyclones, and work on similar principles, but are much smaller, thanks to the much smaller volumes of liquids handled. Inlet velocity is usually within the range of 3–5 m/s. Replaceable machined nozzles may be fitted at the discharge, to allow handing of varying quantity of solids. Hydroclones may remove fines larger than 120–150 micron; smaller fines would escape with the outlet miscella. Pressure drops across miscella hydroclones are typically around 1–1.5 Bar. High-pressure Hydroclones may be designed for removal of fines up to 80–100 micron, but the high pressure drop increases the pumping cost.

15.3.1.2 Miscella Filter

The equipment widely used for this purpose is, in fact, a screener, and not really a filter. A true filter allows deposition of solids, on a support medium such as a screen or a cloth, which then act as the filter bed to trap all fines. A Screener, as used in miscella filtration, on the other hand, has a scraper to scrape the screen continuously; since solids bed is not allowed to form, some fines go through the screen, depending on the opening in the screen. It is this scraping action, which makes the ‘filter’ self-cleaning, which may be operated in continuous mode, without the need for intermittent stoppage for cleaning.

The miscella filter has a cylindrical wedge-wire screen element, mounted in a casing vertically. The clearance between adjacent wires, assembled radially, may be in the range of 30–80 micron. For requirement of low ‘insolubles’ content in crude oil, to meet subsequent quality standard of lecithin, screen size of 30 or 50 micron is necessary.

The screen is continuously wiped by brushes fitted on a rotating shaft (Figure 15.23). The wiped solids collect in the conical bottom. Bottom valve is opened with a timer, which regulates both the ’open’ and ‘close’ times. These time durations typically may be 2–5 seconds and 2–3 minutes, respectively. The slurry of solids-in-miscella is dumped on extractor bed.

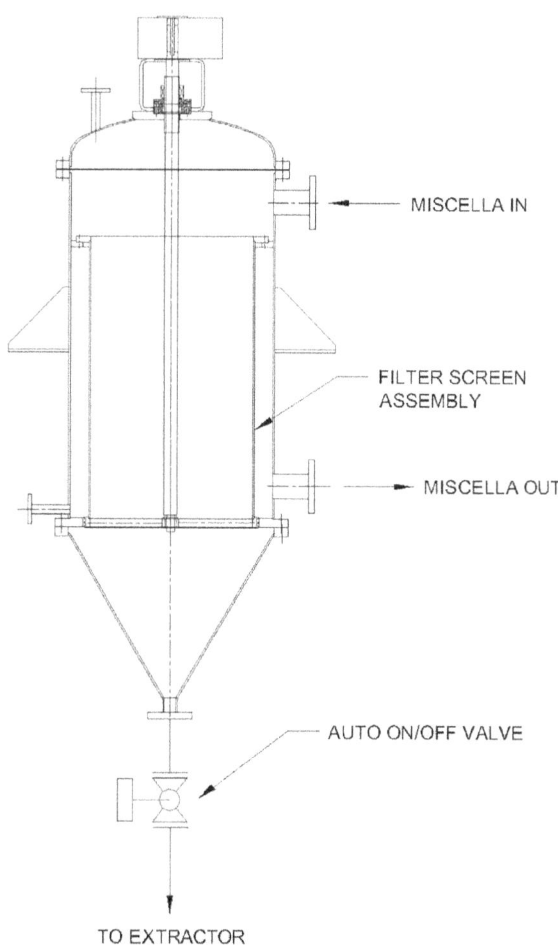

FIGURE 15.23 Miscella Filter.

A back-flush arrangement is provided to clean the screen with a blast of solvent from outer surface, in case of screen fouling. However, this is a rare occurrence and may indicate mis-operation of the brush.

For a 50 micron screen, screen area requirement may be determined, with respect to miscella flow of soybean, as $40\,m^3/h/m^2$. For a 30 micron screen, the flow is lower @ $30\,m^3/h/m^2$, thus area requirement is higher. This area requirement may, of course, vary depending on the screen opening in the extractor. For other seeds, especially cakes and rice bran, which may contain much fines, the area requirement may be higher. At full flow, pressure drop across a miscella filter may be around 3–3.5 Bar.

15.3.2 Miscella Tank

A buffer tank is provided, downstream of miscella filter, to allow for uniform feed to distillation line. The tank is usually sized for 10 minutes hold up in large plants (more than 1,000 Tpd). For smaller plants, hold up may be up to 20–30 min. Apart from the main inlet from the miscella outlet pump of extractor, another inlet is provided from overflow of miscella hoppers of extractor. In case the tank becomes full, an overflow line takes extra miscella to an underground tank.

In some plants, where miscella filtration is not practised, water level is maintained at the tank bottom. Miscella is forced through the water layer, in the hope that fines will absorb water, become heavy and settle down. These are removed by opening bottom drain valve, say, once in a day. It should be noted that this is not a good idea, because contact with water allows gums in the oil also to absorb water. These hydrated gums remain suspended in oil and separate out at elevated temperatures. The separated gums stick to tube surface in Evaporator and also on Stripper surfaces. Over time, evaporator tubes may choke with gums. Plant has to be then stopped to clear the tubes. The deposited gums also impart red colour to oil, as they get charred over time with prolonged heating by steam. Thus, water addition to miscella tank should not be practised; installation of a miscella filter should be the preferred option.

Since the overflow from extractor miscella hoppers, which is not screened, also gets into the tank, it contaminates all the miscella. It is recommended, therefore, to have a separate compartment, say an extension of the miscella tank, to hold the overflow stream. This miscella may then be pumped back, with a separate pump installed for the purpose, to the extractor when the level in miscella hoppers is back to normal. Such a two-compartment tank is especially useful in countries where frequent power trips may cause frequent miscella overflow from extractor hoppers.

The miscella tank may be fitted with high-level and low-level switches. The low-level switch may be used to stop the pump, to prevent it from running dry. The high-level switch may serve to warn that miscella might start overflowing into the underground tank. A continuous level indicator, with its own switch settings, is of course a superior option.

A manhole is provided on top of miscella tank, for maintenance purpose. Vent is connected to the DT condensor, so that any solvent vapours created within the tank are ventilated out.

15.3.3 Economiser

This equipment is the first in the distillation line. As it is also a part of the DT vapour path, it has been covered in the Desolventisation Section 15.2.

15.3.4 Miscella Heater

The enriched miscella at Economiser outlet is still at a low temperature, less than 60C. Therefore, it is a good candidate for heat recovery from finished crude oil exiting from the plant. The crude oil exiting the last equipment in distillation line is at around 100C. Both these streams are contacted in, typically, a plate heat exchanger to recover heat. In the process, the miscella may get heated to around 75–80C, while the crude oil may get cooled to nearly 60–65C. Thus, this equipment saves steam on one hand, and saves cooling water circulation quantity for cooling of oil on the other.

An oil strainer may be fitted upstream of the PHE, to prevent deposition of gums on the plates. A duplex strainer would allow cleaning of a strainer basket without stopping the flow of oil to the PHE.

15.3.5 Evaporator

The Evaporator is, in fact, the second evaporator in the distillation line, the first being the Economiser. However, this is how both these equipment are known in the industry parlance.

The Evaporator is, typically, much smaller than the Economiser. For most oilseeds, and presscakes, miscella of 20–30% concentration is fed to the Economiser and gets enriched to 50–80%. Thus, only a small percentage of the original solvent quantity, just about 10–25%, goes over to the Evaporator. Exceptions may be the plants designed for shea nut cake, where miscella concentration is so low, that it cannot be enriched to a major degree in the Economiser, and a lot of solvent gets carried over to the Evaporator.

Like Economiser, Evaporators are also shell-n-tube equipment, mostly with rising film configuration. Heating medium, steam at low pressure, typically 2–4 BarG, is on the shell side. As the steam gives heat to the miscella and condenses, the condensate flows down by gravity, while the miscella rises within the tubes. Miscella is typically heated to 100–110C, and when flashed under vacuum of 300–350 TorrA (mm Hg absolute), bulk of the solvent evaporates, concentrating the miscella further to 95–97%.

The level of vacuum is decided with respect to temperature of cooling water available. The temperature of water coming from cooling tower depends on the wet-bulb temperature at the site. Should water be available only at 31–32C, the vacuum may only be maintained at 350 TorrA. Higher vacuum (lower absolute pressure) would not allow condensation of solvent vapours. However, should the water be available at 28C instead, higher vacuum of 300 TorrA may be used. It should be noted that at higher vacuum, distillation may be carried out at lower temperature, and with less stripping steam.

When the inlet miscella has very low solvent content, as in miscella of 75% concentration or higher, the inlet velocity in tubes is very small. From this slow-moving miscella, pre-hydrated gums tend to separate out and deposit on tube walls. To avoid such an occurrence, inlet velocity may be increased, either by installing a small (short tubes) pre-heater upstream, or by changing the Evaporator configuration to multi-pass. The latter design is slowly becoming the norm in modern extraction plants (Figure 15.24).

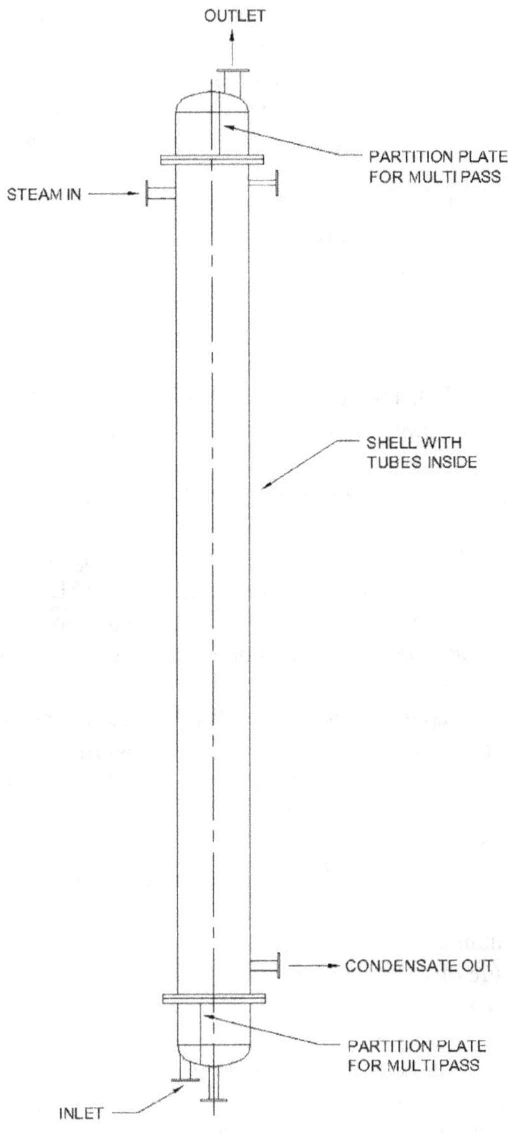

FIGURE 15.24 Evaporator, Multi Pass Design.

Should the Evaporator be of a single-pass configuration, the flasher may be mounted on top, like in case of the Economiser. This type of mounting saves floor space. However, should the evaporator be multi-pass, top mounting of flasher becomes difficult; it may then be mounted on side, with a tangential entry of miscella. To save floor space, the flasher may be mounted on top of the Stripper, which is the immediate equipment downstream.

Since the outlet miscella is highly concentrated, the flasher tends to 'foam', especially with oils of high phosphatides content, e.g. soybean. This issue can be acute when flakes moisture content is high, or at the beginning of soybean season. Tangential entry of miscella, with suitable baffles, can minimise the foam. For safety, a steam nozzle is provided on top of flasher; a blast of steam is able to suppress the foam. Presence of much fines in miscella can also cause excessive foam.

The performance of Evaporator depends on the temperature of miscella at outlet, for a given vacuum. Lower temperature would mean more solvent going over to the stripping stage, and some even escaping with the crude oil, causing off-spec oil quality. Higher temperature, on the other hand, could cause darkening of the oil. Hence, it is important to maintain the temperature constant. This is readily accomplished by an auto-control system, as discussed in later section of this chapter.

15.3.6 Stripper

This is the second, and final, part of distillation, first being the evaporation. Objective is to reduce the solvent content in oil from 3–6% down to 100–700 ppm (0.01–0.07%) as required. It used to be accomplished in two or three stages, operating under increasing vacuum, but during the past two decades, there is a move towards a single Stripper.

Falling film strippers, with multiple vertical plates inside a vessel, were used for many decades. While in principle falling film stripping can achieve high steam efficiency, with a simple construction, in practice it is difficult to maintain a falling film in presence of fine solids.

For the past three decades, a multi-stage **disk-and-donut design** has been widely used. It consists of multiple pairs of tapering disks and donuts, mounted in a vertical cylindrical vessel (Figure 15.25). Steam is injected at the bottom and rises through the disk-donut pairs, counter-current to the falling miscella. The stripper must be built with care, to maintain perfect level in all disk-donut pairs.

In large capacity plants, more elaborate designs, say multi-stage bubble-cap trays column, are often used. These can achieve higher steam efficiency, resulting in net steam saving. However, these face an operation issue when the oil contains high percent of gums, as with soybean, or when the miscella contains much fines. The gums tend to deposit on the trays over time, and affect the stripper performance. Moreover, the equipment cost is much higher than a simple disk-donut column, and may be offset against small steam savings only in large capacity plants, say 1,000 Tpd and bigger.

Strippers, either of disk-donut design, or with bubble-cap trays, are narrow and tall towers. Typical diameter for a 1,000 Tpd soybean plant may be 800 mm; and the tower may be up to 8–10 m high.

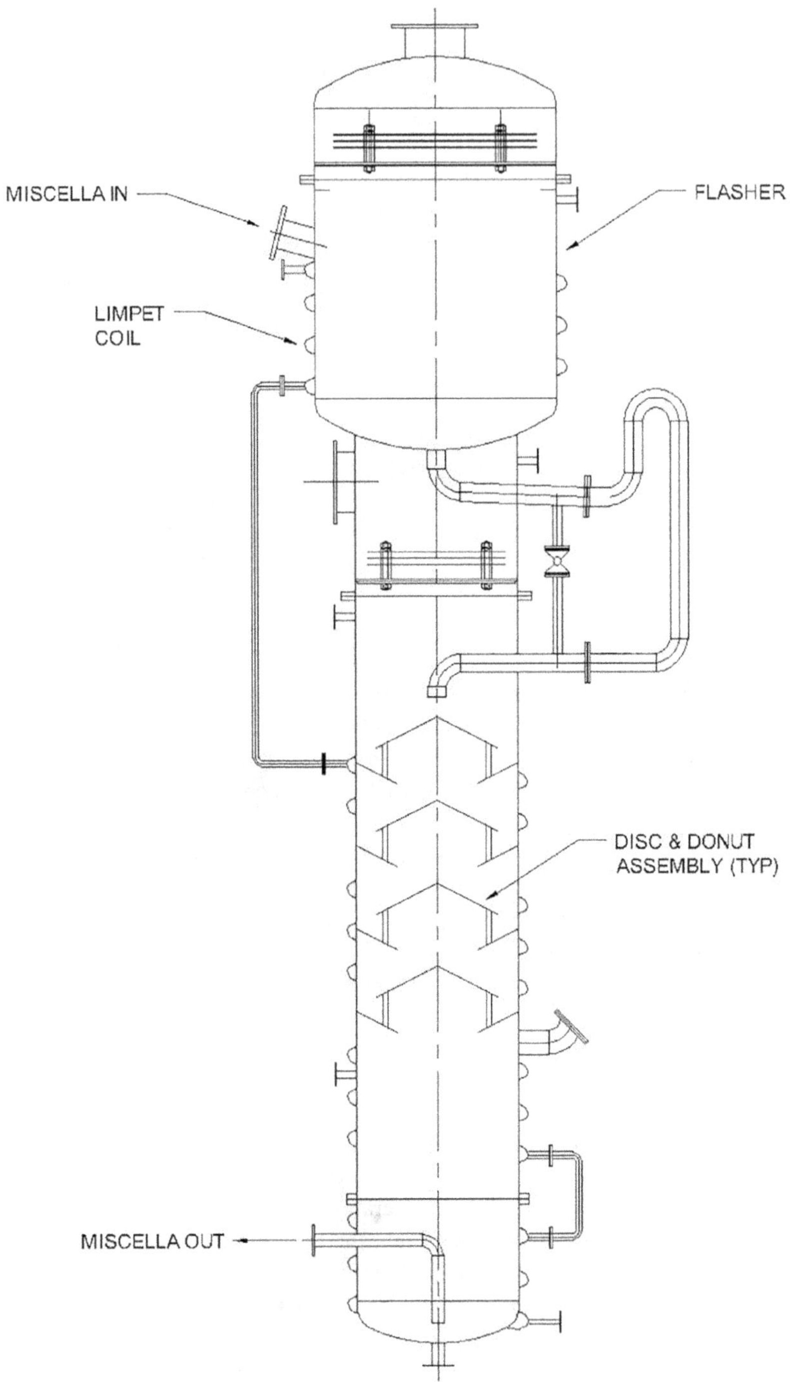

FIGURE 15.25 Miscella Stripper, Disc and Donut Design (With Flasher of Evaporator on Top).

Care must be taken to ensure dry steam for the stripping operation. Wetness in steam can cause hydration of gums, which then tend to stick to stripper walls as also on contacting surfaces. Presence of gums may also lead to foaming in the Stripper. In case of gums deposition in Stripper, these may only be cleaned manually. Multiple hand-holes should be provided to provide access to all the surfaces for annual cleaning.

When multi-stage stripping is practised, the first stripper is usually operated at the same vacuum as the evaporator. This allows the use of a common vapour condensor. The later stages of stripper may be under higher vacuum. The vapours from the high-vacuum stripper(s) may either be condensed in a separate condensor, or may be compressed with a steam ejector and let into the first stage Stripper. In the latter case, the steam from the ejector is used as stripping medium in first stage Stripper. Such an arrangement offers significant advantages, including (a) saving of a separate condensor for the ejector steam, (b) steam saving in first stage Stripper and (c) reduction in steam heat load on condensor system.

When the stripping operation is completed with a single stripper, there is a need to further flash the oil at a higher vacuum to remove excess moisture. The flasher is commonly called the Oil Dryer. The motive steam of the ejector, used to create high vacuum in the Dryer, is profitably injected into the Stripper to serve as the stripping steam (Figure 15.26).

However, care must be taken to ensure dry steam for the ejector. Wet steam can cause deposition of gums in stripper. A dedicated moisture separator in the motive steam line helps in the matter.

The Oil Dryer may be fitted with a large bulb at bottom, to serve as an oil buffer, to ensure that suction of discharge pump is always flooded. The pump is the seal between Dryer which are under high vacuum, and a storage tank which is at atmospheric pressure.

A small oil heater may be placed in between the Stripper and the Oil Dryer, to make up for the temperature drop due to evaporation of solvent in Stripper. This drop may be nearly 8–10C. The heater also helps adjust the final temperature to ensure low moisture in oil. The small heater may just be an in-line unit of shell-n-tube design. Alternatively, the Stripper shell may be fitted with limpet coils in which L.P. steam may be circulated.

Strippers shell may be built in carbon steel, but all internals are preferably made out of stainless steel. For corrosive oils, say the high-ffa oils of rice bran or palm kernel, Stripper, as also Dryer, construction may fully be in SS.

As with the Evaporator, there is a chance of 'foaming' in the stripper with high-phosphatides oils. Hence, a nozzle for foam-break steam should be provided on top of stripper(s).

15.3.7 Oil Cooler

The oil at Stripper outlet is at a temperature of around 100C. For safe storage, it should be cooled to 40–45C. Part of the cooling is accomplished in the Miscella Heater, discussed at the beginning of this section. Further cooling from 65C to 40–45C must be accomplished using cooling water in a plate heat exchanger. Some

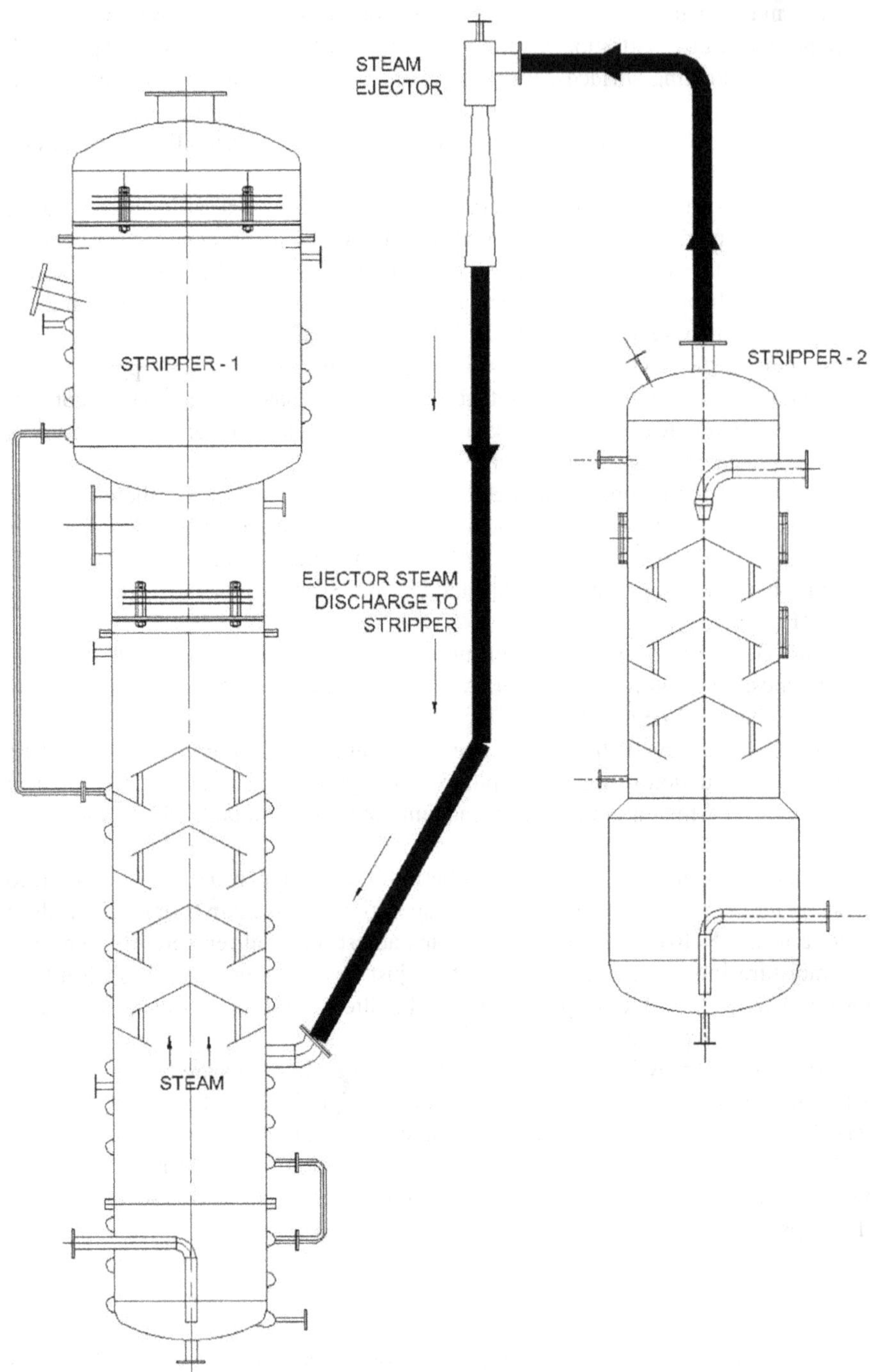

FIGURE 15.26 Single-stage Stripper with Oil Dryer – Flow of Steam.

oils, which may become viscous at 45 C, for example, palm kernel oil, rice bran oil, and shea butter, may be cooled to just 50 C. In fact, a temperature controller need be fitted to prevent further cooling, to avoid difficulties in oil pumping.

15.3.8 Distillation Condensor

All the solvent vapours from distillation line, namely, Economiser, Evaporator and Stripper(s), are condensed in this unit. Thus, this is a large condensor, usually much bigger than the DT condensor. Along with the solvent vapours, all the stripping steam is also condensed here, and so are the solvent vapours from the Vent Absorption unit.

Construction of this condensor is quite similar to that of the DT condensor. As discussed, a central baffle along the axis, allows perfect counter-current configuration, with water flowing in exact opposite direction to the flow of vapours in a two-pass configuration. As mentioned, special care must be taken to minimise the clearance between the baffle and the condensor shell, to eliminate the risk of vapour by-pass.

The vapours coming to this condensor are from miscella distillation, and hence are free of fines. The tube bundle, therefore, does not have to be 'floating' type, unlike in DT condensor. Condensor with fixed tube bundle is not only less expensive, but also can achieve better efficiency with better contact of vapours with tubes, thanks to the small clearance between the bundle and the shell. Better contact results in less tubes surface area requirement.

The water velocity is maintained around 2 m/s, for reasons explained for DT condensor.

Unlike the DT condensor, the Distillation (DS) Condensor operates under vacuum. The vacuum is created and maintained by either a vacuum pump or a steam ejector. Vacuum pumps used to be quite common in older times. These have been replaced with steam ejectors, because of savings in energy and maintenance costs. Also, with process improvements over the years, the ejector steam is used in downstream processes, thus net steam consumption is zero. In the case of the DS Condensor, the steam of vacuum ejector is used for heating the outgoing water in the Clean Water Heater.

Until about two decades back, multiple condensors was the norm. One or two separate condensor(s) used to be provided for the vapours from Stripper(s). Advantages of having a single condensor have been enumerated in section on DT Condensor.

The DS condensor, which is under vacuum, is usually placed at a higher level compared to the DT condensor in the solvent plant. The extra height provides a reasonable positive suction to the condensate pump. Condensate from DT condensor flows by gravity to the receiving tank, namely, the solvent-water separator. The distillation section flow sheet (Figure 14.8) shows the flow patterns of all the streams.

15.3.9 Water Circulation in Condensor System

Cooling water is required to be circulated in the two condensors. In older times, separate lines of cooling water, from the main header, were connected to different condensors. With this arrangement, large quantity of cooling water was required. Significant flow reduction may be achieved if the entire water flow is routed through

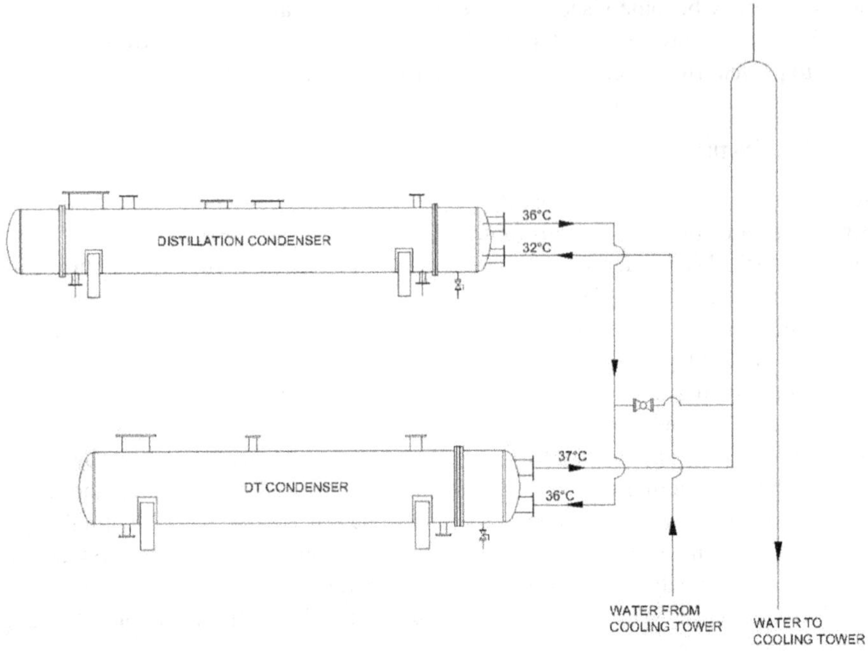

FIGURE 15.27 Condenser Assembly – Water Circulation and Typical Temp. Profile.

the condensors serially. This is how the water circulation systems are designed in modern plants. The logic is explained below.

Out of the two, the Distillation (DS) condensor operates under vacuum, and hence must operate at lower condensing temperature. Then, to maintain a reasonable temperature difference between condensing vapours and water, the water must be available at a low temperature. For this reason, the cooling water is first circulated in the DS condensor. As the water picks up temperature, it is circulated through the DT condensor, and is then let out back to the cooling tower (CT).

Typical temperature profiles are shown in Figure 15.27. DS condensor may typically be designed for nearly 3–4 C increase in water temperature. So, for water inlet temp of 32 C from CT, water temp out from DS Cond may be around 35–36 C. Since the heat load is usually much less on the DT Cond, the water temp rise may just be 1–2 C; thus, final water outlet may be between 36 and 38 C. In this case, the cooling tower has to cool the water by nearly 5–6 C to get back to 32 C. However, the CT may be ordered for a 7 C temp reduction, for safety.

Since the heat load on DT condensor is usually much less than on DS condensor, only a part of water flow from DS condensor may go to the DT condensor and part may be by-passed. If both the condensors are designed for water velocity around 2 m/s, the total pressure drop in the system may be within 10–12 mWc. Considering the height head and frictional losses in lines, it is possible to keep total head on the pump within 30 mWc. At this head, the electric energy requirement for water pumping may be below 2 kWh per T of flakes intake to plant.

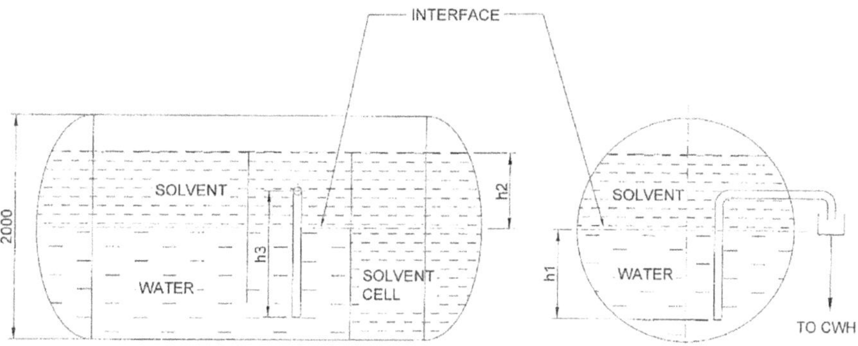

FIGURE 15.28 Solvent-Water Separator.

15.3.10 Solvent-Water Separator

The vapours condensed in both the condensors contain water along with solvent. The water is from two sources: one, a part of the live steam injected into spent flakes in DT, and two, stripping steam as also ejector motive steam from distillation. Since the hexane solvent and water are not miscible, these may be separated by gravity separation, i.e., decantation. In industry parlance, the decanter is commonly known as the Solvent-Water Separator (SWS).

The SWS is typically a horizontal cylindrical tank. The condensate is fed to the SWS at one end, and as it travels through the length, solvent rises to top and water dips to bottom. As the separation is accomplished, water is let out by gravity. In fact, the water outlet pipe is raised to a height so that the water level rises to the centre of the tank (Figure 15.28). The overflow height is decided by following equation:

Head of combined column of water and solvent at point A
= head of water column in overflow pipe

where, point A is at the bottom of overflow pipe.

So,

$$\left(h_1 * \text{water density} + h_2 * \text{solvent density}\right) * g = h_3 * \text{water density} * g \quad (15.2)$$

where, h_1, h_2, h_3 are heights as shown in Figure 15.28.

For example, if the SWS diameter were considered as 2,000 mm, h_1 may be 850, and h_2 may be 800 (point A at 200 mm above bottom of tank). Using these values, and hexane density at 40C as 0.65 kg/L, Eq. 15.2 gives –

$$850 * 1 + 800 * 0.65 = h_3 * 1$$

$$\text{or, } h_3 = 1{,}370$$

Thus, height of the overflow pipe should be 1,370, if the interface level is to be maintained at centre of the SWS. Any reduction in height of the pipe would result in lowering of the interface, and increase in height would raise the interface. It may be noted that the overflow level is 1,570 mm from the bottom of tank, since the overflow pipe started 200 mm above the bottom.

Sight glasses are fitted at the interface level along the length of tank, so that the clarity of separation may be monitored. Multiple circular glasses, or a single vertical glass, are advisable, as the interface may move slightly due to various reasons, such as (a) accumulation of interlayer of oil and gums, or (b) insufficient separation of layers, etc.

Even if the phases are clearly separated, there is always an interlayer, consisting of fatty matter together with water and solvent. If this layer is not removed at regular intervals, the layer can grow thicker. Since the separation of phases has to occur through the interface, a thick interlayer obstructs the separation. The interlayer, therefore, must be removed at regular intervals, say once in a day. An Interlayer Remover device, consisting of a rotating pipe, open at the interface, may be fitted for

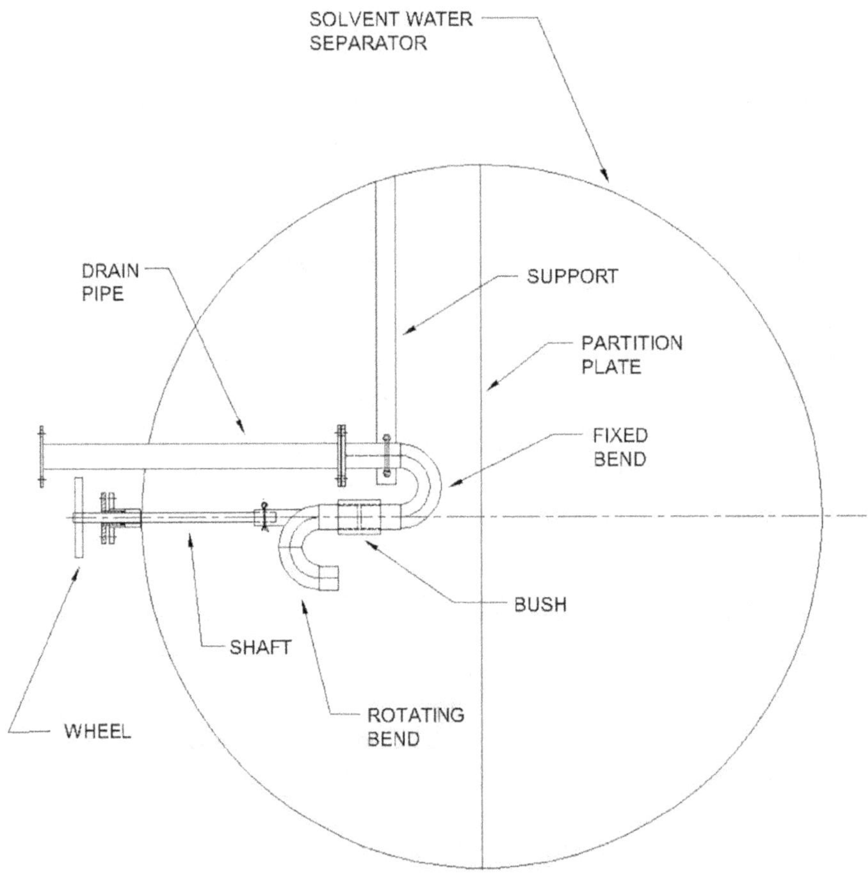

FIGURE 15.29 Interlayer Remover in SWS.

the purpose (Figure 15.29). As the valve on the drain pipe is opened, the liquid in interlayer should come out. A sight glass in the line is used to confirm the presence of the 'blackish' interlayer liquids. During the process, the level of the interlayer may change by a few centimeters. The pipe, therefore, may be rotated, and colour of outgoing liquid observed, to ensure that the interlayer is flushed out. The interlayer liquid is let out to the Dirty Water Heater, discussed in the section on desolventising equipment.

A superior design of SWS, relative to the one shown in Figure 15.28, involves the use of vertical plates along the direction of flow (Figure 15.30). The vertical surfaces help speed up the phase separation. If the plate(s) is fitted along the axis, the flow path also becomes longer, helping the separation further.

The size of the SWS may be determined by the principle of the surface area of the interface. For the separation of hexane and water, in presence of a thin interlayer, the interface area may be such that the solvent velocity through the interface may not exceed 8 m/h.

A solvent buffer cell is maintained at the extreme end of the SWS. The solvent buffer in this cell ensures a uniform flow of solvent to the extractor. HL ad LL switches

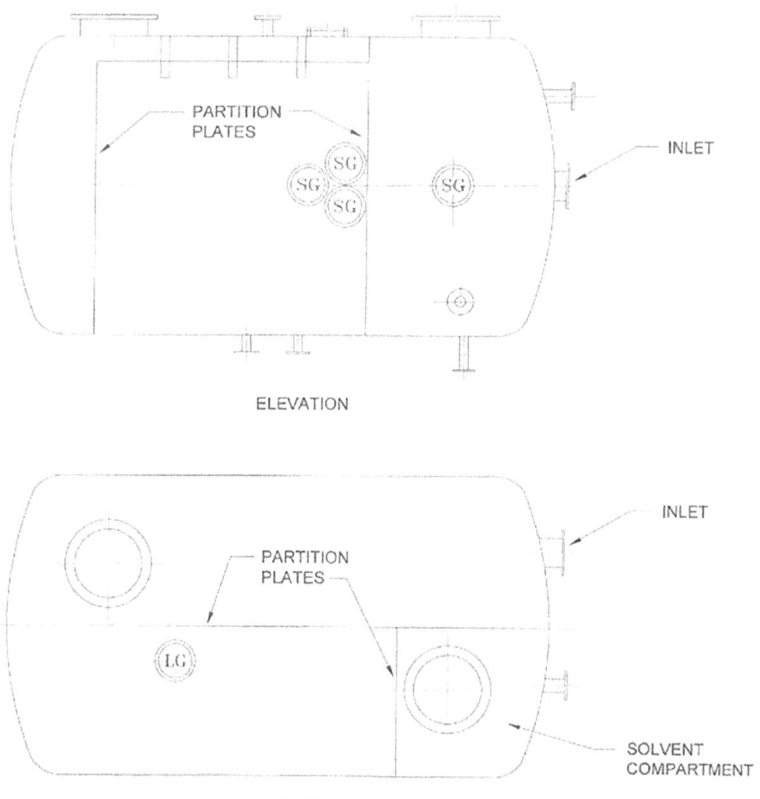

FIGURE 15.30 SWS with Longitudinal Partition Plate.

may be provided on the cell; LL as a warning that the pump may run dry, and HL to warn of the possibility of overflow to an underground tank.

Manholes are provided on top of SWS, for maintenance purpose. Vent is connected to the DT condensor, so that any solvent vapours created within SWS are ventilated out.

15.3.11 Clean Water Heater

Water separated in the SWS must be heated to 85 °C, for the removal of any solvent traces, before being discharged. A small tank, sized for around 30 min hold up, is used for the purpose. Heating is accomplished by motive steam of ejector on Distillation condensor, which is discharged into the tank. Thus, the motive steam is beneficially used here. As a safety, a small steam coil is usually inserted through a manhole. Should the steam coils be used for final heating, the temperature may be controlled within +–2 C through a self-acting steam control valve.

If the interlayer is removed regularly from the SWS, the outgoing water is reasonably clean. After removal of solvent traces, the water may be reused in the plant. One option is to use it as make-up water in cooling tower. This may be accomplished with a small pump to feed the inlet water header of CT.

When the clean water is used within the plant, the effluent water is nearly zero. The small bleed-off from the water used for scrubbing of DT vapours may just be let on to sand pits, as discussed in Section 15.2.

15.4 VENT ABSORPTION UNIT

This section is designed for recovery of solvent vapours from vent air. The equipment in this section are air cooler, absorber, oil stripper, oil heater and cooler, and oil–oil heat exchanger. Two pumps keep circulating the absorption oil through the system in a closed loop. The recovered solvent vapours are sent to Distillation Condensor. In some countries, this process section is known as the 'Recuperation' section.

As will be seen with the discussion below, this is a relatively small section, with two narrow towers of Absorber and Stripper, with a small heater, and two PHEs; this section commonly sits in a corner of the solvent extraction plant.

15.4.1 Air Cooler

Air containing some solvent vapours, from DT condensor, is usually received at around 40–42 C. If cooled to 35–36 C, using cooling water at 30–32 C, part of the solvent can be condensed, reducing the load on the absorption system. Air quantity coming in with flakes may be considered, for a well-maintained plant with a good RAL over the extractor, as approximately 1 m^3 per ton of incoming flakes, with additional leakage flow of 10–20 m^3/h. Thus, for a plant sized at 1,000 Tpd (42 Tph) soybean, air quantity coming with flakes may be considered as 60 m^3/h. Vapour pressure calculations will show that quantity of hexane vapours in the air at 40 C is 2.4 kg/m^3, which at 35 C reduces to 1.8 kg/m^3. Thus, nearly 25% of the solvent may be recovered just by air cooling. The condensed solvent will flow by gravity to the SWS.

Air cooling may be achieved either by direct contact with cooling water or by indirect cooling. Both methods are prevalent in the industry. While the direct contact unit is compact and hence cheaper, the indirect contact type may be preferred as it saves on effluent water.

Direct contact coolers are vertical vessels with a segment packed with packing rings, such as pall rings or raschig rings. Pall rings 1 inch size, in SS 304, are commonly used for the purpose. The diameter and height of packing may be designed for the cooling duty for a given air flow quantity. Air is sucked from bottom to top, while water is sprayed on top of the packing and flows down by gravity. Such a cooler should be made in SS to prevent corrosion by water.

Indirect cooler may be built as a shell-n-tube unit. Air may flow on the shell side downwards, while water may be pumped up through the tubes. In this case, the tubes must be in SS, but shell may be made in carbon steel. The heat transfer coefficient on the air side is low, hence such a unit is bigger than a direct contact cooler.

At times, **chilled water**, say at a temperature of 15C, may be used for cooling purpose. Should the air be cooled to 20C, solvent vapours content in air at this temperature is just about 0.7 kg/m^3. Thus, nearly 70% of the original solvent maybe recovered in the cooler itself. This would not only enable compact design of absorption system, but also would save on steam required for stripping of solvent from oil. However, the saving in steam, and in pumping costs of oil, do not offset the energy requirement and maintenance costs of the chiller unit. Hence, this is not a common feature in modern plants.

15.4.2 The Absorber

The cooled air is contacted with a mineral oil in a packed absorption column. Columns are built of small diameter, to ensure adequate wetting of the packing by oil, and tall, to provide for adequate contact time. The construction tends to be modular, with each contact stage provided with flanges, which are bolted together (Figure 15.31). A perforated plate is fitted in the flange to provide support for the packing. The number of stages and stage height may be calculated for the air flow quantity and for the residual solvent level desired. In most cases, the targeted residual solvent in air is 50 g/m^3, referred to as 50 'ppm' w/v.

The mineral oil quantity may be determined as follows. The solvent content in incoming mineral oil is allowed to be 0.5% max., in order to achieve the desired residual solvent of 50 ppm. This is so because, as the solvent concentration in oil increases down the column to about 5%, we may expect the outgoing air to be in equilibrium with oil at twice the incoming concentration, say at 1% solvent. The vapour pressure at 1% at 36C is nearly 8 Torr. Principles of physical chemistry will show that at this pressure, the solvent vapour in air will be nearly 40 g/m^3. This is a safe figure for the target of 50 g/m^3.

The concentration of solvent in rich oil at outlet of absorber is allowed to be 5% max; this is because, at higher solvent content, there is a chance of oil carryover with

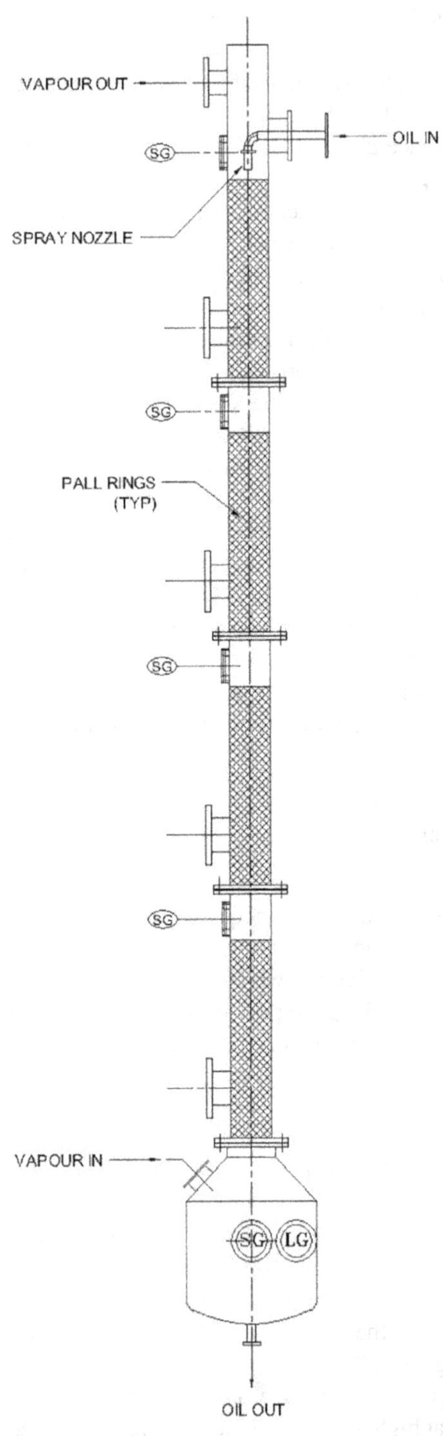

FIGURE 15.31 Vent Absorber.

solvent vapours as the oil is flashed on Oil Stripper downstream. Thus, the oil quantity may be calculated approximately as:

$$\text{Mineral oil flow} = \text{Solvent in air}/(5\text{–}0.5\%)$$

For a 1,000 T soybean plant, air quantity may be considered at 60 m^3/h, as discussed earlier. At 35 C, solvent in air would be 108 kg/h (60 m^3 * 1.8 kg/m^3). Then the mineral oil flow required would be 2,400 kg/h. For mineral oil density of say 0.83 kg/L, flowrate of 2,900 L/h would suffice. This may be adjusted a bit to meet the condition of minimum wetting velocity, as discussed below.

The Absorber diameter may be calculated for the minimum wetting velocity of 0.6 m/min (for the case of 1 inch pall rings). Thus, for the oil flow of 2,900 L/h, dia may be calculated as 320 mm. If we select the standard pipe size of 300 mm, the ID may be 310 mm, which is just perfect for the purpose.

Should the next bigger size, nominal 350 mm, be selected, the oil flow then would have to be increased to 3,660 L/h (Dia 360, v 0.6 m/min). The typical size, for a 1,000 Tpd soybean plant may be diameter 300, with 4–5 stages of packing height 1–1.2 m each. A bulb is built at the bottom, to ensure uniform feed of oil to the pump below.

For adequate wetting of packing in the top stage, cold mineral oil is sprayed with a full-cone nozzle, to cover exactly the entire cross-section of the column. Cone angle and height from top of packing may be carefully selected to avoid much oil being sprayed on the wall; that part of oil would flow mostly along the wall, and not be available for contact with air in the top contact stage. This stage is of critical importance as it primarily determines the residual solvent in vent air. To ensure uniform wetting in lower stages, well-perforated plates should be fitted in between the stages, supported by the connecting flanges.

The absorption column may safely be constructed of carbon steel, with all internals in SS. The packing rings may be in SS or ceramic. Pall rings 1 inch size, SS 304, are commonly used worldwide. The absorption oil may be selected for low viscosity (around 20 cP at 35 C) and high flash point (above 170 C). Mid-range high-speed lubricating oils are suitable for the purpose.

In recent times, there is a move towards minimising the **residual solvent** escaping in vent air. In fact, columns have already been designed for 10 ppm, in place of 50 ppm (Extech, 2018). The requirement, apart from taller column, is that the lean oil which is spread on top of column, should have solvent content below 0.1%. This is to be achieved in Oil Stripper, discussed below.

15.4.3 Oil Stripper

The construction, and operating principles, of this Stripper is similar to that of the Miscella Stripper, discussed in Distillation section. Simple disk-donut design, with three or four pairs, may serve the purpose.

Should the desired residual solvent in oil be 0.5%, as discussed in section on Absorber, a single Stripper, operating at the same vacuum as the Evaporator in distillation section, would suffice. However, if a lower residual of 0.1% is desired, for the purpose of reducing solvent in air below 10 ppm, a second Stripper, operating under higher vacuum, would be preferable; a single stripper for the entire duty would entail somewhat higher steam requirement. An intermediate steam ejector may be used for maintenance of higher vacuum in the second Stripper; the motive steam of the ejector would be beneficially used as stripping steam for the first Stripper. Operating temperature of both the strippers may be 110–115 C. Here too, as in miscella stripping, the ejector steam should be dry; a dedicated moisture separator in the line would help. Wet steam would interfere with the vacuum.

If the second stripper is mounted below the first stripper, care should be taken to leave sufficient clear height from the pump, to maintain NPSH above the pump requirement. Since the stripping steam requirement for the second stripper is very small, the flow may be regulated with an orifice plate.

Foaming is not an issue with mineral oil, unlike in the Miscella Stripper.

Since there are no corrosive liquids, the shell of the Stripper may be made of carbon steel. Internals including the disk-donut pairs, however, should be built in SS, for durability.

15.4.4 Oil Heater and Cooler

The cold mineral oil, which is sprayed on top of Absorber, is maintained at around 35–36 C, the same as the incoming air. In fact, the temperature of oil should not be lower than that of the air. The reason for this is, cooler oil might absorb moisture from the air; too much moisture in oil can create emulsion.

As the oil flows down the packing rings, it absorbs solvent, and gains temperature. Typically the gain would be around 3–4 C, if solvent content is maintained below 5%. This oil is pumped through an oil-oil heat exchanger to gain heat from hot oil coming from the Stripper. The exchanger is typically a plate heat exchanger (PHE), which may be designed to heat up the cold oil (with 5% solvent) from 37 to 90 C, and to cool the hot oil from 115 C to 60 C.

The oil at 90 C is then heated in Oil Heater to the required temperature for Stripper, say 115 C. This is typically a shell-n-tube unit, using L.P. steam as the heating medium.

The hot oil, cooled to 60 C in the Exchanger is further cooled in another PHE, the Oil Cooler, to the required temperature of 35–36 C, using cooling water as the cooling medium (32 C max). Should cooling water of lower temperature be available, air and oil temperatures of 32–33 C may be achieved. Lower the temperature of mineral oil, lower is the vapour pressure of solvent in oil, and lower is the residual solvent in vent air. A reduction of 2C will save nearly 5 g solvent per cubic metre of air.

15.4.5 Vent Fan (or, Ejector)

A small fan maintains slight suction in the Absorber. The fan is designed to create around 10–20 mmWc suction n the DT duct. The corresponding suction at fan inlet may be 80–120 mmWc, depending on the system design. The suction may be maintained with a VFD to vary fan speed. In large plants, this regulation may be automated, with a pressure transducer.

The fan may be replaced by a steam ejector; here, the motive steam pressure may be regulated to maintain the pressure. Since the ejector requires small quantity of motive steam, the steam line may encounter some condensation; the condensate may interfere with the performance of the ejector. Hence, a moisture separator may be fitted in the motive steam line. For auto-control purpose, however, a fan performs better than an ejector, whose vacuum-steam pressure relationship is not linear.

15.5 AUXILIARY EQUIPMENT AND UTILITIES

Apart from the process lines and equipment discussed in previous sections, there are supporting equipment and utilities that form part of the solvent extraction plant. These include an overhead safety water tank, sprinkler system, solvent tanks, and utilities such as steam piping and fittings, cooling tower, compressed air, and electric supply and motor control panels. Let us briefly look at each of these units.

15.5.1 Motor Control Centre (MCC) and Control Panel

The MCC is not built with flameproof fittings. Instead, it is housed in a 'safe room' within the plant. The safe panel room has two-door system, wherein one door is always closed even during movement of personnel to and from the room. A glass window, provided for view of the control panel, with a plant MIMIC board, is fitted with double-glass (toughened) for safety. The room is kept slightly pressurised to eliminate the risk of entry of solvent vapours. The suction to the pressurising fan is taken from outside the plant at a height above the roof of the plant.

A likely place for entry of plant air into the panel room is through the bottom of the MCC at entry of all cables. A perforated steel plate is provided for the cables entry, and the plate filled with a gelling compound, to make it air-tight.

The panel room is located on the main operation floor. A large MIMIC screen, showing the condition of all motors can be seen through the glass window. The panel also indication metres for critical amperages, temperatures, pressures/vacuum, and levels.

When the plant operation is automated, the MIMIC screen may be replaced by a computer screen.

15.5.2 Overhead Safety Water Tank

This is a safety equipment provided to help condense solvent vapours and cool down the plant in case of power failure. As power fails, water circulation pump stops. Thus, condensors and oil coolers are devoid of the cooling water. While steam flow to heaters will also stop, the existing steam in shells of the DT, evaporator and heaters keeps on transferring heat to the meal or miscella, and forming solvent vapours. If these vapours are not condensed, the plant would get pressurised. Excessive pressure might cause leakage of vapours to the atmosphere, and in the extreme, cause failure of vessels designed to operate under vacuum, leading to possible explosions. To prevent such a possibility, water must flow through the system, albeit at a lower flow rate. This is achieved with the overhead water tank. As soon as the power fails, pressure in water pipelines subsides, and a uni-direction valve below the overhead tank opens. Water flow starts thru condensors and cooler system by gravity.

The water tank capacity may be determined for at least 10 min water hold-up at the normal water pumping rate. Since water will flow under gravity, with much less pressure compared to that developed by water pump, flow rate would be much less, and the water quantity in tank would easily last over an hour. That is sufficient to condense all the vapours and allow all the equipment to cool down to a safe level.

Full level of water must be ensured in the tank at all times. This may be achieved by a suitable water inlet valve, such as float valve. Confirmation of level should also be made, either by visual means, such as observation of level with a sight-glass or by observation of a small continuous overflow to a funnel; or by a level sensor showing the level in plant control panel.

The operation of the system, and its adequacy, should be tested periodically, say once in six months, to ensure a fail-safe system.

Until about three to four decades back, the water tank used to be supported on a side bay built in RCC, adjacent to the main plant which was built with steel structure. The RCC bay would also house the electric panel at the main working floor. In recent times, the expensive RCC bay has been discarded. Instead, the water tank is supported on steel structure. For plants with a Linear extractor, the tank may be accommodated in the empty space above extractor.

15.5.3 Water Sprinkler System

Overhead sprinklers are a safety device to douse a fire in the plant. Fusible glass bulbs melt under the heat, and start water spray. The sprinklers are distributed over entire plant operation area (Figure 15.32) and should be connected to high-pressure hydrant system in the factory.

15.5.4 Cooling Tower

The cooling tower (CT) is located within the safe zone around the SEP. CTs are either forced air circulation type or fan-less natural ventilation type. Forced circulation tower, powered by an induced-draft fan, is compact, and is much more common than the fan-less type. The tower packing is either the traditional planks assembly or structured packing. The structured packing is even more compact, but is prone to fouling.

Among fan-less towers, which work on principle of natural ventilation of air, nozzle assemblies with downward spray are more common. However, these require a large area. A recent design consists of nozzles spraying upwards at pressure and creating a mist of fine water droplets (Figure 15.33). Since the surface area of the mist is very large, and heat is exchanged with air on upward path as also during downward fall, these towers are compact compared to the downward spray assemblies. Area requirement for the 'Mist' tower is nearly 0.25–0.40 $m^2/m^3/h$ water flow, depending on the tower capacity. Head required at the nozzle inlet is 8–10 mWc. The head requirement does not put any energy demand on water pump, since the head is naturally available from the condensors located at 10 m or above in the SEP. The saving of

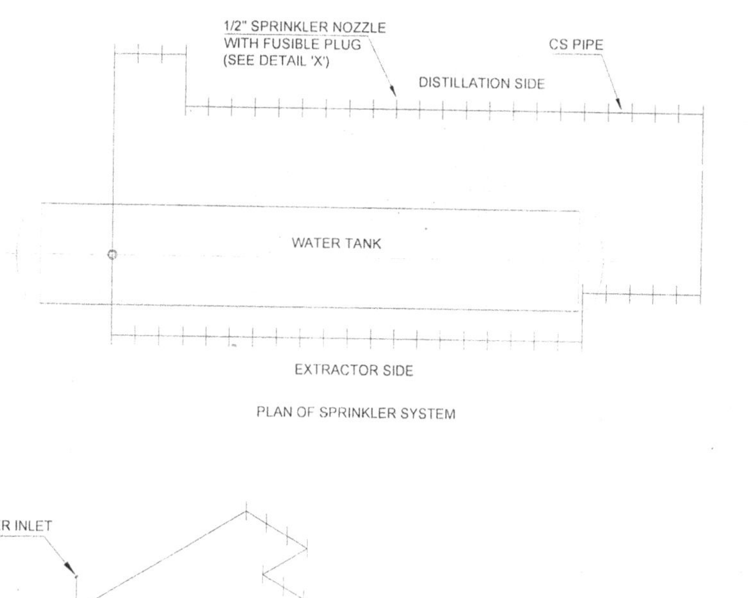

FIGURE 15.32 Sprinkler System.

fan power can pay for the higher cost of civil works, compared to a forced ventilation tower, within a year. Maintenance costs are negligible for the Mist-type towers.

Both the type of CTs, forced ventilation or Mist type, are usually designed for an 'approach' of 3–4C towards the wet-bulb temperature. Thus, for wet-bulb

FIGURE 15.33 Mist Cooling Tower.

temperature of 27C, water temperature may be 30–31 C. For special requirements, CTs may be designed for 2-degrees 'approach'; but that entails much larger CTs, and is not common in vegetable oil industry.

Soft water should be used in CT, to maintain low hardness of the water in sump. As water is evaporated in the CT, the hardness of sump water increases. Soft water with hardness below 5 ppm can help maintain sump water hardness below 100 ppm, with a minimum bleed-off. Should the hardness increase beyond 100 ppm, part of the water should be bled off, and fresh water added. Beyond 100 ppm, chances of scale formation in condensor tubes, and on cooler surfaces, increase.

Direct sunrays on the sump may cause the growth of algae. To control the growth, either algaecide chemicals may be added, or the sump may be covered suitably. In the latter case, the water level in the spray area should be maintained very low, and side sump be built for required water hold up; this sum area may then be suitably covered.

In areas with dusty atmosphere, the CT water may be subjected to filtration. A parallel filtration unit, sized for 1/10th of water pumping rate to plant, should normally be adequate for the purpose. While sand filters have commonly been used, self-cleaning filters as used for miscella filtration, are now making their way into the industry, for the purpose.

15.5.5 Solvent Tanks

In tropical regions, solvent tanks are placed underground. In cold countries, these may be placed over-ground, under a shade.

At least two tanks are installed, one for fresh solvent, and the other a 'working' tank, to receive miscella overflow from extractor or miscella tank or solvent overflow from the SWS. For plants which crush multiple seeds, one more 'working' tank is advisable, to save on 'change-over' time.

The size of the tanks should be decided from the solvent hold-up in a running plant. For a 600 Tpd soybean (flakes) plant, with a Linear extractor, the solvent hold-up may be nearly 40–45 kL. So, the tank should have a working volume of at least 50 kL. Typically, this would be a dia 3 m × 7 m long tank, with flat ends. Flat ends are preferred for the ease of calibration of solvent volume with solvent depth. For two tanks, the pit size may be 8 m × 8 m.

For underground installations, the pit, where the tanks are tied on to a strong foundation, is filled up with sand to keep the tanks cool. A small 'well' is kept at a corner to collect rain water, which is pumped out. The well is lined by loose brick-work, to allow water seepage from the sand pit. It should be noted that if well fills up with water, it means the entire sand pit is also saturated with water; in such a scenario, should any of the tanks be empty, it might float up due to buoyancy. Hence, attention must be paid to water level in the well.

The tanks are placed close to the SEP building, so as not to be too far away from the solvent pump in the plant. Since the pump has to lift solvent from a depth, and due to volatility of the solvent, a self-priming pump should be used for the duty.

Solvent tanks' vents may be connected to DT condensor. Also, a safety valve may be fitted on the vent. The safety valve should be sensitive to small pressures and should be double-acting, to open-out in case of pressure, and to 'open-in' in case of suction; these are commonly known as the 'Tyrose' valves, and may be fitted individually on each tank, on top of a pipe 4–6 m tall, to disperse solvent vapours wide.

Daily solvent losses from plant operation are estimated by measuring the volume of solvent in the underground tank. A dip rod is used for the purpose, and volume is estimated from the depth of solvent in the tank. Tanks are commonly constructed in mild steel, with the flat ends suitably stiffened to help prevent deformation over years.

15.5.6 Steam Piping and Fittings

The boiler operation and precautions to be taken were discussed briefly in Part 1. Those precautions are even more important for the operation of SEP, since fluctuations in steam pressure, as also wetness in steam, can hamper not only the heating operations, but more importantly, maintenance of the vacuum in distillation section. Moisture separator must be installed on the inlet header, to remove moisture condensed during flow from boiler to the plant.

An auto On/Off valve should be installed on the steam inlet header. This is a safety feature, to help cut-off the steam supply in case of power failure. A strainer on the header helps protect downstream valves from metallic and other impurities. Quite often, globe or piston-type valves are used for isolation duty for control valves. These entail significant pressure drop even in fully open position. Gate or sluice-type valves offer negligible pressure drops and should be preferred for isolation duty.

Should inlet steam pressure be in the range of 13–15 BarG, three separate pressure regulation valves (PRVs) are recommended. One for DT (10–10.5 BarG), second for distillation section (3–4 BarG), and third for the vacuum ejectors, set at 8–10 BarG (Figure 15.34). Although the pressure requirement for DT and ejectors may be the same, a separate PRV for ejectors helps avoid fluctuations in vacuum, with fluctuations in steam consumption in the DT, especially as the temperature auto-control systems may throttle or open the steam valves suddenly.

Should the inlet steam be received at 3–4 BarG, from a back-pressure turbine of a co-generation plant, PRVs are not required. In such a case, the DT and the ejectors

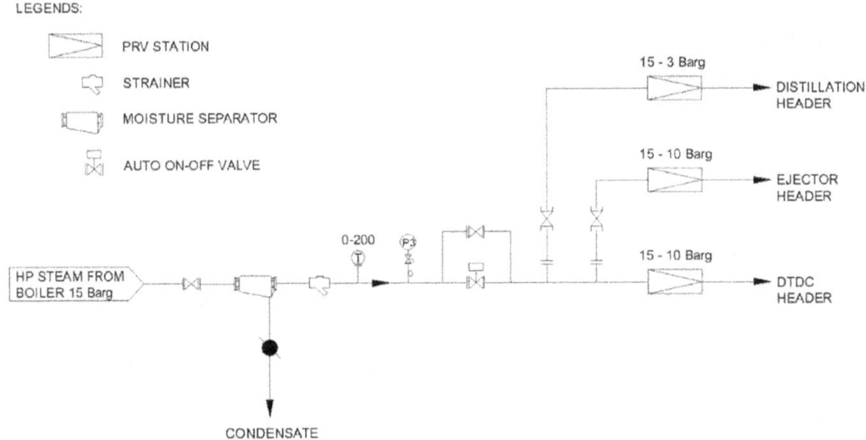

FIGURE 15.34 SEP Inlet Steam Header and PRV Stations.

are to be designed for low steam pressure. However, three separate headers, as above, should be maintained, to minimise the effect of fluctuation in one on the others.

Suitable steam traps help evacuate water condensate from heating equipment, without wasting much steam. Thermodynamic traps (TD) are preferably fitted on steam headers to quickly flush out all condensate, even with a bit of steam. Float-type traps are fitted on heaters to evacuate the condensate without letting out any steam. Air vents may be fitted on headers and heat jackets, as also on DBs and steam chest of DT, to let out air trapped in the system. New type of traps, may come combined with the air vent port, so separate air vents fittings may not be required.

15.5.7 Compressed Air

Compressed air is required to operate pneumatic control systems. Most of the air pressure sensors, as also actuation valves or cylinders are designed to operate with supply air pressure of 3–4 BarG. To maintain pressure at inlet of all the components at 4 BarG or higher, pressure at the Receiver in plant is maintained at 7–8 BarG. Mostly, air compressor is located either in Seed Preparation Plant, or at a common utilities facility within the factory, away from SEP to avoid flameproof motor and fittings.

The air supply pipe, connected from the Compressor to Receiver, is designed for low air velocity, to minimise pressure drop. An air header may be run along the length of plant, from where tappings may be given to individual pneumatic component. The compressed air lines should be either SS or GI (galvanised iron) to prevent rust getting into the sensitive pneumatic components. The tubing to individual components may be nylon or other plastics.

Screw compressors are preferred over piston type, to have oil-free compressed air. Compressed air should be well-dried, preferably in a refrigerated dryer, to dew point of –20C, so that internals of instruments are not exposed to corrosion. Should the compressor unit be located in PREP, it is desirable to have an acoustic enclosure to shut off the noise.

For safety, oil and moisture filters should be provided upstream of each instrument.

15.6 SOLVENT PLANT OPERATION, TROUBLE-SHOOTING AND MAINTENANCE

Operation of a preparation plant or an oil mill, as covered in Part 1 of this book, is quite simple, since the material flow is linear, from one equipment to the next, up to the end of line. So, the start-up procedure is quite straight forward. That is not the case with a solvent extraction plant, since the operation of one section has impact on another and vice versa. For example, operation of desolventising section has a major impact on distillation section, and the two together determine the flow of fresh solvent to the extractor. Thus, the procedures for plant start-up, as also shut-down, has to be in a particular sequence, not necessarily in the direction of flow of solids. These procedures are discussed in Appendix 2.4.

Also, in view of the interdependence of the operation of the three sections, plant tuning and **trouble-shooting** is also a matter of deep analysis and experience. Several trouble-shooting techniques are also discussed in the appendix.

Luckily, probability of machine breakdown in solvent plant is much less compared to that in PREP or Oil Mill. First, there are not many conveying elements. These are the components which are prone to maximum break-downs in the PREP and OM. Only the Marc Conveyor may need some attention. Second, there are no high-speed heavy-duty machines. The DT, which is driven by a relatively large motor, rotates at a slow speed. Hence, once aligned properly, and if metallic impurities are avoided, the DT can run continuously for months without any issue. The mechanical seals of pumps and components of chain in extractor (especially the Linear extractor) do need periodical attention and replacement.

While the likelihood of break-down is less, any break-down in SEP may lead to significant losses of solvent. It is, therefore, best to make a 'preventive maintenance' plan for the plant. If components are attended to, and replaced as necessary, as per the plan, risk of break-downs is minimised. A preventive maintenance is always better than break-down maintenance, as the latter may cause unplanned loss of production, as also heavy losses of solvent at times.

A planned shutdown of one or two days every two or three months normally suffices for preventive maintenance. Additionally, an annual maintenance of 10–15 days may be necessary to carry out some hot work, cutting and welding. This requires that the plant be emptied and drained fully, and de-solventised carefully. This procedure is discussed in Appendix 2.4.

15.7 PROCESS CONTROL AND AUTOMATION

As discussed in Part 1 of this book, control of critical process parameters allows consistency of production and product quality. Automation of the operation allows, not only remote operation, but also better monitoring of the process parameters which can boost process efficiencies. Let us discuss both these aspects of solvent extraction plant.

15.7.1 Process Control Parameters

When we discuss the regulation of process parameters, usually the first focus is on various temperatures and pressures (vacuum). These parameters which are discussed prominently, are relatively easy to control. However, regulation of material flow, especially the flow of solids, may often be of critical value and is also difficult to control. All the important parameters are discussed below. The control parameters for the Seed Preparation are nod included here, as all of those have been covered in the relevant chapter in Part 1.

15.7.1.1 Feed to Plant

Feed to solvent plant is received from the preparation plant, with or without oil recovery from oilseeds by mechanical expression. Although there are feed regulation devices in the preparation plant, as also on Presses, there is always a possibility of some fluctuation in feed rate to the solvent plant. It is not advisable to create a buffer of the flakes/cakes, to ensure uniform feed to the plant, unlike for flow of liquids. So, the feed at varying rate has to be accepted into the plant, and arrangements are required to be made for regulation thereafter.

The first action, for feed regulation is taken in the Feed Hopper of extractor. A buffer is built in the hopper, and level is sensed by an electronic (contact admittance type, or contact-less UV rays type) sensor. Level is then maintained constant by slight variations in the output flow. So, the objective here is not to maintain uniform outflow, but to maintain the level constant. This is so, as to avoid having to stop preparation plant machines if high level in Feed Hopper were reached. If level starts to rise above a set point, speed of extractor is increased slightly; and vice versa. The variation of extractor speed may be restricted to a narrow range of +–3% of a nominal value.

Should high level on Feed Hopper be reached, in spite of the control action, preparation machinery has to be stopped, in the sequence of interlock scheme. Should low level be reached instead, extractor movement is stopped, so as to maintain the minimum level in Feed Hopper to act as a barrier between the solvent atmosphere of extractor and air atmosphere at the feed.

15.7.1.2 Feed to DT

Discharge of spent flakes from the extractor bed is never uniform. Although the degree of non-uniformity may be different for the different types of extractor, as discussed in the section on extractors, there is a need for regulation of the feed to DT in all cases.

A screw is fitted at the bottom of discharge hopper of extractor, for the purpose. Speed of the screw is set in a manner that the sudden discharge from the extractor bed is stopped in the hopper for a while, which is emptied slowly just in time for the next major discharge. This method ensures uniform flow to the DT for most of the time, with less flow for a short duration. The speed of screw may be varied in tandem with that of the extractor, so that the hopper never fills up, while ensuring near-uniform flow to DT at all times.

15.7.1.3 Level Regulation in DT Stages

As discussed before, the DT operation primarily determines the solvent loss from the process. The effectiveness of solvent removal depends majorly on bed levels in each stage of DT.

The level control in PD stages may be accomplished with simple direct-acting control with pneumatic valves. Higher level would increase the valve opening, and hence opening the discharge gate more, and lower level would throttle the gate. When tuned well, the system is able to maintain the level within 50–70 mm of the set point. Thus, the level may be monitored visually in a sight glass, of 200 mm view, at all times.

Levels in upper DS stages may also be controlled in a similar manner as for PD stages. Alternatively, level signal from the top DS stage may be used to regulate speed of discharge RAL (rotary airlock) below the bottom stage; and intermediate stages may be maintained at full bed height with reverse-acting pneumatic cylinders. This method has been discussed in detail in the section on DT operation.

Level in the 'Flash' stage, as also in the DC stages, may be maintained by direct-acting control of speed of discharge RAL.

15.7.1.4 Temperature Control in DT

Temperature control in DT helps complete desolventisation and ensures optimum steam consumption. Two temperature control systems may be operated in tandem for DT. One, the temperature of DT vapours to be controlled by regulation of 'open' steam to the steam chest below the bottom DS stage; and two, the temperature of desolventised meal to be controlled with indirect heating steam to all the double bottoms.

In the second control, manual valves to individual stages may be kept fully open, and the inlet auto valve be regulated to maintain meal temperature between 102 and 104 C. It should be noted that the feedback to both the control system is slow, hence the PID parameters should be set to work accordingly. The second temperature control may be achieved through a pressure control instead, for a quick response; this has also been discussed earlier, in section on DT operation.

15.7.1.5 Temperature Control in Extractor

Temperature of miscella should be maintained in the range of 58–60 C for high rate of extraction. Miscella heaters, in the form of double-pipe, are provided in pump discharge lines. Manual valves to individual heaters may be kept fully open, and auto valve on the common header be regulated to maintain the temperature in, say, second-last miscella spray. Steam valves to sprays towards feed end may have to be somewhat throttled to avoid overheating in those sprays.

15.7.1.6 Miscella Flow to Distillation

Uniform flow to distillation helps in the steady performance of the entire distillation line. A volumetric flow metre/transducer is placed in the line from miscella pump to Economiser. Signal is conditioned and output signal is sent to the auto valve in the line.

It is important that the level in MT is also maintained within a reasonable range. Too low a level would not only affect the discharge flow of the pump, but might also cause air suction into distillation equipment which operates under vacuum. Low-level and high-level switches may be fitted on MT for the purpose. Should low-level be reached, fresh hexane pump would add solvent to the tank. And if high-level be reached, an alarm would be raised. Frequent alarms may either indicate the need for checking the operation of flow control valve, or for increasing the distillation flow.

15.7.1.7 Temperature of Miscella in Evaporator

This is the most important control parameter in distillation section. Variations in this temperature can affect not only the residual solvent but also oil colour. Both are critical aspects of oil quality. Standard control system is a feedback loop from sensing of temperature of outlet miscella to regulate the steam valve; this loop works fine in most cases to keep miscella temperature within 2 degrees of the set value. However, sudden changes in either the temperature of incoming miscella, or its flow rate, or sometimes even change in vacuum, can render this loop inadequate. A superior, and fast, control can be effected by control of pressure of steam in shell side of Evaporator. Pressure change is much faster than that in temperature, and thus feedback is very quick. However, pressure transducers are much more expensive, and also require frequent calibration; hence, temperature feedback loops are very common.

15.7.1.8 Temperature of Hot Oil in Vent Recovery Unit

Mineral oil temperature is typically maintained around 110–115C. Higher temperature, say 120–125C, is not preferred, mostly to safeguard the gaskets of downstream oil-oil heat exchanger. Also, the possible incremental benefit with 10 degrees may instead be achieved with slightly more live steam injection in the Stripper.

The temperature control is via a feedback loop, as for the Evaporator.

15.7.1.9 Pressure in Dryer Stage of DTDC

The air pressure above meal bed in dryer stage should be maintained close to atmospheric to prevent either (a) air entry to DT stages or (b) suction of vapours from DT stages. This is achieved by auto-regulation of speed of the ID fan based on the pressure signal from transmitter fitted in air duct close to exit from the stage.

15.7.1.10 Temperature of Hot Air in Dryer Stage of DTDC

Hot air temperature is maintained within 135–140C for meals with more than 17% moisture. For lower moisture levels, temperature may be maintained lower. The control system is a feedback loop to regulate steam valve. Here, the response is quite fast, hence the loop works satisfactorily.

15.7.1.11 Temperature of Cooled Meal

Temperature of cooled meal is required be 45C max, so that meal quality remains stable in storage. A temperature probe placed in the chute below the RAL of Cooler stage is utilised to regulate the speed of cold air fan. This control achieves two objectives, one to maintain the cooled meal temperature, and two, to optimize the power input to the Cooler Fan.

15.7.1.12 Temperature of Clean Water

The water discharged from the solvent-water separator is referred to as the 'clean' water, as opposed to the 'dirty' water circulated in Dust Scrubber. It has to be heated to around 85 C, for removal of solvent traces, before discharge from the plant. Some fluctuations in the temperature are allowed, in view of low solubility of hexane in water. Hence, a simple, self-acting control valve, is used for the purpose. This is inexpensive, and does not require much maintenance. It consists of a temperature probe filled with a liquid, or gas, which expands or contracts with temperature variation, a capillary tube which transmits the pressure to a valve which then opens accordingly. No external signal, either pneumatic or electronic, is involved, nor is a PID controller required.

15.7.2 Process Automation

As with the oil mills, there is a growing tendency to opt for automation of solvent plant operation. Plant automation allows for remote operation, saving on manpower, as also may help achieve higher production consistency and process efficiencies. Plant automation may be done at different levels, starting from partial automation just for data collection, to full automation and even auto-diagnostics. Since a solvent extraction facility consists of two separate buildings, housing the Preparation plant and the solvent plant proper, usually two separate automation systems are built, and operated from a common control room, mostly situated in the Preparation building.

The different levels of automation have been discussed in detail in Part 1. We revisit those briefly below.

At a primary level (Level 1), a PLC would incorporate all the control loops as also capture other critical process data, send it to a computer for storage and display, in form of data values on process flow sheets, as also as charts of individual parameters over time. Such a level of automation allows for better control of critical plant parameters, by the plant engineer.

The next level (Level 2) of automation includes, in addition to the scope of Level 1, collection of data on more parameters, which were otherwise monitored by field gauges, and incorporate operation of all drives from the screen. This level of automation allows remote-controlled monitoring, and start-stop, of the plant. However, some manual presence is still required, especially for changing/ adjustment of valves on oil/miscella and steam lines, with respect to levels in, and flow between, various equipment.

The next level goes (Level 3) towards full plant automation. With addition of auto On/off valves on pipelines, and level sensors on various equipment, manual intervention becomes redundant. Installation of the level sensors and flow-meters for liquids, allows real time material balance. This level of automation requires significant additional investment and is cost-effective mainly for large plants. Most large capacity plants, more than 1,000 Tpd, have already attained all the three levels of automation.

There can still be a higher level of automation, with 'supervisory controls', which would allow intelligent diagnosis of plant disturbance and take auto-corrective actions. With the advent of artificial intelligence technologies, supervisory control systems may be seen in near future.

15.8 FUTURE TECHNOLOGICAL POSSIBILITIES

Three possibilities are discussed here, one each in the three sections of a solvent extraction plant. A suitable extractor for rice bran, and may be for other oilseeds which face the problem of poor solvent percolation rates; improvement in desolventising operation, so that very low residual solvent may be achieved for all oilseeds; and low-temperature distillation.

15.8.1 New Extractor for Rice Bran (and Low-Protein Seeds)

As discussed earlier in this chapter, and the chapter before, oil extracted from rice bran is very dark, as high as 3–5 Red units (as measured in Lovibond ¼ inch cell). It has also been discussed that the colour is not native to RB, but is generated mostly during pelleting operation, and increased during distillation.

RB as it is received from rice mills is in fine powdery form. It is easily extractable by solvent in a stirred vessel. But the stirring action is not possible in a 'percolation extractor'. When solvent extraction of RB was started, during the 1960s, almost all the extractors in use were of percolation type. The RB powder would not lend itself to extraction in a percolation extractor, because of poor percolation rates through the powder. So it was thought to make the RB powder into pellets to enhance the percolation rates.

Should we not rethink the use of a percolation extractor for rice bran, which is easily extractable in its native form? Can the stirring action, employed in a lab extractor, be replicated on industrial scale? Can the 'immersion' extractors of yore be utilised for RB oil extraction?

The immersion-type extractors had two main issues: one, they were very tall, and two, the miscella contained too much fines and had to be clarified with an elaborate process line. The third issue, of course, was the risk of solvent leakages from the flooded extractor.

If we are able to overcome these issues, it would provide a very suitable solution to RB oil extraction. The oil could be as light coloured as say soybean oil. Refining would become simple, refining losses would reduce, and refined oil colour could be light yellow, in place of reddish, as preferred by consumers.

Should such an extractor be successful on RB, could the application be extended to other oilseeds whose cake is so fragile, that it crumbles under solvent sprays, creating fines which severely restrict the percolation rates? A prime example is the cake of shea nut. In fact, as a general rule, cakes of all the oilseeds which are low in protein content, as sheanut, palm kernel as also 'neem', are fragile. So, can the new extractor be useful to all the low-protein oilseeds too?

Alternatively, can the low-protein, high-carb seeds be prepared differently? If cooking is performed with high moisture, say 15%, will the carbohydrates gelatinise, to form a strong matrix, and hold flakes together in solvent? For Palm Kernel and Sheanut, it might be preferable to increase moisture during pre-heating, and then reduce the moisture to optimum press requirement during cooking. The high moisture during pre-heat may help achieve better softening of cell walls, leading to wall damage on flaking; it may also reduce energy requirement for flaking.

15.8.2 Desolventising under Vacuum

As discussed in the section on DTDC, a 'flash' chamber, operating under vacuum, has been introduced by leading plant suppliers during the past decade. At present, the 'vacuum' is low, just about 10–20 Torr. Under this vacuum, the flash chamber may at best act as a safety against mis-operation of the DT, and not as a final desolventising stage. If at present, meals from most oilseeds (other than soybean) contain nearly 300 ppm solvent or more, it would require exposure to much higher vacuum to reduce the residual solvent further.

May we consider a separate unit, to act as the last stage of desolventiser, designed to operate under high vacuum, as in case of distillation where the last equipment does operate under high vacuum? What kind of equipment that might be?

15.8.3 Mild Distillation

Some oils contain colour pigments, which tend to combine with oil molecules during final stages of distillation, above say 80 C. Examples are cotton seed oil, and RBO produced from pellets. Such association creates what is commonly called 'colour fixation'. These 'fixed' colours are difficult to be removed by adsorption on bleaching earth during the bleaching operation of oil refining. If distillation temperature were restricted to 80 C, this fixation of colour could be minimised. Such a redesign of distillation operation would, of course, require higher vacuum for the last stage of distillation. Quite likely, the extra steam required to create higher vacuum may be balanced by lesser heat required to heat the oil.

Another aspect of 'mild' distillation is low dwell-time of oil at elevated temperatures. The lower the dwell time, the lower will be red colour development. As seen in the chapter on soybean extraction, recent advances have enabled the production of soy oil with low red colour, less than 1.0 units (in ¼ inch cell, Lovibond scale), compared to 1.1–1.2 units in most modern plants. Our aim should be to reach 0.6 units of Red, as this is the colour of chemically neutralised and washed oil during oil refining. If this low Red level could be achieved, it might help simplify the entire refining process.

16 Meal Pelleting and Edible Products of Soybean

Toasted meal of soybean may be bagged and used in feed formulations. Untoasted meal, on the other hand, may be utilised to make different edible products, such as soybean flour, soybean nuggets (meat analogues), soy protein concentrates and soy protein isolates. The concentrates may further be utilised to produce several other food products. Isolates are further used in food and pharmaceutical product formulations.

Cottonseed meal may directly be used, along with hulls and soapstock, to produce feed pellets. Enhanced meal of cottonseed, high-pro and very low on free gossypol, is also entering the market.

These products and their production processes are discussed briefly below.

16.1 SOYBEAN FLOUR

White flakes may be ground to -100 mesh size and bagged as soy flour. Grinders used for the purpose are usually either hammer mills, or the 'universal' mills, with two grinding wheels with spikes. The mill must be properly ventilated to aid product discharge and also to prevent dusty surrounding.

In some operations, the flakes are graded by size, and only the fines, passing through 40 mesh sieves are used to produce flour. The 'overs' may be used to produce soy nuggets.

16.2 SOY NUGGETS (TEXTURISED)

White flakes are usually broken with light hammering. The product is graded by size, with fines going to make flour, and grits of size −20 + 40 mesh used for the production of nuggets.

The soy grits are metered into a twin-screw mixer, where water is added to make moisture content up to 25%. The mixed wet flour is fed to an extruder. These are typically short barrel extruders, with high shaft speed of nearly 300 rpm, and worm overall compression ratio of nearly 3:1. The high speed is critical for the shear required to make meat analogue structure. Extruder barrel consists of several sections bolted together. Each section is provided with a jacket for water cooling. Discharge temperature is nearly 110C. The noodles coming out of the die-plate are cut to about 15 mm size, with a cutter blade rotating close to the die-plate. The nuggets are cooled slowly over 30 minutes on a slow-moving horizontal screen.

Premium quality nuggets (meat analogues) may be produced from soy protein concentrates; these have not only higher protein levels, but are free from the native 'beany' flavour. Also, the protein level of nearly 70% matches with those in meat products.

DOI: 10.1201/9781003309475-18

16.3 SOY PROTEIN CONCENTRATE

White flakes typically contain about 47–48% protein (those from Asian and African beans contain typically 49–51%). For some food applications, higher protein content is required, as also the content of soluble sugars should be low since the latter interferes with protein functionality. This is achieved by the removal of soluble carbohydrates by aqueous ethanol extraction, to boost the protein content to nearly 65–70%. At protein concentration of 70%, most of the soluble sugars have been removed, with the balance 30% consisting mainly of insoluble fibre and minerals. Protein concentrates are free of the 'beany' flavour and attract much premium than the increase in protein content alone can justify. SPCs may also be produced by the dispersion of flakes in hot water or acidic water, to solubilise soluble sugars, which are then removed by centrifugation. The wet solids may again be dispersed in water and spray-dried. However, the ethanol extraction process removes more sugars than the other two processes.

Percolation extractor is used for the ethanol extraction process. The bed has to be shallow to achieve a reasonable percolation rate of ethanol. Because of low percolation rates, the extraction time is long, and the extractor size is more than twice of soybean oil extractor of similar tonnage capacity. Desolventising methods are the same as those for white flakes, namely, the vertical desolventiser with indirect heat or the FDS. Live steam cannot be injected, even for finishing step, since it is soluble with ethanol. The miscella is concentrated to contain 50% sugars, rest being mainly water, in a multi-stage column (a 'Beer' column), to what is known as 'soybean molasses' and is used mostly as binding agent in animal feed pellets.

Typical capacities of plants for soy protein concentrates are between 10 and 100 Tpd. Smaller plants, up to 20 Tpd, may also be of batch design.

16.4 SOY PROTEIN ISOLATE

Spy protein isolate (SPI) is almost a pure protein product, with protein content near 90%. It is almost totally free of carbohydrates, including those sugars causing flatulence.

SPI is produced by dispersion of untoasted soybean flakes, of high PDI, in alkaline water, solubilizing the protein at alkaline condition, filtering out the solids and precipitation of the protein from the filtrate at pH of 4.5 (isoelectric pH). The precipitated product is spray-dried to obtain SPI powder.

16.5 COTTON SEED MEAL PELLETS

As seen in the section on cottonseed processing, five products are obtained from the solvent extraction of cotton seed. These include the main products, namely, oil and meal, and the by-products, namely, the hulls, lint and soapstock. Oil and lint have ready markets. The meal has to be sold in competition with soybean meal, usually at a heavy discount due to lower protein content (38%), dark colour and the presence of free gossypol. Disposal of hulls and soapstock is often difficult, as the markets are limited. A solution to these issues is to make pellets using all the three products, namely the meal, hulls and soapstock, and sell directly as cattle feed. The pellets may contain around 25% protein and 2–3% fat, which is acceptable for cattle feed.

Production process for the pellets is quite simple. Meal and hulls are mixed in a mixer screw (paddle screw, ribbon mixer, twin-screw), soapstock is heated and sprayed onto the mass in the mixer, and some steam injected to increase moisture to 11–12%. The screw may be jacketed to increase the temperature to 55–60C. If the meal is taken online from the DT, without cooling, the jacket steam need not be used. Higher moisture level helps in binding of components in pellets. The mixture is fed to a pellet mill, usually a vertical mill, to produce pellets of dia 10–15 mm. The hot pellets may be cooled quickly in a vertical Cooler, a zig-zag channel, as shown in Figure 5.2 (Chapter 5), with cross-current or counter-current flow of ambient air.

There is also a market for Hi-pro pellets. These may be made with meal and soapstock, without mixing the hulls. Intermediate grades may be made with partial mixing of hulls.

16.6 DE-GOSSYPOLISED COTTONSEED MEAL

One of the reasons for the lower value of cottonseed meal is the presence of free gossypol, apart from lower protein content compared to soybean meal. Extraction with aqueous ethanol, as for the production of soy protein concentrate, can increase the protein content up to 50–52%, thanks to the removal of soluble sugars and also reduce free gossypol to less than 100 ppm. The product is a hi-pro, very low in free gossypol, and lighter in colour, cottonseed meal. Plants with up to 100 Tpd product capacity have recently gone into production.

17 Bagging and Storage of De-oiled Meal

Soybean meal consists of some quantity of lumps, formed by condensation of steam on flakes surfaces. Meals from most cakes (and pellets/collets) also contain pieces of broken cake along with powdery mass. These may be required to be ground to produce mass of near-uniform size. For this purpose, the lumps, or pieces of cake, may be separated and ground, before mixing with the rest of meal for bagging.

17.1 LUMP SEPARATION AND GRINDING

Lumps may simply be separated using a rotary screen. Vibrating screens are not required in view of the big size difference between the lumps and the powdery mass. The screen may be a perforated cylinder with 8 mm dia holes.

Lumps discharged at the end of screen are fed to a hammer mill. This typically is a low-speed operation, in view of the moderately soft lumps or cakes. Screen perforations may be 8–10 mm dia. Should the lumps, or cake pieces be 15–25% of the meal, the energy consumption in grinder may be around 0.7–1 kWh/T meal.

The ground product flows into the conveyor which carries the screened meal to bagging machines. A typical flow sheet for the section is shown in Figure 17.1.

17.2 MEAL BAGGING

In small plants, capacity of 500 Tpd or less, meal may be bagged manually, especially in industrially developing countries. For larger plants, however, auto-bagging machines are recommended, and these are the norm.

An auto-bagging machine is basically the same as an auto-batching machine, as discussed in the section on online weighing, but with a clamp attachment for bags at the bottom (Figure 17.2). Typically such a machine can help fill 5–6 bags per minute. If a 50 kg meal is filled per bag, this translates to nearly 360–430 Tpd meal. Thus, a single machine may suffice for a 500 Tpd soybean crushing plant. Bag size may vary, up to as much as 1,000 kg (Jumbo bag).

A small buffer is provided in a Feed Hopper mounted above each bagging machine. The hopper may be sized for 30–50 bags weight of meal. When multiple bagging machines are used, a distribution conveyor is mounted overhead to distribute the meal to all hoppers and take any overflow to an overflow bin.

Bags released from the bagging machine are conveyed by a slat conveyor to a stitching machine mounted at a distance of say 2–3 m. Stitched bags may then either be taken away for stacking manually or handled on a conveyor system including Stacker conveyors.

DOI: 10.1201/9781003309475-19

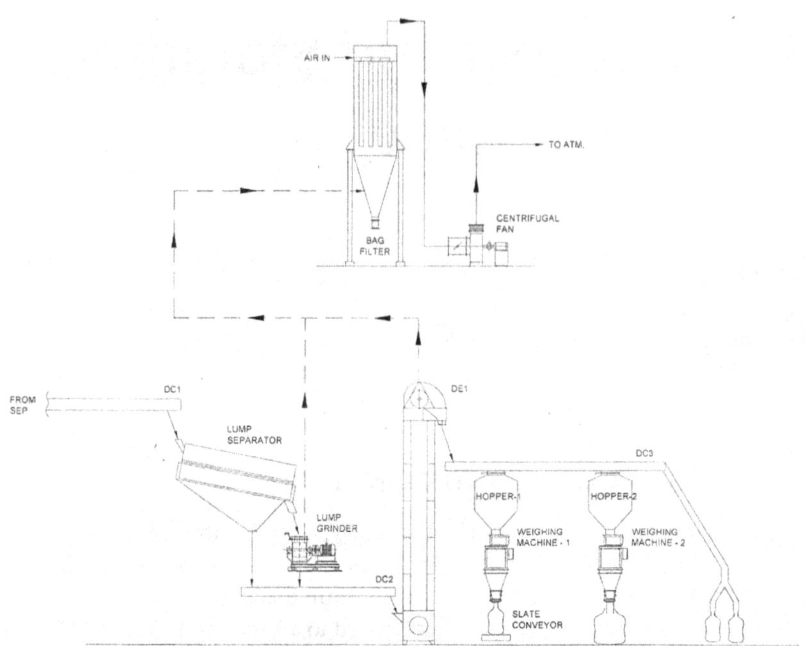

FIGURE 17.1 Flow Sheet for Meal Sizing and Bagging.

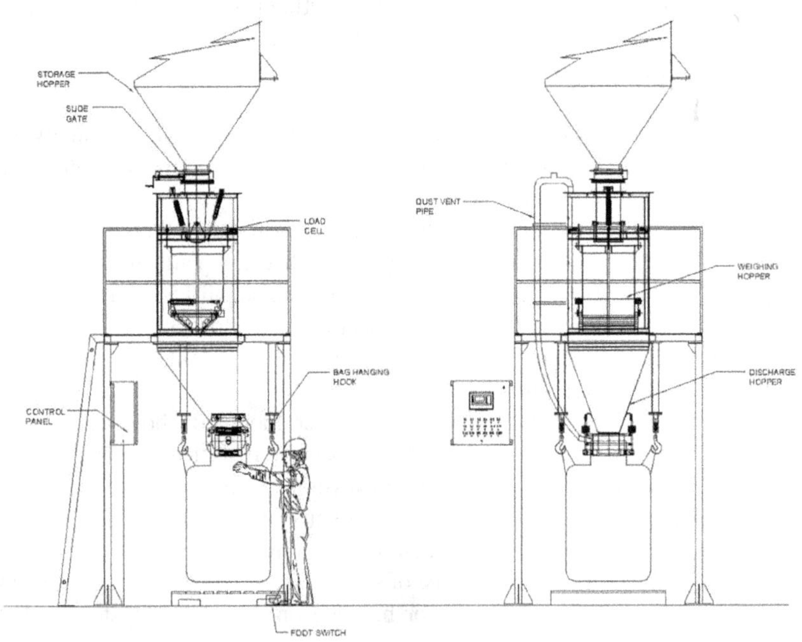

FIGURE 17.2 Auto Bagging Machine. Courtesy: Precia Molen India Pvt. Ltd.

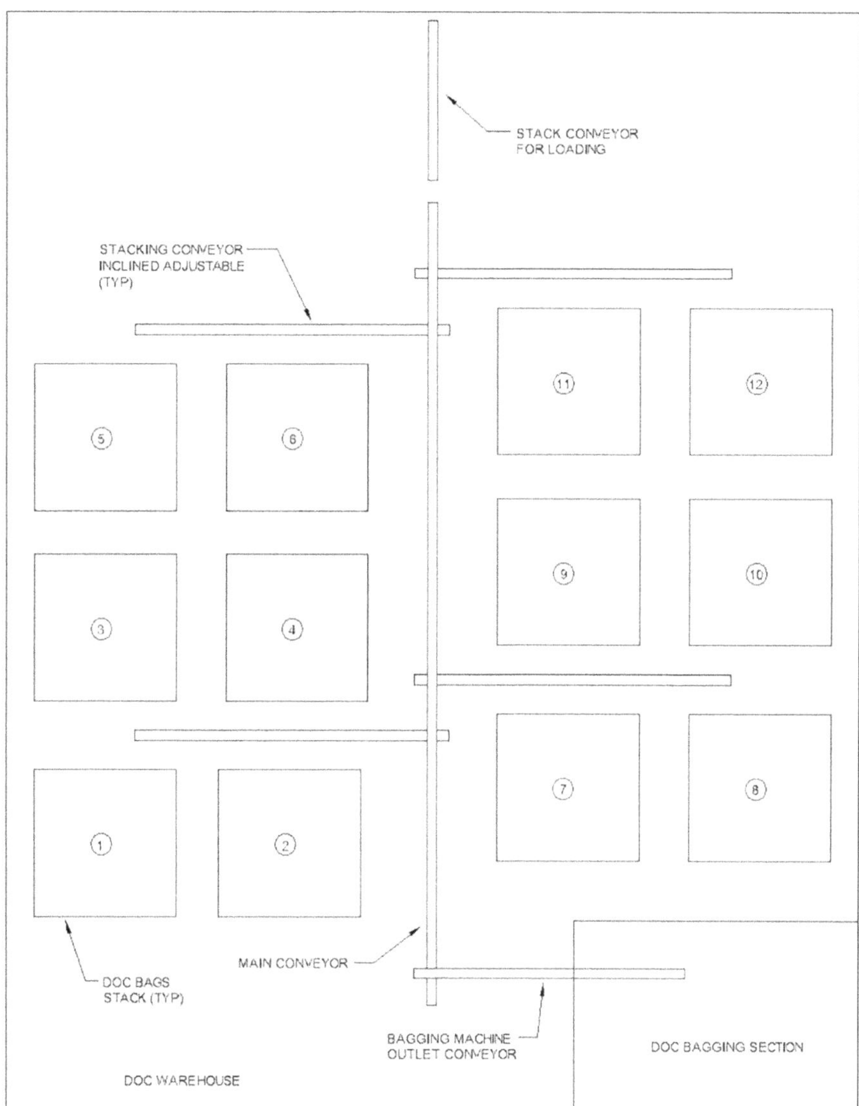

FIGURE 17.3 Conveyors Arrangement in Meal Warehouse.

A small buffer storage, for a few hours, may be provided upstream of the bagging section. This would provide for sufficient storage for 3–4 hours, to cover for the lower productivity of workmen at night. To avoid bridging, and ensure free discharge, this storage may be divided into multiple bins, sized for say 1 hour each. Each of these would be fitted with a sliding gate at bottom. A common discharge conveyor, of tubular screw design, driven through a variable speed drive, serves to ensure uniform supply of meal to the bagging section.

17.3 MEAL STORAGE AND DISPATCH

Bulk storage of meal is not common, unlike seed storage, since meal tends to compact, and at times cake up, when stored in bulk for a length of time, especially at moisture levels above 12%. Even when held in bulk for a day, it may bridge and hamper free discharge. Thus, while a short storage of 3–4 hours may be provided upstream of bagging, as discussed above, only bagged meal is stored in warehouses.

While in small warehouses, and for small plant capacities, bags may be moved, stacked and shifted for dispatch, manually; for plants of capacity of more than 500 Tpd meal, a conveyor arrangement should be provided for the purpose. The most common conveyor type for bag handling is belt conveyor.

A common arrangement of conveyors is shown below (Figure 17.3). The main conveyor runs the entire length of warehouse starting from the line of stitching machines. The bags are carried over the length, and may be routed on any of the 'stacking' conveyors placed perpendicular to the main conveyor, by means of diverters. Bags are stacked in rows in between the cross conveyors. Movable stacking conveyors, whose angle can be adjusted, are used for stacking. Stacking to a height of about 4–5 m, in a warehouse 7 m high, is quite common.

Alternatively, an overhead conveyor may run in a circular route, and chutes may be used to slide down the bags at desired locations to mount the stacks.

Retrieval of bags from stacks for dispatch is achieved using the same conveyor system. The stack conveyor is run reverse to get bags to ground level, and bags are shifted by the main conveyor up to the loading point. Here again, another stack conveyor may be used to move up the bags to the truck or container.

The use of the conveyor system helps reduce manpower requirements to a minimum, for medium and large-capacity plants.

18 Effect of Processing on Quality of Oil and Meal

Plant operations are set, and machines are tuned, so as to give optimum production and consistent quality products. The impact of slight mis-operation of a machine may not be obvious at the particular stage of operation, but it may show in the final product quality. Such issues have been mentioned in the chapters on operation and design aspects of equipment. However, it is worthwhile to discuss these together, to bring these into focus.

18.1 OIL QUALITY

The quality of solvent-extracted oil is affected by several factors, critical among these are: overheating in cooker during seed preparation, high shear on oilseed mass as during press operation or pelleting, presence of solids in miscella, separation of gums during distillation, and overheating in distillation (Vadke, 2015b).

It is quite straightforward to control the overall temperature in cooker closely. However, variations in heating of parts of seed grits may happen because of two mis-operations. One, if fines are created during the cracking operation, these could get overheated even while the temperature of the mass is within the desired range. Second, should there be uneven clearance between sweep arms and the double-bottom plate, of a vertical stack cooker, it might create pockets where grits are held much longer than the average dwell-time. **Uneven heating** is reflected in the uneven colour of cooked mass. Both these pitfalls should be avoided by proper monitoring; regular sieve analysis of cracked grits can point to the presence of fines. Such an analysis also helps identify the presence of un-cracked seed or splits, which would cause an increase in ROC.

The effects of excessive shear within a press, and ways to control it, have been discussed in Part 1 of this book. Effects of pelleting of rice bran, and ways to overcome it, will be discussed separately in the following section.

Should fine **solid particles**, part of seed, get carried over in miscella, these may deposit inside heater/evaporator tubes, get charred over time, and impart dark colour to oil. Solids must be filtered out before the miscella is sent for distillation. Miscella filter, discussed before, is a good solution to this problem. A good standard screen opening is 50 micron, although 30 micron may be even better. It should be noted that the red colour in crude oil is indicative of the damage to oil during processing. In most oils, red is not a native colour, yellow is. Exceptions are: one, palm oil, which contains red carotenoids, but that oil is not extracted by solvent; and two, cottonseed oil, which contains gossypol the red pigment.

Gums, or phosphatides, are present in every oil, in less or more proportion. Soybean has high gums content, as much as 2.5–2.8%. On the other extreme,

DOI: 10.1201/9781003309475-20

lauric oils such as coconut oil and palm kernel oil have very low levels, less than 0.2%. When high-phosphatide oils are extracted, and if miscella contains moisture, the gums may separate out in the distillation section. The moisture may either come from high-moisture beans, or when water is added in process, say in miscella tank for settling fines, or from wet 'live' steam.

The separation of gums is enhanced at higher temperatures, above 90C. Should the gums separate out of miscella, those would deposit in evaporator tubes or on surfaces of stripper; they would get charred over time and impart dark colour to oil. Thus, distillation temperatures should be maintained low as possible. Also, shorter dwell times in Evaporator and Stripper help minimise the problem.

Overheating in distillation is a sure recipe for the darkening of oil. Temperature of miscella at Evaporator is auto-controlled to avoid this risk. Quite often, should the vacuum be low than desired, for whatever reason, the tendency of plant operator is to increase the miscella temperature in Evaporator (changing the set point), so as not to have 'flash' in extracted oil. While this is a correct immediate remedy, attention must be paid to improve the vacuum and reduce the temperature, in the shortest time possible. At times, the flash at lower temperature, than say 150C, is a result of less stripping steam. Stripping steam flow may be monitored via steam pressure upstream of an orifice plate fitted in the stripping steam line.

18.2 QUALITY OF RICE BRAN OIL

Crude rice bran oil (RBO) is quite dark, with red colour between 3 and 5 units (on ¼ inch cell Lovibond). Much of this red colour is produced during RB pelleting, by reactions between proteins and carbohydrates. During distillation, this colour gets 'fixed' and is very difficult to remove during the refining of RBO. Thus, even after double-bleaching process, refined RBO is still dark.

The 'browning' reactions' are aided by shear, temperature, moisture level and exposure to air. These are also time dependent. Thus, the red colour generation may be reduced by (a) reducing the shear, say by chamfering the holes in die-plate of the pellet mill, for smooth entry; (b) conditioning the RB at lower temperature, may be for a longer duration, and (c) minimising the hold-up of RB in the pelleting mill. However, while all these steps may help reduce the generation of red colour, it would still continue to be a formidable problem. The real solution may lie in changing the preparation method altogether; this issue has been discussed in Section 15.8.

As discussed before, part of the problem, the 'colour fixation', may be minimised with low-temperature distillation.

A question may be asked, if the bran is not pelletised and is extracted in its powdery form itself, would the problem of dark colour oil be solved? This matter has been discussed in the section on 'future innovative technologies' (Section 15.8).

18.3 QUALITY OF LECITHIN

The quality of lecithin may suffer generally on two counts, due to mis-operation in seed preparation and solvent extraction; one, the insolubles content, and two, the colour of lecithin.

The (hexane) insolubles content is presently required to be within 0.3% for food-grade lecithin. It is not only the solids from seed that affect the insolubles content but also soil particles coming with seed. While most of the solids from seed may be filtered out at miscella stage using a proper Miscella Filter, very fine soil particles may escape the MF. The soil particles must be removed at the stage of cleaning of seed. Soil particles may escape with 'clean' seed if the Cleaner were overloaded, or if the opening in the lower screen in the screen deck were too fine, or if air aspiration were not effective. The insolubles content in soybean lecithin may be controlled within 0.3%, if those were less than 0.04% in crude oil.

Colour is required to be 14 max., on Gardener scale, but the lower colour of 12 units is preferred for food-grade lecithin. The colour may suffer either if colour of crude oil were high, or due to overheating in the lecithin production process. The issue of oil colour has been discussed in the section on oil quality. Overheating in the lecithin-making process may be avoided by using a good scraped surface heater, lower heater temperature and good temperature control. Lower heater temperature is possible, even while ensuring low moisture in lecithin, by maintaining high vacuum. While standard lecithin drying process employs vacuum at 50 TorrA in most plants, a higher vacuum at 20 TorrA would enable adequate drying at 85–88 C temperature.

It should be noted that colour generation reactions are time-dependent and require exposure to air. In many plants, the gums are pre-heated to 70–75 C in 'blenders' which operate at atmospheric pressure. This is a sure recipe for the generation of red colour and should be avoided. The vessel may only be utilised to mix and maintain the temperature of gums at around 55–60 C. All the further heating and evaporation, should be accomplished under vacuum.

18.4 QUALITY OF SOYBEAN MEAL

The main quality parameters sought in toasted soybean meal are protein content and urease activity. These may readily be maintained with steady operations of the dehulling section in Beans Preparation plant, and by control of DT operation, respectively. However, sometimes, there is a problem of the uneven colour of meal. Some portions of the meal appear darker than the rest.

Overheating may occur if the top plate of the double bottom develops un-evenness, or if the clearance between the plate and sweep arm is large. Pockets of particles may remain un-swept on the plate for longer durations and would get overheated. These would normally show up as only small dis-coloured portions.

At times, major portions of the meal may appear dis-coloured. Should the sweep arms undergo deformation, and the clearance from the plate increase overall, bigger portions of the meal may get overheated, and the darker part may not remain as just a few particles.

The urease activity (U.A.) can be controlled by the manipulation of open steam quantity into the DT. The lower the live steam, the higher the U.A. and vice versa. The easiest way to regulate the live steam quantity is to set the gas temperature (DT duct) as required.

18.5 COTTON SEED MEAL

The main quality requirement for cotton seed meal is the level of 'free gossypol'. For some applications, a level below 500 ppm (mg/kg) is required, while for some other, even as low as 200 ppm is sought. However, in most meals the level is between 300 and 600 ppm.

There exist methods for gossypol removal from meal, and these are now starting to gain economic viability. A few plants have recently started production of de-gossypolised meal, as discussed in Section 16.6. There have also been efforts, over the past few decades, to grow cotton varieties whose seed would be almost free of gossypol. However, these have not been commercialised yet.

A simpler way, to achieve low free gossypol (FG) would be to target lower residual oil in meal, since much of the FG is associated with oil. For example, meal produced by 'un-drained expander' method, with 0.9–1, 0% ROC would typically have over 500 ppm FG, whereas meal produced by 'drained expander' method, with 0.6–0.7 ROC, would have less than 400 ppm FG. If, by better preparation of cotton seed grits, ROC less than 0.5% be ensured, we may expect to reach FG level closer to 250–300 ppm. This may be achieved with better conditioning of flakes, specifically, with increasing the time of conditioning from 4-5 minutes, common at present, to 20 min. Conditioning temperature may also be raised to 80–85 C, similar to that used in preparation prior to pre-pressing. Slight darkening of meal, due to higher temperature, is not a worry, since cotton seed meal is quite dark anyway.

It may be noted that the de-gossypolisation of meal, with ethanol/methanol extraction, not only reduces the FG content, but also helps produce lighter meal, thanks to the removal of the red gossypol pigment.

19 Utility Requirement of Oil Extraction Plants

Utility requirement, **steam and electricity**, may be calculated for each equipment, for each seed, with knowledge of the operating parameters. These figures for some of the main machines and equipment have been discussed in earlier chapters. However, for the benefit of plant engineers and managers, it would be useful to know the benchmarks for 'standard' utility requirements.

It may be noted that the steam and electricity requirements depend not only on the oilseed being crushed but also on plant capacity. The electrical energy input especially varies significantly with plant capacity. Typical figures are mentioned for various seeds, for 'common' plant capacities mentioned for each. For a plant double the size, one would consider nearly 15% reduction in specific electricity requirement and 5% in steam, as also a 5% reduction in solvent losses (Table 19.1).

Soybean is the easiest material to extract, and also easiest to desolventise, which results in low steam consumption, and low solvent losses. Shea nut is the most difficult, as discussed before, and requires high energy inputs, both electricity and steam.

While comparing the figures for various oilseeds, the respective plant capacity specified must be considered. As mentioned above, larger capacity plants require less specific utilities, especially electricity.

It may be noted that the steam pressure required in seed preparation and in SEP ranges from 3 to 10 BarG. Solid heating equipment such as the Cooker and the DT are normally designed for steam at 10 BarG, whereas heating of miscella is with L.P. steam @ 3–4 BarG. Air heaters, for DC, are better provided with M.P. steam @ 10 BarG, while vacuum steam ejectors may be designed with motive steam pressure of 8–9 BarG. However, it is possible to design all the equipment, including DT, air-heaters and ejectors, to operate at 3 BarG; this flexibility opens up the possibility of own generation of electric power by means of co-generation.

Since solvent plants require both steam and electricity on continuous basis, it is worthwhile to consider the option of **co-generation of power**, to produce electricity from high pressure steam, and utilise the L.P. exhaust steam in the plant. This enables the production of electricity in-house, and at low rates. Such a system is especially good for areas where either the power supply from utility companies may be erratic, or rates of electric energy may be high. In fact, solvent extraction plants are almost ideal candidates for co-generation systems, since the ratio of steam and electric energy requirements tends to be in ideal range to meet all the demand for the latter (Vadke, 2015a). Applicability of Co-generation systems for solvent extraction plants is discussed in Appendix 2.7.

Apart from the two main utilities and solvent requirement, there is a **daily water requirement** to operate the plant. Water is required not only for steam generation but also to make up the evaporation losses in cooling tower. Part of the make-up water

DOI: 10.1201/9781003309475-21

TABLE 19.1
Utility Requirements for Solvent Extraction of Various Oilseeds

	Steam, kg/T			Electricity, kWh/T			Solvent
Oilseed / Cake	**PREP**	**SEP**	**Total**	**PREP**	**SEP**	**Total**	**Loss, kg/T**
Soybean 800 T	80	190	270	15	11	26	1.2
Rice Bran 400 T	70	230	300	12	12	24	1.8
Cottonseed Meat[a] 400 T (WCS 600 T)	100	230	330	24	15	39	1.8
Rapeseed cake 500 T[b]		200	200	2	14	16	1.8
Sunseed Cake 500 T[b]		180	180	2	13	15	1.5
Palm Kernel Cake 300 T[b]		220	220		17	16	4
Sheanut Cake 200 T[b]		250	250		20	20	4

[a] Figures are on the basis of cotton seed meat; so, consumptions per T of WCS would be just 65% of these.
[b] Figures are on the basis of Cake tonnage; so, consumptions per T of seed would be just about 0.7 or 0.5 times these (please refer the material balance in Part 1 – Appendix 4).

TABLE 19.2
Utility Requirement for Prepress Plus Solvent Extraction of Rapeseed and Sunflower Seed

	Steam, kg/T			Electricity, kWh/T			Solvent
Oilseed	**Prepress**	**SEP**	**Total**	**Prepress**	**SEP**	**Total**	**Loss, kg/T**
Rapeseed[a]	90	140	230	25	10	35	1.3
Sunseed (whole)[b]	90	115	205	50	8	58	1.0
Sunseed (dehulled)[c]	80	85	165	40	6	46	0.7

[a, b] Utility figures in solvent plant (SEP) are calculated with quantities of presscake derived from 700 Tpd rapeseed and sunflower seed, with oil contents 40% and 45%, respectively.
[c] Utility figures in solvent plant (SEP) for dehulled sunflower seed are calculated considering 15% hulls removed.

requirement may be met from the 'clean' water recovered from the solvent-water separator. Should this water be free from impurities, as from the interlayer emulsion, it may be circulated to the cooling tower as make-up water. Such water-saving methods may become critical in areas with water scarcity.

We have discussed the utility figures encountered in pre-pressing in Part 1 of this book. Let us now combine the figures for oil mill and solvent extraction, to get an overall picture. Table 19.2 shows utility figures for rapeseed and sunflower seed, as these are the two most important oilseeds which are subjected to the prepress plus solvent extraction process worldwide. Capacity considered is 700 Tpd on seed basis. The corresponding quantity of presscake entering the solvent plant would be nearly 500 Tpd, 330–350 Tpd, and 440–470 Tpd on rapeseed, sunflower seed (whole) and sunflower seed (dehulled), respectively. The quantities of presscake may vary depending on oil content and moisture of seed – please see material balance calculations in Appendix 4 in Part 1. The utility figures in the solvent plant have been presented on the quantity of seed basis, instead of presscake, so that the numbers may be added to those in prepress.

It may be noted that the above figures are average figures for modern mid-size plants. Actual figures may somewhat vary depending not only on plant size, but also on consistency of plant operation and technology employed.

20 Environmental and Safety Issues

20.1 SAFETY ISSUES AND PRECAUTIONS

The importance of safety practices can never be overstated. Safety is not only to be followed during the erection of a process plant, but is a continuous process. If a plant operation is trouble-free, it can sometimes lead to complacency and can seed habits which may cause risks in future. Many factories, therefore, have ongoing safety training programs.

20.1.1 COMMON SAFETY ISSUES

Common safety issues faced by most factories, including mechanical and electrical issues, have been listed in Part 1. Apart from those, several specific safety hazards may be encountered in solvent plants.

20.1.2 FLAMMABLE AND EXPLOSION-PRONE SOLVENT

The first and foremost hazard is the flammable nature of the solvent used. Hexane is categorized as Flammable Class IB solvent, which means the hazard risk is 'moderately high'. In case of fire, large quantities of water may be pumped to flood the fire area. However, water spray should not be used, as it may cause spread of fire. Fire extinguishers, containing foam, dry chemicals or carbon dioxide, may be used on small localized fires. For the purpose of water flooding, fire hydrants are installed all around the solvent plant, just outside the boundary wall of the safe zone.

Since hexane at low concentrations in the air may cause explosion, all electric fittings have to be flame-proof and spark-proof. The range of concentrations, for the explosion to happen, is 1–8% v/v in air. This range may be reached during the maintenance of plant; thus, it is extremely important that all implements and tools used in plant maintenance, unless the plant has been de-solventised, should be spark-proof.

Short exposure to hexane vapours may cause drowsiness, while chronic exposure may cause serious problems of neuro-systems. Also, exposure to the skin over prolonged duration may cause skin irritation. Therefore, if in case a workman has to enter an equipment, say the Extractor, he must use full-body suite with oxygen supply. In case of eye contact, eyes must be rinsed with water immediately.

The precautions in plant design, on account of explosion and fire hazards, such as safe area, boundary wall and sprinkler system, etc., have been listed in the discussion of plant equipment and plant layout.

DOI: 10.1201/9781003309475-22

20.1.3 Fire Hazards

In a solvent plant, there are two specific areas which are subject to fire hazard. Both these are around the DTDC. One is the Dryer stage itself, and the other is the Air Heater.

The dryer stage of the DTDC is equipped with an Air Chest below. Air rises from the chest through holes in the upper plate and then contacts the wet warm meal in the stage. Due to mis-operation, fines may drop through the holes and collect in the chest. As hot air, at 130–140C is blown in, temperature of the fines rises, these become bone-dry over a period; and may catch fire. A live steam line is provided to the chest, to douse fire. However, if the smoke, which emerges prior to fire, is not noted, and steam valve is not opened, the fines may catch fire. Fire may spread into the dryer stage, and then in to the exhaust air duct.

This hazard may be prevented by two corrections. First, the hole size should be small, not more than 5mm. Second, and the most important, the DTDC shaft and the sweep arms must never rotate unless there is air pressure in the chest. This may be achieved by simply interlocking the two drives, DTDC and Hot Air FD Fan. An additional safety may be to fit a pressure-switch on the air chest. In case of low pressure, below a set value, the DTDC drive may stop.

The second location is the Air Heater. Should the inlet air be dusty, the dust would accumulate on the fins that are fitted on the outside of steam tubes for increased heating surface area. Over time, the dust fines would become bone-dry, get overheated, and catch fire. This again is noted first as smoke emerging with exhaust air of the cyclone. At the first sight of smoke, live steam should be opened in air chest. If the smoke does not subside even with live steam, it is an indication that the location of fire is not the air chest, but is the air heater. In that case, heating steam to the air-heater must be closed.

To prevent the risk of fire on air heater, air suction should be from dust-free area. If air cyclones are located near the air heater, the surrounding atmosphere may be dusty. Dust from the cyclones may be minimised by (a) using high-efficiency cyclones and (b) preventing dust at cyclone bottom discharge. As discussed earlier, the cold air cyclone is better replaced with a bag filter. It is a good practice to inspect the fins of air heater, for fines deposition, on regular basis. Normally, an inspection once a month, on the monthly preventive maintenance day, should suffice.

In the past, there have been fire accidents in meal bagging area. If inadequately desolventised meal were stored, the solvent vapours in some pockets could reach the lower explosion limit. However, with superior controls on DT operation, followed by air blowing through the meal for (drying and) cooling, this possibility is now minimal. However, hexane sensors should be placed, with audible alarm for high level, at different locations in the warehouse.

20.2 ENVIRONMENT ISSUES

There is no solid waste emerging from the solvent plant. However, issues of water and air pollution need to be tackled.

20.2.1 Water Pollution

In many plants, there is a single waste water tank. Thus the condensed water received from the solvent-water separator is used for scrubbing of dust from DT gases. This makes the entire water stream dirty, which must be treated in an effluent treatment facility. Typically, for a 600 Tpd soybean plant, water outlet flow may be as much as $2\,m^3/h$.

However, if the condensed water is heated separately, and a separate water tank is used for scrubbing water, the condensed water can remain clean. As discussed in chapter on equipment operation, if the interlayer between solvent and water (in the Solvent Water Separator) is removed on a regular basis, the water may be clean enough to be used as make-up water in cooling tower. Thus, the entire water stream may be reused, and environmental issue be avoided.

The scrubber water, in the second tank, does become concentrated with fines over time. Hence, a part of the water should be bled off, and fresh make up water added, to maintain the fines concentration. This bleed-off quantity is much lower than the condensed water flow rate. Typically, for a 600 Tpd soybean plant, bleed off may be less than $0.5\,m^3$ per day. This quantity is so small, it may easily be sent to a sand pit. If multiple sand pits, 3 or 4, are maintained, to be used sequentially, each for a day's quantity, the first pit would dry off before the last pit is filled. Thus, no water has to go out of the plant area.

Should sand pits not be allowed, by law, a small effluent treatment unit, basically a flocculation tank, and a clarifier, would suffice. Since the quantity of water is so small, the unit may be operated for a single shift, just once a week. Alternatively, the water may be evaporated with a fountain.

20.2.2 Air Pollution

Air that is vented out of the solvent plant contains some amount of solvent vapours. As discussed in the chapter on Vent Absorption system, vent air from most plants may contain up to 100 g solvent vapours per m^3. For a 600 Tpd plant, vent air quantity may be $30–40\,m^3/h$. Thus, solvent vapours escaping the plant may be up to 3–4 kg/h. While this much emission quantity may be within allowable limits in many countries, it is possible to reduce the emission to just $1/5^{th}$ or even less. The extra cost of operation to achieve this low level of emission is quite low, and is, in fact, more than offset by the cost of extra solvent recovered, especially since no additional chilling units are now required for the purpose.

So, better absorption systems, to reduce the solvent in vent air to $10–20\,g/m^3$, should be installed in solvent plants. These have been discussed in the relevant sections.

Apart from the emission of solvent vapours, there is the matter of dust in air, in seed cleaning section of the preparation plant, and in meal Dryer-Cooler in solvent plant, as also at meal bagging section. At all these locations, except for the Dryer, air bag filters should be installed; these are preferable to air cyclones, which cannot remove all the dust in air.

One location where the operation of air bag filter is difficult is the meal Dryer. The exhaust air from the Dryer is very moist and can foul the filter elements. In this particular case, an air cyclone may be installed; however, it must be a high-efficiency (high-pressure type) cyclone, to minimise the fines in outlet air, as discussed in the section on meal drying.

Appendices

APPENDIX 1.1 TECHNOLOGY OF PALM OIL PRODUCTION

Palm Oil: Palm oil is the largest produced vegetable oil in the world and is, by far, the most traded internationally. The main producers are the equatorial nations of Indonesia and Malaysia, which together contribute over 80% of the production. Other equatorial countries around the globe, including African countries of Nigeria, Ghana and Ivory Coast; Latin American countries of Columbia and Costa Rica, and several others, contribute the remaining 20%.

Fruit Bunches: Palm oil is obtained from the fruits of oil palm. The seed, within the fruit, also has edible oil; the technology of recovery of oil from palm kernel is covered in the main text of this book. Palm fruit bunches are plucked from the trees and brought to a mill located usually on the plantation itself. The key to good oil quality is to crush the fruits as early as possible, certainly within 24 hours of plucking. Longer the interval, the longer the action of lipase enzyme present in the fruit, and the higher the percentage of free fatty acids (FFA). The oil content, on 'fresh fruit bunch' (FFB) weight basis, is nearly 22–24%.

Oil Mill Sizes: Palm oil mills are sized based on the mass of FFBs processed per hour. Typically, process lines are designed for 10 T FFB/h to 40 T FFB/h. Several mills have multiple process lines within the same premises.

Reception and Sterilisation: At the palm mill, The FFBs are loaded onto cages made of perforated side plates. The cages are moved on rails into 'Sterilisers', which are pressure vessels where the fruit bunches are sterilised with open steam for nearly 2 hours. Objective of the sterilisation operation is twofold: one to arrest the enzyme activity; and two, to loosen the fruits on the bunch.

Threshing: At the end of the sterilisation cycle, cages are removed, lifted one by one, and fruit bunches are dumped into a hopper, from which these flow to a 'Thresher'. This is a rotary drum, wherein the bunches are lifted and dropped repeatedly, to separate the fruits on impact; fruits then pass through the perforated drum, and the empty bunches are collected at the discharge end (Figure A1.1.1). The fruits are taken for further processing. The empty bunches may either be sent back to plantation as natural soil conditioner; or pressed for de-watering, and burnt in boiler for steam production.

Digestion: Fruits separated in a thresher are continuously conveyed into a 'Digester'. This is a cooking operation, wherein the fruits are macerated into pulp, the pulp is heated to 85–90 C and digested for nearly 10–15 min. In a vertical digester, horizontal beater arms macerate the fruits into pulp, and open steam is injected to heat up the pulp. Proper digestion helps weaken the cell walls and helps recover oil in the press.

Press: The palm oil press is a low-pressure press, with a compressing screw, single or twin, rotating within a cage made of perforated steel sheet, supported by a light frame. Twin-screw design is more common (Figure A1.1.2). The screw is made of a

FIGURE A1.1.1 Thresher. Courtesy: Henan Pand Machinery Equipment Co.

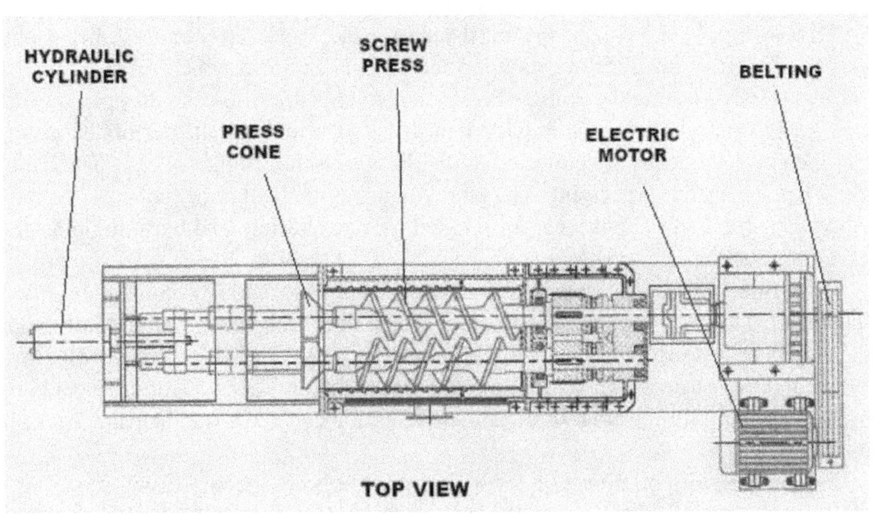

FIGURE A1.1.2 Oil Palm Press (Firdaus et al., 2017).

single stock, with machined flight profile. The reducing flight pitch and depth progressively increase the pressure on the pulp, squeezing out the oil and water, which then flows out of the cage perforations. The de-oiled and de-watered fibre mass, together with seed, is further processed for recovery of seed. Nearly 90% of the oil is recovered, along with water, leaving 7–8% oil-in-fibre.

There may be a scope to improve the design of press flight profile. The principle of 'reducing compression ratio', explained in the section on design of press wormshaft, may be utilised for the purpose.

Oil Clarification: The oil-water mixture is passed onto a vibrating screen for removal of fibre and is then subjected to gravity settling to remove sand/soil particles. Steam is injected into the settling tank, to help settle 'foots' by denaturation of proteins, which in native form cause foots to remain in suspension in water. A hydro-cyclone may also be used for separation of sandy particles. Since the sand is erosive, the cyclone should be made of wear-resistant steels. Once the proteins in foots are denatured, their emulsifying power is gone, and water and oil are ready for decantation.

The clarified oil-water mixture is then pumped into a high-speed decanter for separation of the two phases. Three-phase decanters are sometimes used, to separate out the remaining solids with a single operation. In a three-phase decanter, a screw is added to remove the solid phase.

Oil Drying and Cooling: The decanted oil must be dried to remove traces of water. Drying is typically accomplished by heating the oil to 90–95C and flashing under vacuum. The dried oil should be cooled to below 45C, prior to storage.

Nuts Recovery: The fibre-seed mass is subjected to multiple operations of low-speed beating and polishing to separate the fibre from seed (or 'nuts'). The separated fibre is removed by aspiration and collected under a cyclone.

Kernel Recovery: Nuts, free of fibre, are dried to reduce the moisture down to below 10%. This is typically done in vertical silos, with hot gases generated by burning of fibre. Most dryers are batch-operated, although some mills utilise continuous gravity flow dryers. Dry nuts are broken in a 'ripple mill', by crushing between a rotating roll made of rods of circular cross-section and static corrugated shell plates. The advantage of the ripple mill is that it can handle seed of different sizes, without damaging the kernel. The broken pieces of shell may be separated by aspiration. The kernel, containing some attached shell, is further dried and then crushed for oil recovery, in screw presses.

Flow sheet of all unit operations in a palm oil mill is given in Figure A1.1.3.

Fibre in Steam Production and Co-generation: The dry fibre is a convenient fuel for steam boiler. A palm oil mill is always self-sufficient in steam production. Should a part of the de-watered empty bunches be used for steam production, a co-generation system may be installed to produce both steam as also electricity required for the mill operation. The co-generation system is very beneficial to mills that are located on plantations, away from towns.

Oil Recovery from Fibre: Dry fibre contains 7–9% oil. Efforts are underway to recover this oil by solvent extraction. The main challenge here is the bulky nature of fibre. Bulk density is just about 0.15–0.2 kg/L. Unless the fibre is compacted, the solvent would just pass through without contacting much of the fibre cells, in a

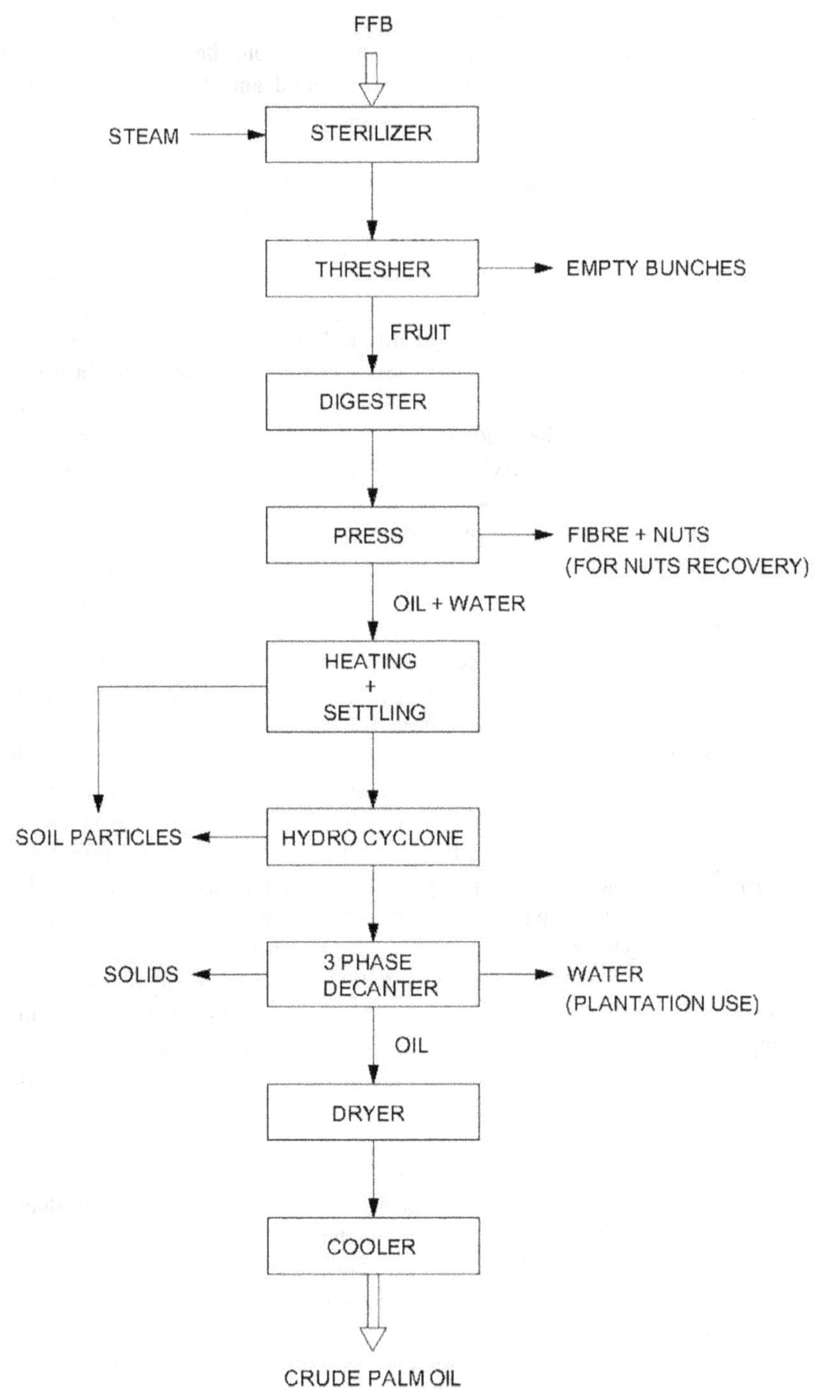

FIGURE A1.1.3 Flow Sheet of Palm Oil Production.

percolation extractor. A couple of plants are reported to have started operation on briquettes made of the fibre. Another way would be to add a binding agent, like 'gums' from a vegetable oil such as soybean, and pass through an extruder to form pellets. Still another way could be to cut the fibre and extract using an 'immersion' extractor.

Continuous Sterilisation: A major issue faced by the millers is the manpower required for sterilisation operation. This is a batch operation. The cages have to be pulled into the sterilisers, covers closed, covers to be re-opened after the completion of the steaming cycle, cages pulled out, and cages have to be lifted and tilted to load fruit bunches into the Thresher. All these are manual operations.

Continuous sterilisers have recently been designed in Malaysia and are being adopted in palm oil mills. In this design, bunches go in continuously, and sterilised bunches are conveyed to the Thresher continuously. This process eliminates much of the manual work as involved in batch operation.

One of the challenges in the design of a continuous steriliser is the steam-locking at FFB inlet and outlet. This may be overcome by means of a double rotary air locks system at both ends.

Another challenge is to reduce the sterilisation time. If the bunches could be poked, to make clear passages for steam to the base of fruit, sterilisation time could be reduced. Passage of the FFBs through a pair of rolls, with spikes (somewhat similar to those on a Cake Breaker) could serve the purpose. Problem of handling bunches of different sizes would still have to be overcome.

APPENDIX 1.2 OLIVE/AVOCADO OIL PRODUCTION

A1.2.1 Olive Oil

Olive oil is an important oil in the world edible oil basket and is ranked tenth in terms of production tonnage. It is known for its distinctive aroma and is the most pricy edible oil today. The olive trees grow mainly in the Mediterranean basin, and olive oil is the traditional oil of the region.

The oil is obtained from the fruit of olive. The fruit must be processed within 24 hours of harvesting, to prevent the excessive rise in free fatty acids. Fruit contains around 20–25% oil, and has moisture content of 45–50%.

As the oil is known for its aroma, it cannot be subjected to heat processing. In fact, 'cold-pressed' olive oils are mandated to be extracted at temperatures below 27C. Grades of cold-pressed oil depend on the FFA content, 'extra virgin oil' having less than 0.8% (or 0.5% in USA) FFA, and 'virgin oil' with up to 1.5% FFA. Oil produced with some amount of heating may be refined, to produce 'refined olive oil'.

Traditionally, olive oil used to be extracted in vertical batch presses. Modern mills use high-speed decanters for oil extraction, and operate on a continuous mode (Figure A1.2.1).

As the olive fruits are received at the factory, they are first washed with water to remove any soil impurities. Water is sprayed onto a slow-moving belt conveyor carrying the fruit. Water is drained as the fruit is passed over a screen. The washed fruit is conveyed to a grinder to make pulp. The grinder may be a slow-rotation disk mixer, or a metal-tooth mixer, or a simple hammer mill. The rotation speed of the

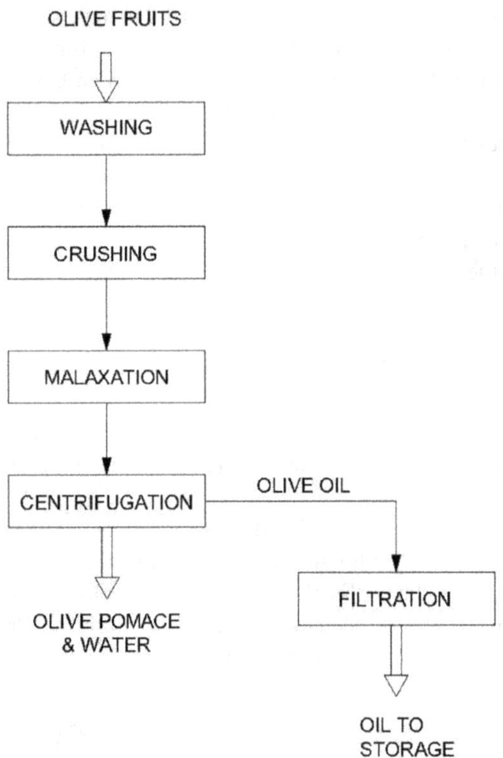

FIGURE A1.2.1 Flow Sheet for Olive Oil Production.

grinding element is low, and so is the energy requirement, thanks to the 'soft' texture of the fruit.

The pulp is gently mixed ('malaxed') for 30–60 min, to allow oil globules to aggregate, as also to help in aroma generation. Longer malaxing time improves oil extraction, but may result in some oxidation, and can lower the shelf life. Typical malaxing equipment is a slow moving disk or ribbon mixer. No heating is allowed during the process. Modern malaxers are nitrogen blanketed, to produce better quality oil with improved shelf life. For the production of 'refined' grade, malaxers may come with steam jackets.

The malaxed pulp is then sent to the oil extraction operation. The extraction is accomplished by means of high-speed centrifugation, in a horizontal Decanter. These are either two-phase or three-phase decanters. For a three-phase decantation, water is added to the pulp, and then oil, water, and dry 'pomace' phases are separated. The advantage of a three-phase decanter is a single-step operation; however, a major disadvantage is the production of a voluminous water phase which must be treated to prevent pollution.

Alternatively, a two-phase decanter separates the oil from a wet pomace phase. This may be followed by second extraction, after addition of water to the wet pomace, in another two-phase decanter. The second extraction can add as much as 6% of the fruit oil to the 90% extracted in the first extraction. Water addition requirement is much lower due to the smaller quantity of 'pomace' in relation to the malaxed fruit.

The oil produced with two-phase decanters is of better quality due to less water addition, which allows better aroma retention, as also retention of anti-oxidant compounds.

The extracted oil phase still has to go through another centrifugation step, with bowl or disk centrifuges running at higher speeds than decanters, to remove water entrained with oil. The dry oil may be passed through a polishing filter before bottling.

The pomace contains 7–8% oil, which may be extracted, after drying, with the solvent extraction process. The oil is called 'pomace oil'.

A1.2.2 Avocado Oil

Avocado oil is primarily used for cosmetics applications, in refined form. The trees are native to Central America and equatorial Africa, but have now spread to far off places like New Zealand. The fruit contains nearly 22–26% oil on wet basis and nearly 65% water.

The processing is similar to that of olive oil, except the additional steps of skin and stones removal, both done manually, prior to pulping and malaxing. Malaxing temperature is higher, nearly 45–50 C, since aroma retention is of no concern, as any aroma would be removed during refining anyway. The downstream process, decanter – centrifuge – polishing filtration, is similar to that for olive oil.

APPENDIX 1.3 MUSTARD (PUNGENT) OIL PRODUCTION

A1.3.1 Kohlu Press

Pungent mustard oil is the cooking oil of choice in eastern parts of India. It is also used in various pickles all over northern India. This has been so for generations over centuries. Presently, nearly 4–5 million ton of mustard seed, half of the mustard crop in India, is crushed to produce pungent oil, mainly in the western Indian state of Rajasthan. The pungent oil attracts a price premium of nearly 10% over the Press oil.

The pungency is created by slow maceration of seed, in presence of moisture, to allow the enzyme myrosinase to act on glucosinolates and produce allyl-isothiocyanates (AITC). No adverse health effects of the compounds, as feared in the west, have ever been found.

Mustard seed is gently macerated in Kohlus (similar to a 'Stump' press), with a conical bowl, full of seed, rotating around a thick centre pole at 18–20 rpm (Figure A1.3.1). Material of construction is carbon steel. A wooden pole, normally from a tamarind tree, is stuck off-centre to create mixing vortex, as also to add 'tinge' to the oil. Water is added to raise moisture content of the batch of seed up to 11%. The slow grinding action creates gentle heat, raising the seed temperature to 40–45 C, which is ideal for the enzyme action. Part of oil is released and is collected from a port at the bottom. Batch time is nearly 45 min. At the end of the cycle, nearly half the oil is recovered, leaving nearly 25% oil in the mash. The moisture content increases to 13–14%, due to the reduction in mass. The concentration of AITC in the oil is about 0.35%, indicated as pungency of '35 points'.

FIGURE A1.3.1 Kohlu. Courtesy: Nirmal Oil Industries, Alwar.

The mash is then pressed multiple times, four or five, in short-cage screw presses. The short-cage design helps in keeping temperature rise low. In the first press, the temperature rise is restricted to around 60 C. This mild temperature rise is due to two reasons, one the high moisture content (nearly 13%), and two, the short cage. The pungency of the oil recovered in the first press is nearly 25–28 points. Oil-in-cake is typically between 13% and 15%.

In the second press, foots from all the presses, first to fourth, are added at inlet. Cake gets further heated to around 70 C, oil-in-cake is typically 12–13%, and the oil has pungency of 17–18 points. It may be noted that the low oil recovery in the second press is mainly on account of the high level of moisture at inlet (13–14%).

The oils recovered from the first and second presses are mixed with the Kohlu oil, to produce oil with the overall pungency of around 30 units.

In successive third and fourth presses, feed is heated with indirect steam in a small conditioner (short holding time). The oil-in-cake reduces to 10–11% and 8–9%, and oil pungency reduces to 12–13 and 6–7 points, respectively. The low recoveries of oil are mainly due to high moisture content. Some mills use even a fifth press, to ensure final OIC at 8%, and to reduce moisture content below 8% for cake stability.

The oils recovered from the third press onwards are very low in pungency, on account of steam heating and heat generation within successive presses. These oils are cooled and sold separately as 'Mustard Press oil'.

Each Kohlu typically can take 20–22 kg of seed per batch. Cycle time being 1 hour, the capacity of each Kohlu is nearly half a ton per day. Thus, for a mill rated nominally at 100 Tpd, the Kohlu room would typically have 192 Kohlus! These are arranged in 16 lines of 12 Kohlus each, with each pair of lines being driven by a common motor of 40 hp, through a long shaft and belts (Figure A1.3.2). The room would have overhead conveyors for the distribution of seed into the Kohlus, and ground-level conveyors to take the mash away to the Press room. Each line of 12 kohlus is operated by two workmen. Thus, a room with 192 Kohlus would have 32 workmen

FIGURE A1.3.2 Line of Kohlus. Courtesy: Nirmal Oil Industries, Alwar.

per shift, apart from four to five men on the presses. Thus the room is worked by nearly 110–120 people working in three shifts.

The pungency in the air in Kohlu room is very strong, which a first-time visitor may not withstand for more than a minute.

At present, the main challenge before the industry is to eliminate the fourth and fifth presses, and still achieve final cake of acceptable OIC and moisture. The key here would be to reduce moisture of cake entering the third press to below 8%.

Another challenge is the difficulty in oil filtration, due to high moisture in foots. The foots block the screen when a PLF is used. Therefore, common filters are the plate type, which involves significant manual labour.

A1.3.2 Screw Press

During the past decade, screw presses have been adopted for pungent oil production, in place of Kohlus. This has allowed continuous process operation, compared to the batch Kohlus, and reduced the manpower requirement significantly. Thanks to the less manpower, and less area requirement, larger capacity mills are easier to manage with the process. A 100 Tpd oil mill typically has five lines of four presses each (cage ID 200 mm), with nearly ten people managing the entire operation every shift. Thus the mill would have 30 people working in three shifts, compared to nearly 120 men on Kohlus.

In this process, the seed is suitably prepared for the production of the pungent aroma. Seed is broken into grits, using fine corrugations (short pitch and depth) on rolls of a Seed Cracker. The grits are then moistened in a rotary drum. The moist grits are gently mixed in a kettle, mounted on top of the press, designed for nearly 45 min hold up.

The presses are designed to operate at low temperature, and also to keep the chambers cool, in order to trap the vapours of AITC, which are all volatile compounds. The cages are water-cooled, and so are the press covers. In fact, chilled water @ 10–12 C is circulated for the cooling operation.

Cake RO at second press is nearly 12–13%. Oils from first and second presses are mixed to produce pungent oil of nearly 28 point pungency. The third and fourth press kettles are steam-heated, and the final RO is around 9–10%.

A1.3.3 Large Screw Press

Similar operation was attempted with a line of larger presses, cage ID 305/280mm, length 2.5m (Vadke, 2014). Seed was flaked, moistened, passed through an expander with drain cage, to recover just 2–3% oil, and gently mixed for 40min prior to pressing. Flaking was introduced to achieve complete breaking of the seed. Moistening of flakes, by water spray, was done in a rotating drum. The purpose of expander treatment was to shear the mass and bring the enzyme in close contact with oil globules. Mixing of flakes was accomplished in a vertical multi-stage cooker, but without any steam to the double-bottoms. Press doors were jacketed for the circulation of chilled water.

Capacity achieved was 120 Tpd, and RO after the third press was 10%. Pungency of oil was 25–26 points, which is about the lower limit accepted by consumer. The mill was operated by five workmen per shift.

A cooking-cum-drying operation was proposed prior to the third press, to try to reduce the RO and moisture of the final cake. However, the modification could not be carried out, as the mill went out of operation for commercial reasons. The main challenge here would be to increase the pungency to 28 points. Could that be achieved with a longer mixing time?

APPENDIX 1.4 TYPICAL MATERIAL BALANCE AROUND MECHANICAL EXPRESSION PROCESS

It is important to understand the material balance calculations around pressing processes, not only to cross-check for material losses, if any, in the oil mill; but also to be able to compute the energy consumptions at different stages of pressing, apart from correct sizing of downstream equipment. Calculations are presented below for some of the oilseeds. These are typical cases, with assumed percentages of oil and moisture in seed. One may conduct the same steps of calculation to arrive at exact figures for their seed composition; and for actual moisture and oil content in presscake.

A1.4.1 Rapeseed Pressing

Basis: 1,000kg rapeseed containing 40% oil, 7% moisture

On **Prepress**, if Cake contains 18% oil and 6% moisture,

$$\text{Mass of Cake} = 1{,}000 * (100\% - 40\% - 7\%) / (100\% - 18\% - 6\%)$$

$$= 697.37 \text{ kg}$$

$$\text{Oil in Cake} = 697.37 * 18\% = 125.53 \text{ kg}$$

$$\text{So, } \mathbf{Press\ Oil} = 1{,}000 * 40\% - 125.53 = \mathbf{274.47} \text{ kg}$$

Overall balance: Cake + Oil = 697.37 + 274.47 = 971.84 kg
So, moisture loss = 1,000 − 971.84 = 28.16 kg
(moisture loss figure may be cross-checked from seed moisture and cake moisture)
On **second Press**, Cake to have, say, 8% oil and 6% moisture

$$\text{Mass of Cake} = 697.37 * (100\% - 18\% - 6\%) / (100\% - 8\% - 6\%)$$

$$= 616.28 \text{ kg}$$

$$\text{Oil in Cake} = 616.28 * 8\% = 49.30$$

So, **second Press oil** = 125.53 − 49.30 = **76.23** kg

$$\text{Overall balance : Cake + first press oil + second press oil}$$

$$= 616.28 + 274.47 + 76.23 = 966.98 \text{ kg}$$

$$\text{So, moisture loss} = 1{,}000 - 966.98 = 33.02 \text{ kg}$$

Thus, the first press recovers nearly 70% of the seed oil, and the second press nearly 19–20%. So, it may be seen that when prepress and solvent extraction process is practised, only about 30% of the oil is left to be extracted by solvent.

A1.4.2 Sunflower Crushing (Partial Dehulling)

Basis: 1,000 kg sunflower seed (Ukrainian hybrid) containing 75% Kernel and 25% Hulls – Kernel containing 60% oil, 20% protein, 5% moisture
and, Hulls containing 2.5% oil, 4.5% protein, 8% moisture
so, Combined seed: 45.5% oil, 16.1% protein, 5.8% moisture
On Dehulling, say, 15% Hulls fraction (14.6% hulls + 0.4% kernel) with 3.55% oil and, 85% Meat fraction containing 52.9% oil, 18.1% protein, 5.4% moisture
(Figures arrived at by considering 0.4% kernel gone with Hulls fraction, and 10.4% hulls retained in meat fraction)
On Prepress of Meat, if Cake contains 18% oil and 7% moisture,

$$\text{Mass of Cake} = 1000 * 85\% * (100\% - 52.9\% - 5.4\%) / (100\% - 18\% - 7\%)$$

$$= 472.6 \text{ kg}$$

$$\text{Oil in Cake} = 472.6 * 18\% = 85.1 \text{ kg}$$

So, **Press oil** = 1000 * 85% * 52.9% − 85.1 = **364.6 kg**

$$\text{Overall balance : Hulls + Cake + Press oil}$$

$$= 150 + 472.6 + 364.6 = 987.2 \text{ kg}$$

$$\text{So, moisture loss} = 1000 - 987.2 = 12.8 \text{ kg}$$

Check protein content of de-oiled meal after solvent extraction –
Meal containing, say 0.7% oil and 11% moisture

$$\text{Meal mass} = 472.6 * (100\% - 18\% - 7\%)/(100\% - 0.7\% - 11\%)$$

$$= 401.4 \text{ kg}$$

So, **Protein content** = $850 * 18.1\%/401.4 * 100 = \mathbf{38.3}\%$

If 20% Hulls fraction were separated, protein content would rise to over 43%. At this level of protein, the meal would compete well with soybean meal, which is the 'meal standard' internationally.

It may be noted that in the case of dehulled sunflower seed, the cake mass, going for solvent extraction, is less than half of the seed.

A1.4.3 Cottonseed Crushing (Delinting, Dehulling)

Basis: 1,000 kg white cotton seed, consisting of:

- 60% kernel, containing 36% oil, 35% protein, 7% moisture
- 30% Hulls, containing 1% oil, 4% protein, 12% moisture
- 10% Lint, containing 0.8% oil, 3.5% protein, 12% moisture
  Combined WCS: 21.9% oil, 22.2% protein, 7.8% moisture

On **Delinting and Dehulling**,

30% lint + hulls fraction separated, containing 2.07% oil
70% Meat fraction, containing 30.4% oil, 30% protein, 7.8% moisture

On **Prepress**, if Cake contains 16% oil and 7% moisture,

$$\text{Mass of Cake} = 1{,}000 * 70\% * (100\% - 30.4\% - 7.8\%)/(100\% - 16\% - 7\%)$$

$$= 561.8 \text{ kg}$$

$$\text{Oil in Cake} = 561.8 * 16\% = 89.9 \text{ kg}$$

So, **Press Oil** = $1{,}000 * 70\% * 30.4\% - 89.9 = \mathbf{122.9\ kg}$

$$\text{Overall balance: Hulls + lint + Cake + Press Oil}$$

$$= 1{,}000 * 30\% + 561.8 + 122.9$$

$$= 984.7 \text{ kg}$$

$$\text{So, moisture loss} = 1{,}000 - 984.7 = 15.3 \text{ kg}$$

Check protein content of de-oiled meal after solvent extraction –

Meal containing, say 0.7% oil and 11% moisture

$$\text{Meal mass} = 561.8 * (100\% - 16\% - 7\%)/(100\% - 0.7\% - 11\%)$$

$$= 489.9 \text{ kg}$$

So, **Protein content** = 700 * 30%/489.9 * 100 = **42.8**%

A1.4.4 Copra Crushing (Two-Stage)

Basis: 1,000 kg Copra cups containing 67% oil, 6% moisture
On **Prepress**, if Cake contains 25% oil and 6% moisture,

$$\text{Mass of Cake} = 1{,}000 * (100\% - 67\% - 6\%)/(100\% - 25\% - 6\%)$$

$$= 391.3 \text{ kg}$$

$$\text{Oil in Cake} = 391.3 * 25\% = 97.8 \text{ kg}$$

So, **Press Oil** = 1,000 * 67% – 97.8 = **572.1** kg

Overall balance: Cake + Oil = 391.3 + 572.1 = 963.4 kg
So, moisture loss = 1,000–963.4 = 36.6 kg
On **second Press**, Cake to have, say, 8% oil and 6% moisture

$$\text{Mass of Cake} = 391.3 * (100\% - 25\% - 6\%)/(100\% - 8\% - 6\%)$$

$$= 313.9 \text{ kg}$$

$$\text{Oil in Cake} = 313.9 * 8\% = 25.1$$

So, **second Press oil** = 97.8 – 25.1 = **72.7** kg

Overall balance: Cake + first press oil + second press oil

$$= 313.9 + 572.1 + 72.7 = 958.7 \text{ kg}$$

So, moisture loss = 1,000 – 958.7 = 41.3 kg

- Note the high moisture loss, at over 4% of the mass of copra.

The total oil recovery is more than 96%, by mechanical expression alone, thanks to the very high oil content of copra.

Out of the total, as much as 85% of oil is recovered in first press itself. Thus, one can see heavy streams of oil coming out of cage segments. Note, however, it is the second press, which recovers just 1/8th of oil compared to the first press, that decides the economics of the entire oil mill operation. A single percentage point increase in RO in final cake can significantly affect the profitability.

APPENDIX 1.5 DETERMINATION OF SIZES OF INTERMEDIATE WORMS

We have seen, in Section 4.5.2, how to determine the sizes of the Feed worm and the last worm on the wormshaft of a Screw Press. We now discuss methods to determine sizes of intermediate worms.

The compression, or the 'volume ratio' between the feed worm and the second worm may be considered @ 1.5. This ratio is actually quite high and might appear to be contrary to the principle cited in Section 4.5.2. However, this high VR is allowed, and in fact necessary, for quick expression of air contained between the flakes. The feed worm and second worm may be designed with the same boss dia, and the pitch of the second worm would be shorter by 1/3rd (VR 1.5). A collar placed after the second worm would ensure that the material is fully packed in the second worm, and all the air is driven out. The third worm onwards, the VR should be progressively reduced. As the material gets compacted, and oil released, viscosity of the material increases. Hence, subsequent compression should proceed slowly, to avoid pressure peaks.

For example, the subsequent VRs may be reduced by 5% progressively. Thus, VR_2 may be 1.425, VR_3 may be 1.354 and so on, until VR_8 is 1.05 (1.5, 1.425, 1.354, 1.286, 1.222, 1.161, 1.103 and 1.047). Multiplication of all these VRs would give the Overall VR as 6.1. Since this is less than the required overall VR value of 6.5, we may tweak the second VR slightly, to say 1.45. The subsequent VRs would then become 1.38, 1.31, and so on (1.243, 1.181, 1.122), until VR_8 is 1.065. Multiplication of all these VRs would give Overall VR equal to 6.88, which is 5% more than the required overall VR. The actual overall VR may be higher than the required overall VR by 5–10%. The intermediate VRs indicate the volume, and hence the size, of successive worms.

The above method may be called a **'constant compression ratio' method**. Here the constant Compression ratio (CR) of 0.95 has been used.

$$CR_n = VR_{n+1}/VR_n$$

An alternative principle, for the determination of intermediate VRs, could be to have **reducing compression ratios**, in place of a constant ratio of 0.95. Thus, while we may start with a higher ratio of say 0.97 between VR_1 and VR_2 ($VR_2 = 1.455$), each successive VR may be determined with reducing ratio, say 0.96, 0.95, 0.94, and so on. The overall VR would be matched as described above, and minor corrections made as required. This is a superior method, as it gives reducing compression rates towards the discharge end, where the pressure development is high due to reducing oil content.

APPENDIX 1.6 TYPICAL LAYOUT OF OIL MILL

A1.6.1 Rapeseed Oil Mill

Figure A1.6.1 shows a typical layout for a small mustard (rapeseed) oil mill of 300 Tpd capacity. The building size is 21 m × 12 m × 15 m high. The mill has two presses of 150 Tpd prepress capacity each.

As can be seen, the main machines, namely presses and flaker, are placed on the main operation floor. The Conditioner and the Cookers are placed on a floor above,

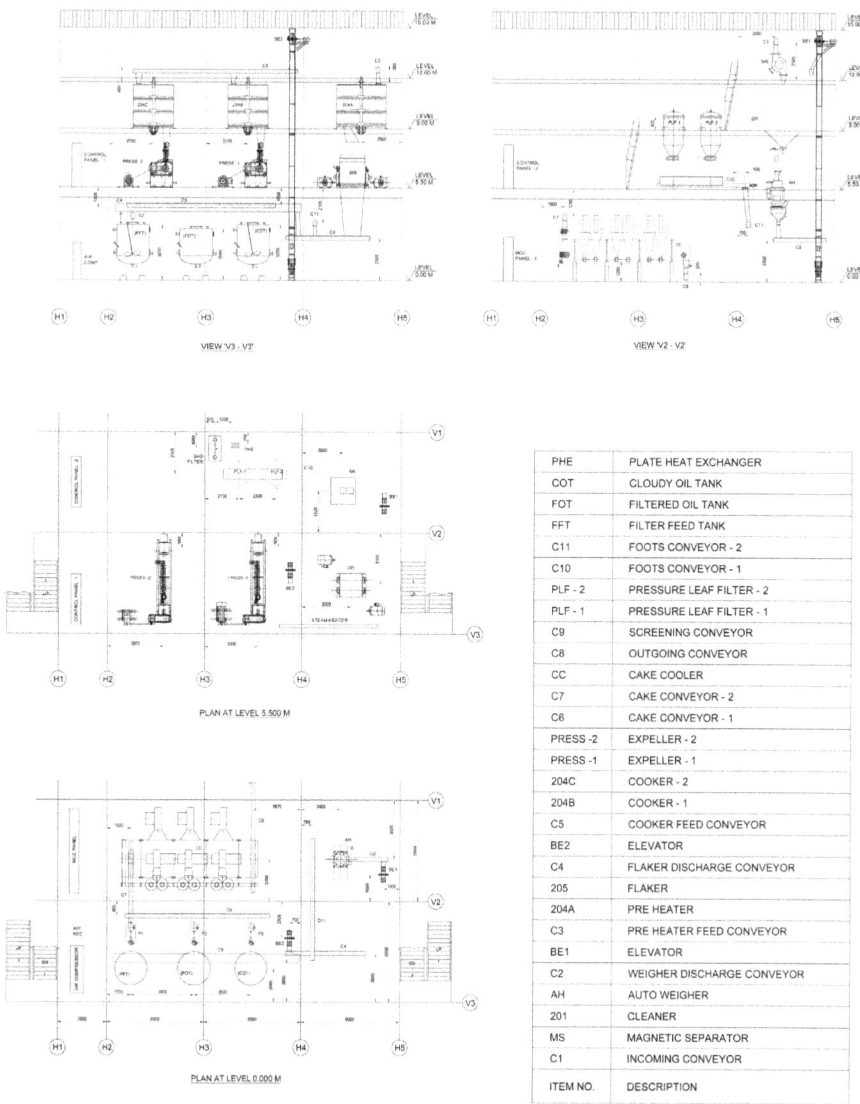

FIGURE A1.6.1 Oil Mill Layout (Rapeseed).

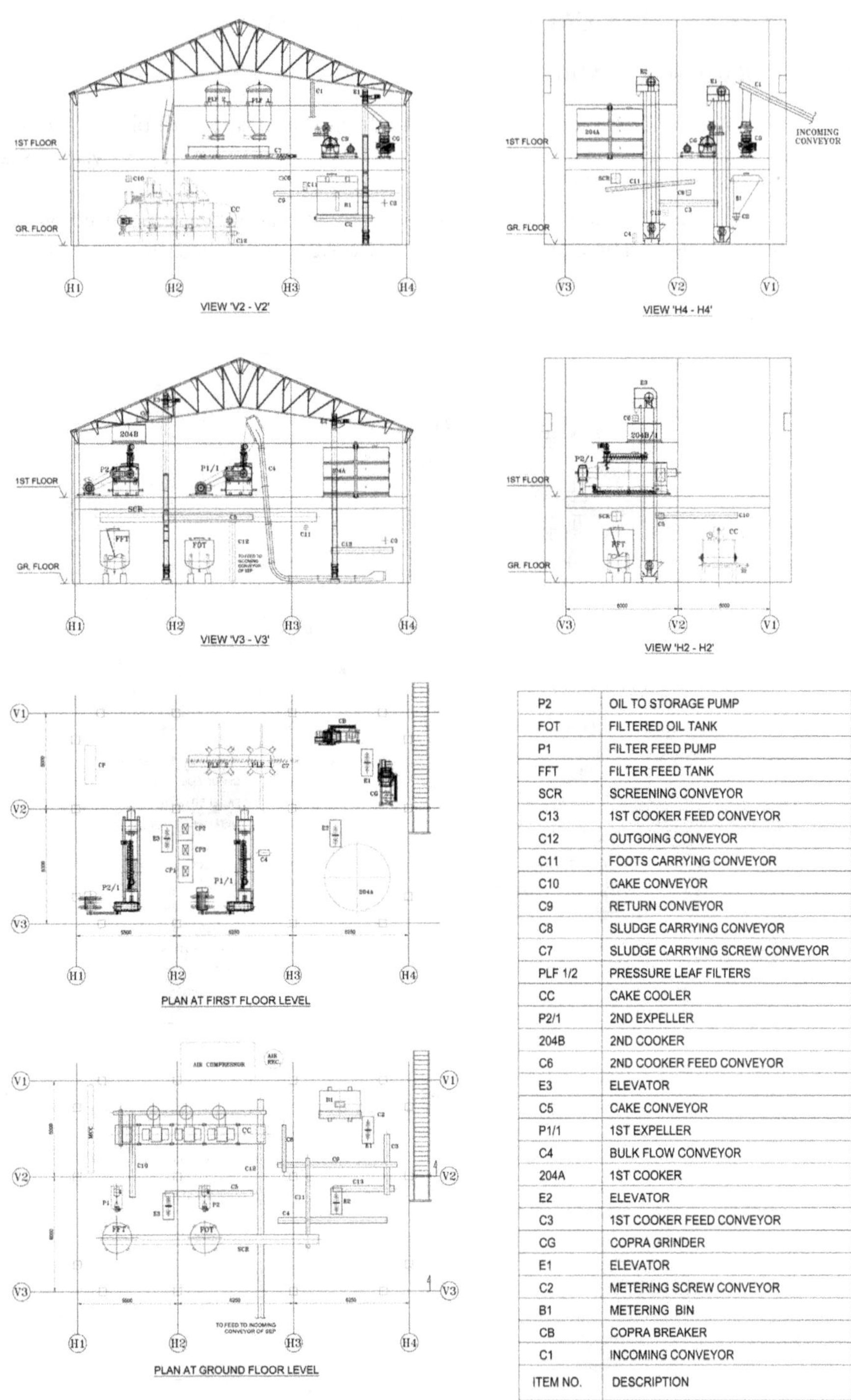

ITEM NO.	DESCRIPTION
P2	OIL TO STORAGE PUMP
FOT	FILTERED OIL TANK
P1	FILTER FEED PUMP
FFT	FILTER FEED TANK
SCR	SCREENING CONVEYOR
C13	1ST COOKER FEED CONVEYOR
C12	OUTGOING CONVEYOR
C11	FOOTS CARRYING CONVEYOR
C10	CAKE CONVEYOR
C9	RETURN CONVEYOR
C8	SLUDGE CARRYING CONVEYOR
C7	SLUDGE CARRYING SCREW CONVEYOR
PLF 1/2	PRESSURE LEAF FILTERS
CC	CAKE COOLER
P2/1	2ND EXPELLER
204B	2ND COOKER
C6	2ND COOKER FEED CONVEYOR
E3	ELEVATOR
C5	CAKE CONVEYOR
P1/1	1ST EXPELLER
C4	BULK FLOW CONVEYOR
204A	1ST COOKER
E2	ELEVATOR
C3	1ST COOKER FEED CONVEYOR
CG	COPRA GRINDER
E1	ELEVATOR
C2	METERING SCREW CONVEYOR
B1	METERING BIN
CB	COPRA BREAKER
C1	INCOMING CONVEYOR

FIGURE A1.6.2 Layout of Oil Mill (Copra).

directly above the flaker and the presses, respectively. Filters are also supported on upper floor, but hang partly below the floor. This placement allows all the valves and sight glasses for filters to be accessible from the main floor. The control panel is also on the main operation floor.

The oil tanks, pumps, oil cooler, and the cake cooler, are all placed on the ground floor. The MCC is also conveniently located on the G.F.As can be seen, all conveyors receiving flakes, cake, or oil-n-foots from machines on first floor, run below the floor, and feed equipment, or elevators placed on G.F.

A1.6.2 Copra Oil Mill

Figure A1.6.2 shows a typical layout of a copra oil mill. This mill has only two Presses, one for the first press and the other for the second press. Mill capacity is 10 Tph (or, 250 Tpd). Both the presses are placed on the first floor. Most of the other major equipment, namely, copra grinders and the main oil filters (vertical PLFs) is also placed on the first floor, making it the main operation floor. Naturally, the Control Panel is located on the same floor. This floor has been built in RCC to support the heavy machinery and absorb vibrations.

The ground floor has accommodated the electric motor starters panel (MCC), which tends to be much bigger than the control panel placed on the first floor. All the elevators are supported on the ground floor. Also the oil tanks and pumps are placed on the ground floor, apart from the Cake Cooler.

The Cooker for the first press is placed by the side of the first press; this was done to accommodate the Cooker in a given shade, with height limitation. In general, Cookers are placed directly above each Press, and a separate work floor is provided for the Cookers.

APPENDIX 2.1 BATCH EXTRACTION

Batch extractors, which were the first extractors during the early 20th century, are still in practice for special applications, with up to 10 T batch capacity. Common applications are in oil extraction from rice bran, for the advantage of unbroken pellets in the meal; and sugars extraction from de-oiled soybean flakes, for the production of soy protein concentrate. Such extractors are quite commonly used for the extraction of flavours ('essential oils') from spices and flowers.

A batch extractor is typically a cylindrical vertical vessel, fully closed, with a screen near the bottom. Filter mats are laid on the screen and tightened with a ring on top. A coil with holes is fitted on top for solvent/miscella spray. The plant may have a single extractor, for very small capacities, or multiple ones. A steam coil may be inserted in the bottom to heat up miscella.

A2.1.1 Single Extractor

Let us first consider the case of a single extractor. This is accompanied with multiple miscella tanks. Prepared oilseed flakes are loaded into the extractor, from top manhole, and the manhole is closed. Miscella is pumped onto the bed of flakes, which then percolates through the bed and collects below the screen. The miscella may be circulated continuously, by means of a pump, for say half an hour. At the end of the circulation cycle, the miscella is pumped out to the last miscella tank, say tank no. 4. Fresh solvent is then pumped again to the extractor and the cycle is repeated. At the end of the second cycle, the miscella is pumped out to MT no. 3. Two more cycles are repeated and miscella are collected in MT nos. 2 and 1, respectively. An additional cycle may be performed, should the miscella in MT 1 show colour of oil (oil percentage more than 0.2). During the extraction cycles, the extractor is ventilated to maintain small suction of 15–20 mmWc.

After the required number of extraction cycles, bed is allowed to drain, and miscella sent to MT 1. Then, the bed is steamed for say 40–50 minutes, vapours are condensed in a water-cooled condensor. When a sample of condensate shows no solvent traces, steaming is stopped. Manholes to the extractor are then opened, and the deoiled desolventised meal is allowed to cool down for about an hour. The meal is then discharged manually and sent for cooling.

Miscella in MT 4 is sent for distillation, which may operate in continuous mode.

When a second batch of flakes is loaded into the extractor, the first wash cycle is with the miscella in tank no. 3. After the cycle, the miscella is forwarded to MT 4, ready for distillation. Similarly in next two cycles, miscella in MT 2 and MT 1 are used for washing, and are forwarded to next tanks, MT 3 and MT 2, respectively. The last wash is with fresh solvent, which at the end of cycle is sent to MT 1.

The entire extraction cycle may take up to 3–4 hours. Desolventising may take an hour. The meal discharge may take another hour. So, the total cycle for each batch of flakes, including the loading time, is around 6 hours. In 24 hours, four batches may be completed. If the batch size were 10 T, plant capacity could be 40 Tpd.

A2.1.2 Multiple Extractors

For larger capacities, more than 30–40 Tpd, multiple extractors may be employed. A typical arrangement, with four extractors, is shown in Figure A2.1.1. Miscella hoppers of the four extractors are connected, with overflow from one going to the next. Here, the extraction mechanism is more of immersion than percolation.

Flakes are loaded into all the extractors by an overhead conveyor. As the extractors are filled, the inlet manholes are closed. Fresh solvent is pumped onto the bed in Extractor 1. As more solvent is pumped into Ext 1, when the bed gets saturated, level in miscella hopper rises. As solvent/miscella fills up the extractor up to a certain level, flow of miscella starts by gravity onto the top of Ext 2. Similarly, as more solvent is continuously pumped into Ext 1, miscella overflows into Ext 3 and then into Ext 4. When miscella level rises in the bottom part of Ext 4, it overflows into Miscella Tank (MT) outside. It may be noted that the miscella inlet levels in successive extractors are lower than the preceding one.

As miscella starts overflowing into the MT, sample of miscella flowing from Ext 1 to Ext 2 is checked for oil content. If it is more than 0.2%, solvent pumping is continued. When the required oil content is reached, pumping of fresh solvent to Ext 1 is stopped, and is diverted to Ext 2. After solvent drain, flakes in Ext 1 are steamed for desolventisation. Likewise, after completion of extraction in succeeding extractors, the respective beds are desolventised.

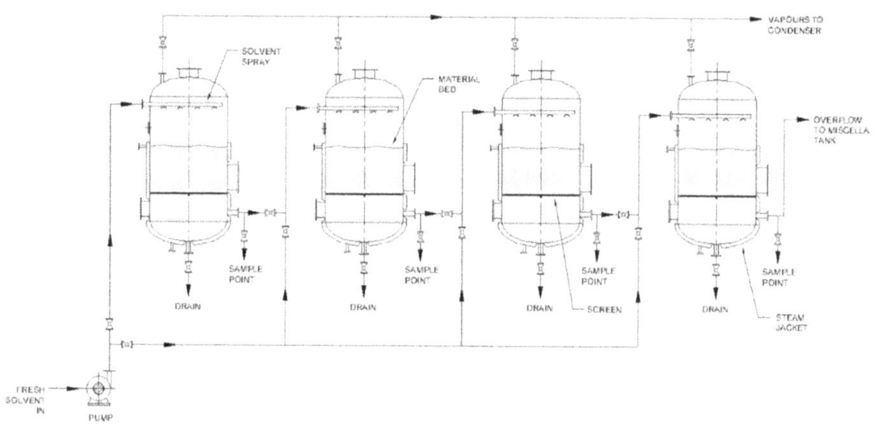

FIGURE A2.1.1 Multiple Batch Extractors.

In this configuration, multiple miscella tanks are not required. A single tank receives overflow miscella from Ext 4, as also drained miscella from all the extractors; this miscella is continuously sent to distillation.

A2.1.3 Utility Consumption in Batch Extraction

As for utility requirements for batch extraction, electric energy requirement per ton of flakes is lower than that for continuous extraction process, steam requirement is much higher, and solvent loss may be only marginally higher. The main problem is a lot of manual work involved in discharge of meal and handling up to bagging.

APPENDIX 2.2 DT CALCULATION FOR MOISTURE-IN-MEAL, AND FOR NO. OF PD STAGES REQUIRED (600 TPD SOY FLAKES)

As seen in Chapter 15, section on Schumacher DT, the desolventising section, for soybean flakes, may be designed for 25 minutes desolventising time, Thus, for a plant capacity 600 Tpd, the size may be determined, as dia 3,000 mm × 4 stages.

A) For this size, moisture content of meal may be computed using mass and heat balance equations.

Heat balance: For steady state, Total heat input = Total heat output

$$\text{Or, } Q_1 + Q_2 + Q_3 = Q_4 + Q_5 \tag{15.1}$$

where,

Q_1 is the heat content of incoming flakes,
Q_2 is the heat input from steam in DBs,
Q_3 is the heat input with live steam
Q_4 is the heat going out with DT gases,
and, Q_5 is the heat content of outgoing meal.

Equation for moisture balance may be written as:

$$\text{Moisture in spent flakes} + \text{Live steam} = \text{Moisture in meal} + \text{moisture in DT gas...} \quad (15.2)$$

Let us compute all the heat content taking temp Zero Celsius as the base point.
Then,

Q_1 = (dry matter in flakes * sp. heat of d.m. + moisture in flakes * sp. heat of moisture + solvent in flakes * sp. heat of solvent) * temp. of spent flakes

Q_2 = No. of DS stages * DS stage heating area* (steam temp – flakes temp)) * Heat transfer coeff.

Q_3 = Live steam qty * sp. heat content of live steam

Q_4 = (Solvent qty + steam qty in DT gas) * average sp. heat content

Q_5 = (d.m. in meal * sp. heat of d.m. + moisture* sp. heat of moisture) * temp of meal

Let us consider following values for various quantities in above equations:

Moisture in flakes = 10% of soybean flakes = 2,500 kg/h

D. M. in flakes = deoiled soybean – moisture = 25,000*82% – 2,500 = 18,000 kg/h

Sp. heat of moisture = 1 kCal/kg/C

Sp. heat of D.M. = 0.35 kCal/kg/C

Sp. heat of solvent = 0.56 kCal/kg/C

Temp of spent flakes, outgoing meal and heating steam, may be considered as 55 C, 105 C, and 185 C (at pressure 10 BarG), respectively.

For each of the 4 DS stages, heating area may be considered as 95% of DB area = 6.7 m^2

Heat transfer coeff. for transfer of heat from condensing steam to wet flakes may be considered as 110 kCal/m^2/C/h

Sp. heat content of live steam, derived from L.P. steam @ 3 Barg= 655 kCal/kg
Solvent qty in DT gas = solvent in spent flakes = 42% of flakes = 8,610 kg/h
Steam going with solvent in DT gas @ 70 C = 9.4% of solvent vapour * = 810 kg/h

Average heat content of DT gas = 158 kCal/kg*

(Note *: the steam content in DT gas, and the sp. heat content, may be computed from vapour pressures and heat contents of solvent vapour and steam at 70C)

From the above data, the heat quantities may be computed as follows:

$Q_1 = 749{,}188$ kCal/h

$Q_2 = 251{,}147$ kCal/h (equivalent heating steam = Q2/475 = 529 kg/h)

$Q_3 = 665$ * Live steam

$Q_4 = 14{,}88{,}256$ kCal/h

$Q_5 = 661{,}500 + 105$ * Moisture in meal

From Eq. (15.1), we get, 10,00,335 + 655 X = 21,49,756 + 105 Y

From Eq. (15.2), we get, 2,500 + X = 810 + Y => Y = X + 1,690

From the above two equations, we get, Live steam qty, X = 13,26,940/550 = 2,413 kg/h

And, moisture in meal, Y = 4,103 kg/h

Meal qty = 4,103 + 18,000 = 22,103 kg/h; So, moisture-in-meal will be 18.6%.

This high level of moisture is not necessary for toasting, and would unnecessarily add to drying load.

B) So, let us repeat the calculation by addition of, say 1 PD stage.

Then, Q_2 = Heat transfer in PD stages + Heat transfer in DS stages

$$\begin{aligned}\text{Heat transfer in PD stages} &= \text{No. of PD stages} * \text{PD stage heating area} \\ &\quad *(\text{steam temp} - \text{flakes temp}) * \text{Heat transfer coeff.} \\ &= 1 * 6.35\ \text{m}^2 (\text{being 90\% of the area}) * (185 - 65) * 110 \\ &= 83{,}820\ \text{kCal/h} \\ \text{So, } Q_2 &= 83{,}820 + 251{,}147 = 334{,}967\ \text{kCal/h} \\ &\quad (\text{eq. heating steam 705 kg/h})\end{aligned}$$

Repeating the above procedure, we find, X = 12,42,965/550 = 2,260 kg/h

Then, Y = 3,950 kg/h; So, meal-in-moisture, in this case, will be 18.0%

This moisture content is within the required range for toasting, 17.5–18.5%.

Therefore, addition of 1 PD stage is a good idea.

So, the selected size of DT is Dia 3,000 × (1 PD + 4 DS stages).

APPENDIX 2.3 TYPICAL MATERIAL BALANCE AROUND SOLVENT EXTRACTION PROCESS

A2.3.1 Soybean Crushing

A2.3.1.1 Whole Soybean (No Dehulling)

Basis: 1,000 kg Soybeans containing 92% Kernel and 8% Hulls –
Kernel containing 20.5% oil, 38% protein, 10% moisture
and, Hulls containing 0.6% oil, 12% protein, 10% moisture
so, Combined beans: 18.9% oil, 35.9% protein, 10% moisture
On **extraction**, if meal contains 0.6% oil and 12% moisture, then

$$\text{Meal} = 1{,}000 * (100\% - 18.9\% - 10\%)/(100\% - 0.6\% - 12\%)$$

$$= 813.4 \text{ kg}$$

$$\text{Oil in meal} = 813.4 * 0.6\% = 4.88 \text{ kg}$$

$$\text{So, } \textbf{Crude Oil} = 1{,}000 * 18.9\% - 4.88 = \textbf{184.2} \text{ kg}$$

Overall balance: Meal + Oil = 813.4 + 184.2 = 997.6 kg
So, moisture loss = 1,000–997.6 = 2.4 kg
(moisture loss figure may be cross-checked from seed moisture and meal moisture)
Check protein content of meal:

$$\text{Protein in meal} = \text{Protein in beans} = 1000 * 35.9\% = 359 \text{ kg}$$

$$\text{so, } \textbf{Protein} \text{ content in meal} = 359/813.4 = \textbf{44.2\%}$$

A2.3.1.2 Dehulled Soybean

Basis: 1,000 kg Soybeans containing 92% Kernel and 8% Hulls:
Kernel containing 20.5% oil, 38% protein, 10% moisture
and, Hulls containing 0.6% oil, 12% protein, 10% moisture
so, Combined beans: 18.9% oil, 35.9% protein, 10% moisture
On Dehulling, say,
7.5% Hulls fraction (7.25% hulls + 0.25% kernel) with 1.4% oil
and, 92.5% Meat fraction containing 20.33% oil, 37.8% protein, 10% moisture
On Extraction of Meat, if Meal contains 0.6% oil and 12% moisture, then,

$$\text{Meal} = 1000 * 92.5\% * (100\% - 20.3\% - 10\%)/(100\% - 0.6\% - 12\%)$$

$$= 737.4 \text{ kg}$$

$$\text{Oil in meal} = 737.4 * 0.6\% = 4.4 \text{ kg}$$

$$\text{So, } \textbf{Crude oil} = 1000 * 92.5\% * 20.33\% - 4.4 = \textbf{183.6 kg}$$

$$\text{Overall balance: Hulls} + \text{Meal} + \text{Crude oil}$$

$$= 75 + 737.4 + 183.6 = 996 \text{ kg}$$

$$\text{So, moisture loss} = 1000 - 996 = 4 \text{ kg}$$

Check protein content of meal –

$$\text{Protein in meal} = \text{Protein in Meat} = 1000 * 92.5\% * 37.8\% = 349.6\,\text{kg}$$

$$\text{So, } \textbf{Protein} \text{ content in meal} = 349.6/737.4 = \mathbf{47.4\%}$$

Thus, the meal qualifies as Hi-Pro meal (more than 47% protein).

A2.3.2 Extraction of Rapeseed Cake

Basis: 1000 kg rapeseed containing 40% oil, 7% moisture

On **Prepress**, if Cake contains 18% oil and 6% moisture,

$$\text{Mass of Cake} = 1{,}000 * (100\% - 40\% - 7\%)/(100\% - 18\% - 6\%)$$

$$= 697.37 \text{ kg}$$

$$\text{Oil in Cake} = 697.37 * 18\% = 125.53 \text{ kg}$$

$$\text{So, } \textbf{Press Oil} = 1{,}000 * 40\% - 125.53 = \mathbf{274.47} \text{ kg}$$

Overall balance: Cake + Oil = 697.37 + 274.47 = 971.84 kg

So, moisture loss = 1,000–971.84 = 28.16 kg

(moisture loss figure may be cross-checked from seed moisture and cake moisture)

On solvent extraction of Cake, if Meal contains 0.8% oil and 10% moisture, then,

$$\text{Meal} = 697.37 * (100\% - 18\% - 7\%)/(100\% - 0.8\% - 10\%)$$

$$= 586.35 \text{ kg}$$

$$\text{so, Oil in meal} = 586.35 * 0.8\% = 4.69 \text{ kg}$$

$$\text{and, } \textbf{Crude} \text{ Oil} = 125.53 - 4.69 = \mathbf{120.84} \text{ kg}$$

Total oil = 274.47+120.84 = 395.31 kg (Press oil 69.4%, Extracted oil 30.6%)

$$\text{Overall balance on extraction: Meal} + \text{Oil} = 586.35 + 120.84 = 707.19\,\text{kg}$$

$$\text{So, moisture loss} = 697.37 - 707.19 = -9.82\,\text{kg}$$

Thus, there is a net gain of 9.82 kg moisture in the process of solvent extraction.

So, the overall loss of moisture on Pressing and extraction is 28.16 + (–9.82) = 18.34 kg.

APPENDIX 2.4 PLANT START-UP, SHUT-DOWN AND TROUBLE-SHOOTING

A2.4.1 Plant Start-Up

It is not the objective here to list out all the action steps for start-up of a solvent extraction plant (SEP). We take an overview of the philosophy of start-up and the precautions that need to be taken.

A solvent plant is not a straightforward 'linear' process plant. A 'linear' process plant may be a seed preparation plant, or an oil mil, where the oilseed material passes from one machine to the next in a single direction. In an SEP, however, while the solids pass in a single direction, the other important stream, the solvent, flows in nearly an opposite direction. This is true not only for the extractor, but also for the overall plant operation. While the oil-rich solvent (miscella) flows from extractor to distillation, and part of solvent goes with spent flakes to the DT, solvent flows from both the sections back to the extractor. Therefore, the operation of the latter sections has a direct impact on operation of extractor. Also, the interplay between the DT gases and the miscella in the Economiser impacts operation of both the sections, DT and distillation.

Due to the interplays mentioned above, the extractor is not the first equipment to be started for start-up of SEP. The condensor system must be activated first, with cooling water circulation, so that no solvent vapours would escape from the plant. After that, the Vent absorption system, with the ventilation fan, should be activated, to prevent pressure build-up in any equipment. Only then, solvent may be pumped into the miscella tank, to the SWS, and to the extractor hoppers. Once the solvent has been taken to fill up the hoppers, the plant is ready to receive flakes (collets/cakes) from the PREP.

As the extractor is started, and miscella starts to flow to miscella tank, distillation may be started. DT is the last section to start, since the vapours from DT must be condensed in Economiser with miscella going for distillation. Extractor start-up has to be slow, to ensure complete extraction of the first material, which has not formed a proper bed. Plant start-up time, before meal starts to exit the DT, is anywhere between 6 and 10 hours, depending on the extraction time required for the feed material.

A2.4.2 Plant Shut-Down

The procedure is nearly the reverse as that for start-up. Material feed to extractor is stopped. The extractor speed is reduced to ensure complete extraction of the last material. After all the material has been discharged, the extractor is stopped. DT is stopped after all the meal is discharged. Slide gates at extractor feed and DT outlet are closed. Feed to distillation is stopped, and each equipment in the line is drained after flow subsides from the respective equipment. All the solvent may be drained to underground tanks. Vent system is continued for a few hours, until the plant has cooled down. Water circulation is the last to be stopped.

If the shut-down is for a short duration, say just a couple of days, the extractor may not be emptied. This is so that the subsequent start-up may be quick. As the feed to extractor is stopped, extractor may be stopped with full bed. DT and distillation sections are emptied, in that order, and the plant is shut down when cool.

For annual shut-downs, when annual maintenance may involve hot-work as cutting and welding, additional steps are necessary. The plant, after emptying, has to be de-solventised. After the solvent has been fully drained, and plant cooled down, steaming of equipment is started for removal of solvent vapours. The condensor and Absorption system are kept on, to condense the steam and solvent vapours. Steaming is continued for 4–6 hours. All the manholes on various equipment are then opened and plant allowed to self-vent for half a day.

In any case, before starting any hot work, careful double-check of any residual solvent vapours, by smell-sense as also with digital monitors (flameproof) is a pre-requisite that cannot be over-emphasised.

A2.4.3 Trouble-Shooting Procedures

Table A2.4.1 lists numerous issues related to process parameters and product quality, and outlines the steps to solve those.

TABLE A2.4.1
Trouble-Shooting in Solvent Plants

Note: **The parameter figures mentioned below pertain specifically to soybean oil extraction process. However, the principles apply uniformly to all other oil-bearing materials.**

				Performance Issues			
A.	High Oil Content in DOC	Check Milling Defect of Extractor Feed	MD > 0.3	Problem of inadequate Preparation	Check cracked soybeans size distribution -see Manual; If too many fines, or too much unbroken beans, correct the setting of rolls.		
					Check Cooker temp and moisture; If moisture less than 10%, increase open steam (or, add water if beans moisture is less than 8%). Check Flake thickness, adjust roll setting and pressure if required. Are roll surfaces smooth? If rough, time for roll grinding. If Expander is used, ensure continuous cake formation. If Dryer-Cooler is used, ensure moisture below 10%.		
		ED = Oil Content at Extractor outlet - MD	For Soy, ED > 0.3	Problem in Extractor	Fines in Flakes?	Screw conveyor in line, with hanger bearing?	Replace with Chain conveyor
						Check for material churning at bend section of chain conveyor	Correct the design of bend to avoid churning
					Extractor Screen choked?	Screen Rinsing device mis-operation?	Check Rinsing pump pressure. Ensure at least 2.0 Barg. Check orientation of spray nozzles.
						Second rinsing spray at feed end, with Miscella, stopped?	Restart; ensure fines-free miscella.
						Water in Solvent?	see below
B.	High moisture in Meal	Check moisture at DT outlet	Higher than design moisture?		Higher carryover of hexane from Extractor	Check for fines in Flakes	See action above

(*Continued*)

TABLE A2.4.1 (*Continued*)
Trouble-Shooting in Solvent Plants

			Check for high moisture in flakes	See action above
			Check for Extractor Screen choking	See action above
			Stop one last spray pump to improve draining	
		DT moisture OK, but D-C action inadequate	Check hot air temp	increase steam pressure to Radiator Check functioning of steam trap
			Check fan motor loads - low amps means low air flow	open damper, or increase speed (in case of a variable speed drive) increase fan speed if required or, increase fan impeller size
C. High Hexane loss	Check 'Flame' in Meal	Positive?	See below	
	Check Flash point of Oil	Lower than 130 C?	See below	
	Hexane in Vent Air	More than 50 ppm?	See below	
	Vent Air quantity	More than 1.5× the equivalent air flow?	Check air tightness of Feed Rotary Valve	
	(measured with Anemometer)	(equivalent air flow in m^3/h numerically equal to material feed rate in Tph; e.g., for a 600 Tpd plant, equivalent air flow would be 25 m^3/h)	Check speed of Feed RV; if high, reduce as per feed rate. Check air tightness of all flanges on Extractor and DT	

(*Continued*)

TABLE A2.4.1 (*Continued*)
Trouble-Shooting in Solvent Plants

D.	High steam consumption	Check DT gas temp	More than 74C?			Reduce open steam in DT, esp. in upper stages	
		Check Economiser miscella outlet conc. - lower than 75 % (soy)?				Check Economiser vacuum - lower than 400 Torr? - see below	
		Final oil temp	Higher than 100C?			Reduce steam to final heater	
				Operational Issues			
A.	Meal 'Flame' test positive	Check DT outlet temp	Less than 103C?	Increase DB steam			
				If DB steam already set to 10 Barg (or, 3.5 Barg for L.P. DT)	Increase open steam at last stage; ensure dry steam.		
		Check DT gas temp	Less than 70C?	Increase open steam			
		Check bed levels in each compartment	If low,	restore the levels			
		Check hexane content in Marc		If high, take action			see above (Meal moisture)
		Check steam holes in last stage	If blocked,	clean the holes.			
			Ensure, at DT start-up, that open steam in last stage is 'ON' before feed conveyor (solvent-tight) is started.				

(*Continued*)

TABLE A2.4.1 (*Continued*)
Trouble-Shooting in Solvent Plants

		DT duct pressure?	Should be -10 to -20 mmWc	Is Extractor vapour line valve fully open?	Throttle, to prevent too much air suction from Extractor	
				If gas temp too high (>75 C), due to open steam in upper stages (classic DT), reduce it to <74 C.		
				Check Economiser operation	If not adequate (outlet conc. > 75% for inlet conc. < 30%), check vacuum. Should be > 400 Torr.	
					If inlet conc. is > 30%, reduce it, to achieve more condensation of DT gases.	
			Ensure suction > 50 mm at Vent Ejector	If lower, increase vent steam (or, speed of vent fan)	If by increasing vent steam, smell in vent increases, check -->	1. Hot oil MV to be < 0.5% 2. Absorber oil inlet Temp 36 C max. 3. Oil flow to be adequate 4. Pall rings adequate in Absorber (4x1 m) 5. Pall rings to be clean
A2	Lumps in DT: Poor discharge from stages	Lumps formation in last stage (just above open steam injection) only?	May be due to condensate of steam		Rectify the moisture separator /trap operation, to ensure dry steam.	
		Check Cracking	Too many uncut beans?	Adjust Cracker rolls		
		Check Cooking	Inadequate cooking?	Set Cooking operation		
		Check fines at Extractor inlet	If much fines, identify and clear the cause			
		Much pressure in lower DT stages?	Too much steam injection? Reduce, in stages, till DT duct temp reduces to 70 C.			

(Continued)

TABLE A2.4.1 (*Continued*)
Trouble-Shooting in Solvent Plants

		DT door operation	If does not open properly, correct the operation			
		DT Door size	If small, will have to increase			
B.	Low flash point in oil	Check distillation vacuum	If low,	clean the ejectors()		
				tighten various flanges on all distillation equipment		
				check condensor temperature	If high, check condensor operation	see below
				Check seal leakages at pumps	If leaking, correct/ replace the seal	
				Check water level in CWH	If high, reduce the level to ensure that the ejector discharge pipe is dipped 200 mm max in water	
				Check for tube leakage in Econom'r/ Heater		
				Finally, drain the plant, and conduct a hydro-fill test	Seal any leakages found	

(*Continued*)

TABLE A2.4.1 (*Continued*)
Trouble-Shooting in Solvent Plants

C.	High hexane in Air Vent	Check DT condensor temp	If high,	check condensor operation (water flow, temp; incoming vapour temp)
				check Economiser operation. If inlet miscella conc is higher than 30%, much of DT vapours will remain uncondensed.
		Check Absorber oil temp	If high,	check oil cooler operation
		Check absorber oil flow	If low, restore the flow	
		Check Absorber Hot Oil MV	If higher than 0.5%	check Rec-Stripper operation
		Check the difference in manometer readings at Absorber inlet and outlet	If less than 20 mmWc	Increase oil flow, to ensure adequate wetting of packing rings
D.	High MV (hexane) in Absorber cold oil	Check Absorber oil flow	If low, restore to required flow	
		Check DT condensor temp	If high, check the Condensor operation	

(Continued)

TABLE A2.4.1 (*Continued*)
Trouble-Shooting in Solvent Plants

			If both above are OK, then air intake into the plant is high		Check 'suction' at Extractor	If more than 2 mmWc, throttle valve in duct going to DT Condensor.
					Check for leakages in flanges on Extractor and DT	Seal any leakages
E.	Condensor temp high	For Distillation Condensor	Check miscella concentration		If low, restore concentration.	
			Check miscella flowrate		If high, restore to required range.	
		For DT Condensor,	Check DT gas temp		If high, reduce open steam	
					If OK, check hexane in Marc. If high, take action to reduce it	see below
		Check water flow	Check pump motor current - flow may be deduced from the Pump Curve.			
			If low,	check if any blockage in suction pipe - if yes, clear the same.		
				check if header pressure is high - if yes, it is time to open the condensor, and clean the tubes.		
		If water flow ok,		there may be shell-side fouling (on DT Condensor). Pull out the bundle, and clean from outside.		

(*Continued*)

TABLE A2.4.1 (*Continued*)
Trouble-Shooting in Solvent Plants

			check water inet temp	If high,	check C.T. operation
F.	Hexane in Marc high	Check moisture in material at Extractor inlet	If high,	check Collet D-C (or, Cake Cooler) operation	
		Check 'fines' content in material	If high,	reduce fines coming from PREP.	
				If fines are generated in Extractor, check collet quality - may require more cooling time.	
			Try stopping last miscella circulation pump		
			Try reducing Extractor speed		
G.	Water in Solvent	Check if vapour ducts of SWS and CWH are connected to the same Condensor (among, Economiser and DT Condensor).			If not, correct the connections.
		Check if 'Interlayer' is being drained regularly. If not, effect the same.			
		Check the NRV in the SWS drain line connected to Condensor pump.			
		Check the height of water overflow - see calculation in the text			

APPENDIX 2.5 MILLING DEFECT TEST

Milling Defect is a measure of inadequacy in preparation, of an oil-bearing material, for solvent extraction. An ideal preparation (say, of Soybean flakes) would render all the oil extractable. The oil content in spent flakes would then be very low, say, less than 0.3%, with a properly designed extractor. However, such a perfect preparation is never possible. What we need to measure is how weak or inadequate is our preparation in comparison with the 'ideal' one. This is done as per the following method:

1. Determine moisture content of sample of flakes (or, Cake or Collets, as the case may be) with oven method (A).
2. Weigh approximately 20 g (weight B) original sample **without grinding (no hand grinding either)**, and extract in Soxhlet apparatus for 3 hours. Distill the miscella and weigh the extracted oil (C).
3. Remove the extracted sample from Soxhlet and keep open to dry out.
4. Grind the entire dry sample in Laboratory Disk Grinder, as is normally done for oil content determination.
5. Collect the entire ground sample carefully, place it in Soxhlet again, and re-extract for 3 hours. Weigh the second-extracted oil (D).
6. Calculate the Milling Defect (MD) as follows:

$$MD = (D/B) \times (100 - 12)/(100 - A) \times 100\%$$

(The factor of (100 – 12)/(100 – A) is used to standardise the MD result on a uniform 12% moisture basis)

Total oil content may be calculated as (C+D)/B × 100%.

Extraction Defect (ED): Is the difference between the MD and the oil content in spent flakes at extractor discharge.

ED = Spent Flakes' Oil Content (on 12% moisture basis) – MD

Note:

The MD test is very useful for analysing reasons for high oil content in meal. For a good preparation, MD should not exceed 0.3%. For soybean flakes, with MD of 0.3%, oil content in meal should be less than 0.6%; so, the ED should be lower than 0.3%. However, the Oil-in-meal (ROC), after the DTDC, may be analysed as 0.9–1.0%, because of the 'DT Jump'.

Thus, if ROC is higher than 1.0% in meal, first analyse the MD. If the MD is higher than 0.3%, should correct the problems in Preparation. If not, then the ED is high, and the focus should then be on the operation of Extractor. For very high ROC, more than 1.2%, both MD and ED may be high, and efforts may then be made to correct the operation in Preparation as well as in Extractor.

APPENDIX 2.6 FOOD GRADE HEXANE SPECIFICATIONS

The latest standard for Food Grade Hexane, issued by the Bureau of Indian Standards, in year 2017, IS: 3470 – 2017, is as follows in Table A2.6.1.

TABLE A2.6.1
Specifications for Food Grafe Hexane (IS: 3470-2017)

Sr. No.	Characteristics	Requirement
1	Density at 20 C, g/mL	0.660–0.687
2	Refractive Index η^{20}D	1.375–1.384
3	Distillation - a) IBP, C min^{m}	63
	b) Distilled between 64 and 70 C, % volume min^{m}	95
4	Non-volatile Residue, g/100 mL, max.	0.0005
5	Reaction of non-volatile residue (pH Indicator)	Natural to Methyl orange
6	Sulphur content, mg/kg, max.	5
7	Lead as Pb, mg/kg, max.	1
8	Benzene content, percent (v/v), max.	0.05
9	Aromatics, percent (v/v), max.	0.07
10	Polycyclic aromatics, UV Absorption Test	To pass the test
	Specific Value at wavelength (mm) range 280–290	0.15 max.
	290–300	0.13 max.
	300–360	0.08 max.
	360–400	0.02 max.
11	Saturates, percent by mass, min^{m}	98.5

APPENDIX 2.7 CO-GENERATION OF POWER

Since a solvent extraction plant requires a steady supply of steam and electricity on a continuous basis, it is a good candidate for co-generation of power. Generally, a steam boiler rated at 17.5 Barg (M.P.) pressure is used in solvent plants, so that, even with fluctuations in boiler pressure, a steady steam pressure of 13–14 BarG is available at the plant header. A PRV then is set to supply steam at a constant pressure of 10–10.5 BarG to the DT. It should be noted that nearly 80% of the steam requirement is at low pressure, up to 3.5 BarG, for the distillation section, as also for the live steam into the DT.

If a higher pressure boiler were used, say rated at 45 BarG, in place of the M.P. boiler, it would then be possible to use a two-stage steam turbine, to produce electricity, while letting out part steam at 11 BarG as required for the DT operation, and balance steam at 4 BarG. For the ratio of DT steam and L.P. steam required, the power generation may be nearly 1 kWh per 14–15 kg steam.

As seen in the earlier discussion on utility requirement for soybean crushing plants, the ratio of steam to electricity requirement is just about 10.5 kg: 1 kWh. Thus, the above ratio of power generation is not sufficient for soybean, i.e. not enough electricity will be produced. However, if all the steam were exhausted at 4 BarG, then the turbine may produce 1 kWh electricity with only 11 kg steam. Thus, if say the soybean unloading section is separately supplied electricity, the entire Prep and SEP load may be connected to the turbine. Cost of electricity then would be much lower @ INR 2.25 (3 cents) per kWh, as against the normal cost of INR 7.50 (10 cents).

When the entire steam supply to crushing plant is at 4 BarG, the DTDC, Cooker and the Vacuum Ejectors would have to be designed to operate with L.P. steam. The steam consumption for ejectors is somewhat higher when motive steam is at low pressure. Steam requirement for DT and Cooker does not increase, but these equipment must be built bigger, with more heat transfer area, and will require somewhat more electricity.

The steam to electricity ratio for other oil-bearing materials is, in fact, quite favourable for co-generation. As seen from Table 19.1, the ratio for Rice Bran as also for presscakes of rapeseed and sunflower seed is about 13:1. Thus, in fact, extra power may be produced than required for the plant. If extra power is not desired, boiler steam pressure may be reduced to produce just enough electricity.

In view of the substantial saving in energy cost, the pay-back period for a co-generation plant may be less than 2 years for medium-sized plants, say 1,000 Tpd. It is almost a blessing for plants operating in areas where electricity supply may be erratic.

APPENDIX 2.8 TYPICAL LAYOUT OF A SOLVENT EXTRACTION PLANT

As discussed earlier, a solvent extraction facility consists of two processing plants, namely, an oil mill (to prepare high-oil content seeds) or a Preparation section (for low-oil content seeds such as soybean or rice bran), and the solvent extraction plant proper. Layout of an oil mill has been discussed in Part 1 of this book. Layout of a soybean preparation plant tends to be similar to that of an oil mill, with one main operation floor with main heavy machines such as flakers and expanders, and one or more other floors to locate other machines. A typical layout of a solvent extraction plant is discussed below.

Figure A2.8.1 shows the layout of a 600 Tpd soybean solvent extraction plant. The building size is 30 m × 12 m × 15 m high. Larger capacity plants would be very similar except for bigger equipment. The main operating floor is at around 6 m level. This plant has a linear extractor, which is on one side, and all other equipment, including the DTDC and distillation line, on the other side. As seen, the plant is rectangular, with the length being much more than the width. There is a long corridor between the two sides. The corridor enables the plant operator to monitor all the main parameters easily, since sight-glasses, and gauges, are mounted on all equipment facing the corridor. Distillation Condensor is placed on the upper floor at 10 m; this floor also has a corridor for monitoring of the operation of Evaporator Flasher and the Absorber. Pumps and vessels such as the miscella tank, solvent-water separator, and water heater, are located on the ground floor. Layouts with Loop extractor, or the Sliding Cell extractor, also tend to be similar, except that the extractor has to be monitored at two different levels.

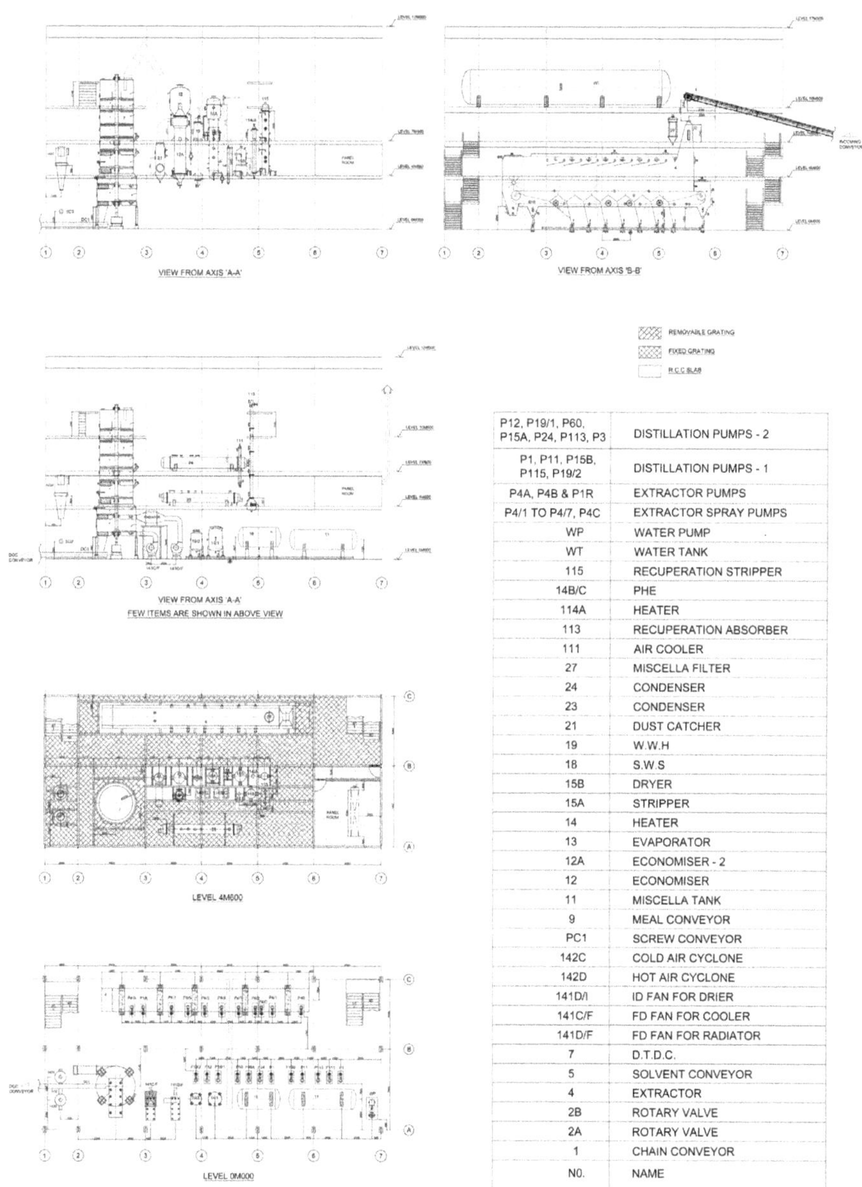

NO.	NAME
P12, P19/1, P60, P15A, P24, P113, P3	DISTILLATION PUMPS - 2
P1, P11, P15B, P115, P19/2	DISTILLATION PUMPS - 1
P4A, P4B & P1R	EXTRACTOR PUMPS
P4/1 TO P4/7, P4C	EXTRACTOR SPRAY PUMPS
WP	WATER PUMP
WT	WATER TANK
115	RECUPERATION STRIPPER
14B/C	PHE
114A	HEATER
113	RECUPERATION ABSORBER
111	AIR COOLER
27	MISCELLA FILTER
24	CONDENSER
23	CONDENSER
21	DUST CATCHER
19	W.W.H
18	S.W.S
15B	DRYER
15A	STRIPPER
14	HEATER
13	EVAPORATOR
12A	ECONOMISER - 2
12	ECONOMISER
11	MISCELLA TANK
9	MEAL CONVEYOR
PC1	SCREW CONVEYOR
142C	COLD AIR CYCLONE
142D	HOT AIR CYCLONE
141D/I	ID FAN FOR DRIER
141C/F	FD FAN FOR COOLER
141D/F	FD FAN FOR RADIATOR
7	D.T.D.C.
5	SOLVENT CONVEYOR
4	EXTRACTOR
2B	ROTARY VALVE
2A	ROTARY VALVE
1	CHAIN CONVEYOR

FIGURE A2.8.1 Typical Layout of a Solvent Extraction Plant.

Layout of a plant, with a Rotary extractor tends to be somewhat different. The extractor and the DTDC are usually placed side by side in a square-shaped plant, and the distillation line with condensors is placed on the other side.

Solvent storage tanks are usually placed underground, by the side of the building, and piping connected to solvent pump placed on ground floor. Cooling tower may be located near the building by the side of distillation line, and water pump may be placed on ground floor.

When plants are automated, they are operated from a control room. The control room may either be located in the solvent plant or the Preparation building. The latter is preferred since it is in a safe area, away from the solvent plant. In such a case, it helps to have a large screen, mimicking the entire process, in the electric panel room of the solvent plant, to help the operator monitor the plant.

References

Anderson, G.E. 2001. Apparatus for enhanced solvent recovery from solvent extracted materials. US patent 6279250.

Bair, C.W. 1979. Microscopy of soybean seeds: cellular and sub-cellular structure during germination, development and processing with emphasis on lipid bodies. Ph.D. Thesis, Iowa State University, USA.

Beach, D. 1983. Rapeseed crushing and extraction. In *High and Low Erucic Acid Rapeseed Oils*, Kramer, J.K.G., Sauer, P.D. and Pigden, W.J. (eds.), Academic Press, Toronto, Canada, pp. 181–197.

Becker, W. 1978. Solvent extraction of soybeans. *JAOCS* 55: 754–761.

Bredeson, D.K. 1983. Mechanical oil extraction. *JAOCS* 60: 211–213.

Brueske, G.D. 1993. Oil/meal separation process. In *Proc. World Conf. on Oilseed Technol. Utilisation.* Applewhite, T.H. (ed.), AOCS Press, Urbana, IL, pp. 126–137.

Bull, W.C. and Hopper, T.H. 1941. The composition and yield Urbana, Illinois of crude lipids obtained from soybeans by successive solvent extractions. JAOCS 18: 219–222.

Cardarelli, D.A. and Crapiste, G.H. 1996. Hexane sorption in oilseed meals. *JAOCS* 73: 1657–1662.

Cardarelli, D.A., Crapiste, G.H. and Mattea M.A. 2002. Modelling and simulation of an oilseed meal desolventizing process. *J. Food Eng.* 52(2): 127–133.

Carelli, A.A., Frizzera, L.M., Forbito, P.R. and Crapiste, G.H. 2002. Wax composition of sunflower seed oils. *JAOCS* 79(8): 763–768.

Coats, H.B. and Karnofsky, G. 1950. Solvent extraction II. The soaking theory of extractions. *JAOCS* 27: 51–53.

Dunning, J.W. 1953. History and latest developments in expeller and screw press operations on cottonseed. *JAOCS* 30: 486–492.

Eggers, R., Sievers, U. and Stein, W. 1985. High pressure extraction of oilseed. *JAOCS* 62: 1222–1230.

Extech, 2016. Redesigning of double-bottoms in Schumacher DT for reduction in pressure drop. Personal notes.

Extech. 2018. Design of solvent absorption system for low residual solvent in vent. Personal notes.

Extech. 2020. Sheanut crushing plant in Nigeria. Personal communication. Extech Process Engineering LLP, Thane, India 400 615.

Firdaus, M, Salleh, S.M., Nazir, I., Ngat, Z., Siswauto W.A. and Yusup, E.M. 2017. Preliminary design of screw press model of palm oil extraction machine. Proc. IOP Conf. Series – Mat. Sc. & Engg. 165-012129, Busan, Korea.

Heinrich, D. 2020. Warm and hot dehulling of soybean. Private communication.

Hutchins, R.P. 1949. Processing of oil seeds and nuts by hydraulic and mechanical screw press methods. *JAOCS* 26: 559–563.

Janssen, L.P.B.M. 1989. Engineering aspects of food extrusion. In *Extrusion Cooking*, Mercier, C., Linko, P. and Harper, J.M. (eds.), Am. Assoc. Cereal Chem., St. Paul, MN. pp. 39–56.

Karnofsky, G. 1949a. The theory of solvent extraction. *JAOCS* 26: 564–569.

Karnofsky, G. 1949b. The mechanics of solvent extraction. *JAOCS* 26: 570–574.

Karnofsky, G. 1987. Design of oilseed extractors I. Oil extraction (supplement). *JAOCS* 64: 1533–1536.

Karnofsky, G. 2005. Some time-tested tips for washing and leaching: a good extractor design will be based on laboratory-test results and the proper selection of the equipment and its configuration. *Chem. Eng*. 112(13): 52–55.

Kemper, T. 2005. Oil extraction. Ch. 2 in *Bailey's Industrial Oils and Fats Products – Vol. 5*, 6th edn, Shahidi, F. (ed.), Wiley-Interscience, Hoboken, NJ, pp. 57–98.

Keshre, P. 2019. Processing of castor beans, a case study. Private communication.

Knott, M. 1991. Design methods of screw presses at DeSmet group. Personal communication.

LeClef, E. 2020. DeSmet Ballestra Group, Belgium. Personal communication.

LeClef, E. and Kemper, T. 2015. Sunflower seed preparation and oil extraction. In *Sunflower – Chemistry, Production, Processing and Utilization*, Martinez-Force, E., Dunford, N.T. and Salas, J.J. (eds.), Science Direct, Elsevier, Amsterdam, pp. 235–242.

Merrikin, E.J. and Ward, J.A. 1981. On-farm production of fuel from vegetable oil. In *Proc. 3rd Conf. on Energy Use Management*. West Berlin, Germany, Fazzalore, R.A. and Smith, C.B. (eds.), Pergamon Press, Oxford, pp. 1697–1705.

Modi, A. 2003. Solvent percolation in rice bran extraction. Private communication.

Modi, A. 2012. Desolventising of rice bran. Private communication.

Mrema, G.C. and McNulty, P.B. 1984. Microstructure of rapeseed and cashew as related to mechanical oil expression. *Ir. J. Food Sci. Technol*. 8: 59–66.

Mrema, G.C. and McNulty, P.B. 1985. Mathematical model of mechanical oil expression from oilseeds. *J. Agri. Eng. Res*. 31: 361–370.

Norris, F.A. 1981. Extraction of fats and oils. Ch. 3 in *Bailey's Industrial Oil and Fat products – Vol. 1*, 4th Ed. Swern, D. (ed.), John Wiley and Sons, New York, pp. 175–246.

Paraiso, P.R., Cauneto, H., Zemp, R.J. and Audrade, C.M.G. 2008. Modelling and simulation of soybean desolventizing-toasting process. *J. Food Eng*. 86(3): 334–341.

Roques, M., Naiha, M. and Briffaud, J. 1984. Hexane sorption and diffusion from rapeseed meals. In *Engineering and Foods*. B.M. McKenna (ed.), Elsevier Applied Science. Vol. 1, pp. 13–21.

Seaman, D.W. and Stidham, W.D. 1991. Method for preparing a high bypass protein product. US Patent 5225230A.

Shirato, M., Murase, T., Hayashi, N., Miki, K., Fukushima, T., Suzuki, T., Sakakibara, N. and Tazima, T. 1978. Fundamental studies on continuous extrusion using a screw press. *Int. Chem. Eng*. 18: 680–688.

Shirato, M., Murase, T. and Iwata, M. 1983. Pressure profile in a power-law fluid in constant-pitch, straight-taper and decreasing-pitch screw extruders. Int. Chem. Eng. 23: 323–331.

Shurtleff, W. and Aoyagi, A. 2016. *History of Soybean Crushing (980–2016)*. Soyinfo Centre, Lafayette, CA, p. 2577.

Terzaghi, K. 1943. *Theoretical Soil Mechanics*. John Wiley and Sons, New York, pp. 265–296.

Tiwari, S.K. 2018. Reduction in red colour in crude soybean oil. Private communication.

Tiwari, S.K. 2019. Solvent extraction of sal seed. Private communication.

Unger, E. 1990. Processing. Chapter 14 in *Canola and Rapeseed – Production Chemistry Nutrition and Processing Technology*. F. Shahidi (ed.), Springer, New York, pp. 235–242.

Vadke, V.S. 1987. Mechanical expression of vegetable oil – technology and modelling. Ph.D. Thesis, University of Saskatchewan, Saskatoon, Canada.

Vadke, V.S. 2007. New generation expellers. In *Proc. 6th Technical Seminar*. Solvent Extractors Association of India, Mumbai.

Vadke, V.S. 2013. Advances in technology for oilseed and vegetable oil processing. In *Vision 2025*, Solvent Extractors Assoc. India, Mumbai, pp. 168–171.

Vadke, V.S. 2014. Pungent mustard oil with large press. Personal notes.

Vadke, V.S. 2015a. Co-generation systems in VegOil plants. In *Proc. 9th Technical Seminar*. Solvent Extractors Association of India, Mumbai.

Vadke, V.S. 2015b. Effects of extraction processes on oil quality. In *Proc. National Conf. on Process and Products Development for Better Economic Benefits to Fats and Oils Industry*, Oil Technol. Assoc. India, Kolkata, p. 51.

Vadke, V.S. 2017. Recent advances in miscella refining of cottonseed oil. Paper presented at 'Cottonseed Oil Conclave', Indian Cottonseed Crushers Association, Mumbai.

Vadke, V.S. and Sosulski, F.W. 1988. Mechanics of oil expression from canola. *JAOCS* 65: 1169–1176.

Vadke, V.S., Sosulski, F.W. and Shook, C.A. 1988. Mathematical simulation of an oilseed press. *JAOCS* 65: 1610–1616.

Van Damme, J. 2009. Device for desolventising under reduced pressure. US Patent 8142178.

Vavpot, V.J., Williams, R.J. and William M. 2012. Extrusion / Expeller pressing as a means of processing green oils and meals. Ch.1 in *Green Vegetable Oil Processing*, Farr W.E. and Proctor, A. (eds.), AOCS Press, Urbana, IL, pp. 1–17.

Ward, J.A. 1976. Processing high oil content seeds in continuous screw presses. *JAOCS* 53: 261–264.

Ward, J.A. 1984. Pre-pressing of oil from rapeseed and sunflower. *JAOCS* 61: 1358–1361.

Yiu, S.H., Altosaar, I. and Fulcher, R.G. 1983. The effects of commercial processing on the structure and micro-chemical organisation of rapeseed. *Food Microstructure* 2: 165–173.

Ziemann, C. 2020. Process of warm dehulling. Private communication.

Index

For Product Safety Concerns and Information please contact our EU representative GPSR@taylorandfrancis.com
Taylor & Francis Verlag GmbH, Kaufingerstraße 24, 80331 München, Germany

www.ingramcontent.com/pod-product-compliance
Lightning Source LLC
Chambersburg PA
CBHW060748220326
41598CB00022B/2362

* 9 7 8 1 0 3 2 3 1 3 8 4 9 *